Teratology in the Twentieth Century Plus Ten

Harold Kalter

Teratology in the Twentieth Century Plus Ten

Springer

Dr. Harold Kalter
Lenox Lane 5
45229 Cincinnati Ohio
USA
kalterh@ucmail.uc.edu

ISBN 978-90-481-8819-2 e-ISBN 978-90-481-8820-8
DOI 10.1007/978-90-481-8820-8
Springer Dordrecht Heidelberg London New York

Library of Congress Control Number: 2010928542

Printed on acid-free paper

Springer is part of Springer Science+Business Media (www.springer.com)

To my wife

Bella Briansky Kalter
with abiding love

Preface

Teratology is among the oldest and youngest of human preoccupations. From the earliest recordings of congenital deformities in humans and animals to modern efforts to understand and control such happenings, abnormality and its meaning have been of fascination. Today, after occupying a lesser place in the halls of philosophical and scientific pursuits, teratology has come to take a position at the hub of a complex crossroads of human medical concerns.

The focus has taken different forms. For academics, clinicians, and experimentalists the quest is the basic one, the search for the causes of prenatal maldevelopment; pursued by disciplines ranging from anatomy to zoology, from embryology to epidemiology, these have given teratology its richness of content and enduring challenge. For others, of diverse purpose, environmentalists, ethicists, the general public, the focus is pragmatic – the determination that tragedies such as that caused by thalidomide may never occur again.

A small minority of human conceptions are morphologically abnormal, and though many of them are sifted out through spontaneous abortion others are born with defects. As over time many other medical problems of infancy declined, attention turned to these malformed newborns and the challenges they present.

The present work focuses on congenital malformations whose origin is the environment, that is, those whose cause comes from outside the embryo and fetus. Summings up of what is at present known of these causes have reckoned that perhaps one-third of all serious congenital malformations are of this origin, a knowledge which has come in bits and pieces from a slowly emerging etiological miscellany – the story of which will be told with the unfolding below. The others – largely outside our purview here – have neither a simple genetic or clear environmental basis, as nowadays defined.

What is known is the portal to the future. This book summarizes the past and the latest findings and opinions about the environmental teratological forces that malform the unborn creature from the moment of conception; assering with Antonio that

> In nature there's no blemish but the mind;
> None can be call'd deform'd but the unkind.

Contents

Introductory Matters

The twentieth century saw the flowering of the science of teratology. The subject of this field of study, physical abnormalities of the newborn – congenital malformations – is an old human concernment, such conditions having afflicted human beings since the dawn of the species; an antiquity attested by prehistoric anthropological evidence and by written records from as early as nearly five thousand years ago. Now, at the outset of the modern third millennium, we note its recent contributions as prelude to continued understanding of these phenomena.

Though this is an old subject objectivity regarding such phenomena, replacing ancient fable and superstition, arrived only with the nineteenth century, when anatomists, embryologists, and pathologists with observational prowess not yet surpassed, meticulously described, classified, and categorized congenitally malformed humans and animals (Wells 1964). And the threshold of rationality was approached when experimentalists, tentatively exploring the how, produced abnormalities in birds and amphibians (Saint-Hilaire 1832–7, Dareste 1877, Martin 1880, Taruffi 1881–94, Ballantyne 1904, Schwalbe 1906–37), thus setting the scene for advancements in later decades.

Coincidental Discoveries

Two unconnected events occurring at the outset of the just past century define the initiating moment for the modern study of malformations: the rediscovery early in 1900 of Mendel's laws of inheritance (see Dunn 1965); and the use of the then recently discovered Roentgen rays to induce malformations in laboratory animals (Hippel 1907). These strands, destined to become intimately entwined, epitomize the dichotomy – heredity and environment – in the ultimate quest in teratology, the search for the causes of human prenatal maldevelopment.

From early on these polarities drove the search for causation. Between them at first there seemed to be an unbridgeable gap; an example of which was reflected in the views of two eminent reproductive pathologists – ironically even studying the same material: early human embryos – when one wrote that "careful study of my [pathological] specimens . . . establishes beyond doubt . . . that all of them . . . are

H. Kalter, *Teratology in the Twentieth Century Plus Ten*,
DOI 10.1007/978-90-481-8820-8_1, © Springer Science+Business Media B.V. 2010

due to external influences" (Mall 1908); while the other believed that the pathologies arose from "intrinsic defects" (Hertig et al. 1959). It may be remarked parenthetically that what is strangest about these two pronouncements is that the former was made at a time when abnormal prenatal development was thought to be predominantly, if not wholly, of genetic origin (Baur et al. 1921, 1931), and the latter, when environmental influences on embryonic development, as shall be seen, had by then been well established.

But in the beginning and for many years the anachronistic former view yielded to the latter, and the main purpose of studying human and animal congenital abnormality was to establish and classify patterns of inheritance (e.g. Pearson 1912, Wright and Eaton 1923, Dobrovolskaïa-Zavadskaïa 1927). Human abnormalities that particularly lent themselves to this pursuit were those of incontestably hereditary origin, specific though individually rare skeletal abnormalities such as brachdactyly and chondrodystrophy, purposes readily addressed when such conditions did not skip generations or occur in overlooked form. Brachydactyly, eminently fitting this prescription, was in fact the first human structural abnormality shown to be inherited in a manner conforming to the rules delineated by Mendel (Farabee 1905).

Early Genetic Studies

It was of such abnormalities that the first genetic studies were made in the early decades of the twentieth century. Vast pedigrees of families containing members with such abnormalities were published by the Galton Laboratory in London in its Treasury of Human Inheritance, edited overall by Karl Pearson (1912).

But while some of these defects were discovered to be inherited in a simple fashion (Bell 1951), others had a more complicated pattern of descent. The latter, in fact, have been the more intellectually challenging, and much study was later to be devoted to the problems they entailed (Carter 1977).

Definition

Introduction: Styles

Teratology, the subject of this history, is the science of congenital malformations, in all its aspects; and as that term is the keystone of the subject we begin with an attempt to convey its meaning. It has been defined in various ways, and as an introduction to the problems encountered in understanding and agreeing about its meaning, we note a few examples of pronouncements regarding it, from times when basic questions were still being worked out.

> Congenital malformations are structural defects present at birth. They may be gross or microscopic, on the surface of the body or within it, familial or sporadic, hereditary or nonhereditary, single or multiple (Warkany 1947).

> A malformation is an abnormality ... caused by antenatal disturbances in development (Potter 1964).
>
> ... a major anomaly is one which has an adverse effect on either the function or the social acceptability of the individual; a minor defect is one which is neither of medical nor cosmetic consequence to the patient (Marden et al. 1964).
>
> ... conditions thought to be of prenatal origin whether or not they were manifest at birth [including] structural defects, functional abnormalities, inborn errors of metabolism, and chromosomal aberrations (Christianson et al. 1981).

Even these few attempts to characterize the term vary in several ways. The first is comprehensive and classical, covering many aspects; others include additional matters – that malformations precede birth, may be manifest or latent, hereditary or not, major or minor, structural or functional – subtle considerations, needing amplification.

Although these views reflect different perspectives and are open to debate, one thing was universally acknowledged: major malformations cause death or serious medical consequences, whereas the so-called minor ones do not. Now let's consider some terms, one by one.

Congenital Means Present at Birth

In medicine it was once the convention to divide disease into two opposing classes, congenital and acquired, the former meaning inborn, i.e. hereditary, and the latter not inherited. For teratology however these two terms need not be contrasting, and as well have taken on specialized signification.

The term congenital has a complex history. A venerable source notes its 1796 provenance with the meaning: existing or dating from one's birth (Anon 1971); which is equivocal since it may or may not exclude presence before birth. Another source removes the ambiguity when it gives the meaning "existing prior to or at birth..." (Anon 1963a). But then adds "but not hereditary," thereby resurrecting a usage that had almost expired.

Cutting through these ambiguities the modern-day scientific meaning of congenital, adopted by the field of teratology, is simply 'present at birth,' whether of endogenous origin, from germ cells, or exogenous, from environment. In this work it is the latter that is the major focus.

Malformations: Abnormalities of Structure

Defining malformation is the greater challenge. Broad definitions, such as "abnormalities attributable to faulty development" (McKeown and Record 1960) or "structural defects present at birth" (Warkany 1971), leave their key words unsettled. Strictly speaking, abnormalities of *structure* may include aberrations ranging from the submicroscopic to the glaringly gross. In practice however such semantic quibbles make for no difficulty, because the malformations dealt with in clinical, epidemiological, and experimental teratology are exclusively those seen by the naked eye or detected by standard clinical instruments and the usual investigational procedures. Thus debarred are not only nonqualifying 'structural' abnormalities, such as molecular and cellular ones, but also those in the categories of isolated metabolic, endocrinologic, functional, and so on; which like everything on earth have a physical basis, but nevertheless do not come into the purview of this work.

The term faulty development can also be dealt with by limiting malformations to irreversible events arising from disturbances of development of primary embryonic structures and organogenesis, which occur in the earliest months of pregnancy; and thus exclude conditions such as tumors, nevi, angiomata, etc. And also excluded, therefore, are conditions arising almost exclusively in the postembryonic period such as those associated with exposure to coumarins (van Driel et al. 2002) and the prenatal growth retarding effect of tobacco smoking occurring by itself (Simpson 1957).

Recognition of Malformations

Just as definition is necessary for agreement of usage, so are uniform criteria of the time of recognition necessary for comparability of observation. It is not surprising that major malformations are predominantly first detected in the neonatal period, when in most parts of the world infants are present in hospital and can be conveniently observed and examined. Thus, the vast record concerning most malformations, not only the conspicuous and medically more demanding ones, rests on observations made during this period. Early examination also allows recording of neonatally lethal conditions which will not yet have been lost. Therefore, with the exception of certain malformations, especially cardiovascular ones, which may not be discovered till some months afterward (Kalter 2007), the great majority of records and reports of what we have agreed to call congenital malformations have pertained to discoveries in the neonatal period, in babies while still in hospital.

Major malformations in experimental teratology, to be discussed below, are defined as gross abnormalities detectable by external observation or special procedure at or preceding birth.

Major and Minor Malformations

Aberrant physical characteristics of prenatal origin are not all of equivalent medical significance. Thus while all may be considered abnormal (the complexities into which this epithet can entangle one will be explored below), they differ in consequences for viability, health, and well being. The conventional distinction is that between major or serious congenital malformations and minor defects and trivial physical variants.

There are pragmatic reasons for this basic distinction. Major congenital malformations are those of such drastic departures from the norm that they cause or are associated with prenatal or perinatal death, require surgical or medical care soon after birth, or are gravely physically handicapping; and, some would add, impose an extreme cosmetic burden; while minor defects and others to be mentioned have no or little medical importance.

Understandably, it is the major malformations that have been the focus of the medical and investigational world, as well as of the lay public; and because of this have been long and widely chronicled and thus form a body of record against which comparison and analysis are made (Warkany and Kalter 1961, Kalter and Warkany 1983).

Ironically, many of these conditions are the most frequently occurring abnormalities of prenatal development, frequent in this context meaning of the order of 20 to 30 per thousand births. They include among them numerous sorts of malformations of every organ and system of the body – central nervous, cardiovascular, orofacial, gastrointestinal, urogenital, skeletal. (Incidentally, considering that medical and surgical advances have made many of these malformations no longer lethal or impair reproduction, what this means as far as evolutionary dynamics is concerned, would make for an interesting topic of discussion, but not here.)

Minor Malformations and Variants

Relatively trivial physical divergences from the typical, commonly known as minor congenital malformations or anomalies, come in many forms, but are usually of little or no medical or cosmetic consequence. Depending on what is considered a minor malformation (about which there has been little consensus) and the assiduity of the search for them, the number a baby may be discovered to possess can vary, as can the frequency of the newborn population so affected.

Let me be clear about one thing. What is being considered here are strictly physical attributes seen in newborn or very young babies, not qualities or features manifested at some later time, whether bodily or psychological.

I will venture briefly into this teratological byway. Some large, but variably estimated, fraction of newborn infants has been said to possess morphological features designated minor malformations or variants – most, probably because of ease of inspection, affecting the external ear, face, and hand – not accompanying major congenital malformations, and thus called isolated (Marden et al. 1964, Méhes 1980, 1988, Holmes et al. 1985, Merlob et al. 1985, Leppig et al. 1987).

Their being of consequence was judged by how often they occurred: only those in fewer than an arbitrary 4% of newborn infants being considered a defect (Smith 1971). But in the absence of agreement as to their meaningfulness for etiological investigation, it is fair to say that interest in them as isolated phenomena seems to have had its day.

For history's sake let's note that many of these features are simply physical or morphometric variants with not the least medical relevance. But without agreeing which of the more frequent of such isolated nonvariants were to be considered defects there could be little progress in determining their heuristic value (Pinsky 1985, Leppig et al. 1987, Merlob 1994). The crucial word here is isolated, since minor defects when appearing together with medically significant malformations may take on a relevance they otherwise lack, e.g. as supposed teratogenic danger signals. That wraps up the topic.

Classification

As important as definition is classification, that is putting entities into an orderly arrangement. In biology the arrangement consists of taxonomy, rules for grouping organisms; and in medicine of nosology, arranging disease according to established criteria. In teratology matters are not that simple. Some years ago James Neel (1958) wrote "no entirely satisfactory classification of congenital abnormalities has yet been devised." And many would say that is still true today.

Teratology being a branch of medicine, schemes for classifying congenital malformations, like those for disease, fall mostly into the categories of causation, pathogenesis, and outcome. Systems founded on causation – etiology – are ultimately oriented toward prevention, in distinction to those based on pathogenesis and outcome, which are directed as much toward theoretical considerations.

Classification by Cause

Congenital malformations have been around for a long time, and over these eons when explanation was attempted they were long attributed to the supernatural and recently to hereditary transmission (e.g. Martin 1880). Systematic efforts at classification by etiology had a late start.

A formulation analogous to Galton's (1889) division into nature and nurture was the partition into genetic and nongenetic offered by Gruenwald (1947). Conforming to the author's customary comprehensive style, the former consisted of spontaneous, induced, and somatic mutations; and the latter, as "agents affecting the phenotype without effect on the genotype," of all imaginable types of environmental agents – mechanical, actinic, chemical, ambient temperature, infectious – almost all known through experiments with birds, amphibians, rodents, and other laboratory species. The later suggested addition of psychological trauma and maternal-fetal interactions, e.g. antigenic incompatibility (Penrose 1951), was never substantiated; but the still later discovered chromosomal aberrations were well proven.

Knowledge of the causes of congenital malformations is limited, as already noted. One summary comprehensively and critically provided a compendium of them, which though compiled some years ago has remained essentially unadvanced to this day (Kalter and Warkany 1983). The summary also provided a novelty, a reasoned calculation of the frequency of each of the identified causal divisions. Thus, of the estimated 3% of the major malformations afflicting newborn children, it was calculated that about 7.5% are caused by single mutant genes; about 6% are associated with chromosomal aberrations; and perhaps 5% are due to or associated with known discrete environmental factors. To these may be added perhaps as much as 20%, the outcome of the combined action of individually obscure environmental and genetic components, i.e. multifactorial situations. Summing these, we see that the etiology of fewer than half of all congenital malformations had been ascertained to one extent or another by the later 1990s; which is not yet bettered.

This has left a vast terra incognita that presents a formidable challenge. How much of this residue has as its basis as yet undiscovered environmental teratogens and the multitude of single genes daily revealed to be responsible for prenatal mishaps is for the future to reveal. Thus, while applauding the discoveries in the just past century, much especially about the interworkings of external and internal causes of abnormal prenatal development remains to be learned.

A final word here. This relative paucity of etiological understanding may be due to the limited final form taken by individual malformations which obscure their diverse developmental pathways. By default therefore the main unambiguous classificatory scheme at present is by abnormality type.

Classification by Type

Achieving the goal of meaningful classification has been problematic. The predominant purpose of putting malformations into an order based on morphological appearance is to facilitate recognition and comparison. The entities, however, that can be included in such schemes are numerous, and discussions about what, according to varying needs, should be included have been controversal (see e.g. Davison 1963, Potter 1964). And it may be added, categorization today has the appearance of a bygone preoccupation, so it may be expected that no further elaboration will be forthcoming – till perhaps molecular studies add their ideas.

The most comprehensive of descriptive classifications of congenital abnormalities is that contained in the ICD, the international classification of diseases (WHO 1992). The ICD is arranged by system, part, and organ, and includes virtually every deviation from normal originating prenatally present at birth or attributable to conditions present at birth, regardless of medical importance or etiology; a inclusiveness in fact that is its fault, in giving undiscriminating equal weight to entities of different prognosis.

Other formulations have attempted to deal with the difficulty of categorizing malformations that often involve several bodily systems (e.g. Neel 1958, Edwards et al. 1964, Leck et al. 1968). A frequent solution of which has been to classify the combination by a given feature, e.g. its most serious component. But this may be at the cost of failing to appreciate that the particular combination is integral to its recognition as a specific constellation or syndrome.

Classification by Pathogenesis

Congenital malformations may also be classified by the abnormal developmental pathways by which they come into being, or are conjectured as coming into being, sometimes called 'mechanism' (a contentious term with several connotations, discussed further below).

One such scheme distinguished between anomalies originating in the organogenetic period and those in the period of fetal growth (Warkany 1947). Another suggested division was by type of abnormal embryological process – developmental arrest, abnormal resorptive events, secondary destruction, etc. (Patten 1957). In this vein were more specific anomalous actions – failure to form or form properly, retrogress, close, open, remain open, unite, etc. (Potter 1964). A later variation on this design differentiated between processes leading to intrinsic and extrinsic congenital abnormalities, between organ and tissue defects, etc. (Spranger et al. 1982). All in all, as it would hardly seem necessary to say, these exercises have been of limited instructive or practical value.

Behind and beyond this question of the processes by which malformations come into being is the ultimate one of the means of their so doing, their 'mechanism.' As Grüneberg and others discovered, going back, back, back, in the analysis of the steps leading to morphological deviation, reached a stage not further analyzable, at least by the tools now available. The complexities and frustrations of this pursuit are discussed further below.

Nomenclature

What we call a thing often determines how we think about it, despite a well-known author's 'what's in a name?' The term 'teratology' long referred to things out of the ordinary, but only fairly recently did it come to refer to studies of anomalies of organization, as the title of Isadore Geoffroy Saint-Hilaire's (1832–7) work indicated; till today it is defined by the medical world as the science that deals with abnormal prenatal development leading to congenital malformations.

Several attempts were made to revise this straightforward definition. One by attributing to teratology a narrow focus, the study solely of gross aberrations, and the neologism 'dysmorphology' introduced to apply to this view. Since this term merely indicates faulty structure it hardly covers the many connotations teratology possesses and yields no improvement in fundamental understanding of prenatal maldevelopment.

Another suggested term is birth defects, a simplistic synonym for congenital malformations; which disguising that malformations originate before birth, is detrimental in that it distracts from their origin and causation. Notwithstanding such objections, the term with its catchiness has become popularized. It will be eschewed here.

Syndromes

How do single etiological factors produce syndromes, characteristic patterns of malformations? Early attention was given to this question by Grüneberg (1938, 1963) in his exploration of the phenomenon of pleiotropism, the multiple morphological effects of mutant genes in mice. He reasoned that multiple abnormalities must be caused by either simultaneous action on different parts of the body, or subordinated ones through "a cascade of secondary and tertiary gene effects..." It was the latter, he concluded, by which teratogens, whether genic, chromosomal, or environmental, produce their course of action. It is to be emphasized that his hypothesis was based on experimental evidence obtained by studying abnormal prenatal development, and could not have been formulated through examination solely of newborn malformed individuals.

A different school of thought held, on the contrary, that causal and pathogenetic relations of anatomically distinct constituents of multiple malformations could be clarified by "careful analysis of the history and physical findings of the patient...," that is, by examining the malformed child (Christiansen 1975, Smith 1975, Spranger et al. 1982). Which was to be accomplished by assigning the proper term to the defects: whether deformation, anomalad, syndrome, sequence and association. Almost needless to say, such thinking has been of no use in analyzing congenital malformations.

Frequency

It is remarkable how few babies are congenitally malformed. Considering the prevalence of faulty genes, the intricate embryonic path traveled, the extrinsic perils potentially bombarding the unborn, it can only be marveled that prenatal development goes seriously awry as seldom as it does.

How often these errors happen began to be looked into during the nineteenth century. First estimates were low, about 0.2–0.9%, while later ones, perhaps more thoroughly pursued, came close to present findings, reaching 1.6–2.8% (Birnbaum 1912). Such variability was not unusual, ranges of frequency continuing to be found even later, from the 1920s through the 1950s (Warkany and Kalter 1961 gave some examples).

It was realized however that such "tremendous differences ... do not reflect biologic differences in the populations studied" (Harris and Steinberg 1954), but other factors instead, including definition. Obviously agreement cannot be reached in the absence of a standard definition, if only tentative and partial of what is to be considered a malformation, and thus to be included and not in the catalogue of such happenings. Differences from study to study in the assortment and frequency of malformations reported have often largely been due to such disagreement. As was astutely observed, way back in the last century, "It is not an easy matter to give correct statistics as to the frequency of malformations. Some figures only include the more marked forms of malformations, while others include even the slightest anomalies. . . ." (Birnbaum 1912).

Responsible as well for the sometimes considerable variation in estimated frequency were differences between studies in their structure, called ascertainment, differences in time, place, and quality of selection and observation. Continuing the quotation from Birnbaum, "many malformations are indeed only discovered for the first time at an autopsy." To expect one frequency, true always and everywhere, is an illusion. Further reasons for this will emerge in later pages.

Difficulties of Establishing Frequency

Older studies of malformations were driven in particular by scholarly purposes, seeking embryological and even evolutionary insights. Later inquiries often turned to estimating frequency with the principal purpose the pragmatic one of forming a basis for comparing findings and charting progress in the prevention of malformation.

Thus the first order of things, temporarily relegating study of causation to the background, was to establish baselines of frequency for different times and places, against which demographic variation and other forms of fluctuation could be mapped and analyzed.

To accomplish this task, wherever and by whomever studies of estimation of frequency would be conducted, common systems of nomenclature, classification, and definition had to be accepted; and procedures of discovery and diagnosis, if not as uniform as possible, at least be made known to colleagues.

Such matters have long been appreciated. Alfred Russel Wallace (1874), in reference to zoologic taxonomic efforts, noted that "one of the first requisites of a good system of nomenclature [is] that the same object shall always be known by the same name." This has obviously been a hard-learned lesson, not yet absorbed 75 years later when Warkany (1947) admonished that "a clear and commonly accepted terminology is indispensable for a satisfactory discussion of a subject. Some of the confusion in the field of congenital malformations is due to the lack of a uniform usage of words." And leaping to the opposite scale of matter, another 50 years later the same need for order was even recognized for protein nomenclature (Anon 1999).

It's clear then that if estimates of malformation frequency were to have validity, a measure of unanimity had to be reached as to the phenomena congenital malformation referred to, and systems of classification formulated as aids to recognition and identification. Only then could the deferred questions of etiology and variability in time and place be satisfactorily addressed.

A Matter of Terms

Before dealing with frequency – i.e. the rate of occurrence of a particular event within a specified interval or among a specified group of individuals – let's recall that even its application has been the subject of disagreement.

This broad meaning was divided by epidemiologists – whose subjects are groups or populations – into concepts of more precise connotation: incidence, to refer to new occurrences of a particular event during a given period; and prevalence, to the totality of such events existing at a given time. Thus, with respect to malformations, incidence is the proportion of abnormal individuals born during a specified interval of time, and prevalence the proportion of individuals of a given age or other characteristic that is abnormal.

Usually much ado about nothing, since a single age group, newborn infants, is almost always the subject of malformation studies. Prevalence, however, is misused when the group being studied consists of different age and survival categories – spontaneous or elective abortuses, stillbirths, neonatal deaths, live births – since the frequency of malformations differs greatly among them (Kalter 1991). But either term is correct respecting observations made over a nonspecified time period confined to a particular age or survival group. Obviously, comparison is clouded when the specifics of these categories are not made explicit. Having acknowledged this matter it can be kept in mind, and frequency continue to be used here, except as necessary for certain distinctions or elegant vatiation.

How Often Do Malformations Happen?

Estimates of how often congenital malformations happen – called here frequency – have varied depending on three basic considerations, the source of the information, the malformations recorded, and the age of the subjects examined (despite the qualification given just above).

The source greatly influences the finding. Records of many studies made over many years in many parts of the world demonstrated this fact well (Kennedy 1967). Data of three sources – public health records, birth and death certificates, and the like; serially registered hospital and clinic records; and data derived from close examination of infants – yielded mean frequencies of congenital malformations respectively of 0.8, 1.3, and 4.5% (the last included studies of lesser malformations and conditions of older children, which when omitted reduced the mean to a more likely 2.9%).

The spread of frequency makes it obvious that the level discovered depends to an enormous extent on the purpose of collecting the information, as well as on the intensity, thoroughness, and competence of the search for it. And thus that in the attempt to understand and compare findings its source cannot be neglected.

The most complete figures have come from hospital records, which have noted that major congenital malformations occur in approximately 3% of liveborn infants (Kalter and Warkany 1983). But even data derived from this source, the most reliable, can vary considerably. For example seven classic studies from hospitals in the US and Europe reported frequencies ranging well over twofold, from 1.43 to 3.30% (Warkany and Kalter 1961), differences that were due to variations in the assortment of malformations included, diligence and eccentricity of examination, and study design; and, additionally, factors beyond human control – demography, geography, temporality.

Studies in more recent years, though not as abundant as previous ones, yielded figures that did not differ importantly from those of older ones; and it can therefore be taken as a given that about 3% of children well examined in the neonatal period will be found to have serious, or, as they are called, major, congenital malformations. Why this particular level of maldevelopment, apparently widespread, is a separate question, well worth cogitation.

It is the proportion of liveborn infants with major malformations found in the neonatal period that gives the customary meaning of the frequency of such conditions. This figure is only negligibly increased when malformations in stillbirths are included, to which another quarter or so may be added by the postneonatal discovery of defects, particularly cardiovascular ones (Kalter 2007).

Sometimes terms like defect or anomaly are used interchangeably with malformation, often with a more inclusive but poorly defined connotation. It should be accepted that these are general terms embracing many types of abnormal conditions of prenatal origin (how many human attributes are not ultimately of prenatal origin), whereas the meaning of malformation is more limited; a delimitation to be grateful for, in an area of sufficient complexity.

An Aside

As yet in this work the subject of what causes malformations – etiology – aside from brief remarks found above, has not been mentioned. This subject per se will be given its due below. Here only its connection with definition will be noted. Congenital malformations as was said are abnormalities present at birth and hence, though it is obviously redundant to say do, originate before birth. At what time during prenatal life they do so it is pertinent to consider. Some congenital abnormalities are endogenous or intrinsic, i.e. are entirely or primarily genetic, and their essential causative factors reside in fertilized ova, even though the abnormalities are not expressed till some time during prenatal life.

This is not true, it would seem at present at any rate, of the great majority of the commonest major malformations, in whose causation genetic influences do not have the greatest part. This large fraction has as its sine qua non environmental or exogenous forces originating outside and acting at some time upon the developing embryo itself. The particulars of the known consequences of these etiological forces are discussed below.

Why Investigate Malformation Frequency

Trustworthy measures of the frequency of congenital malformations, in toto or individually, are of utmost importance for several reasons. Such information is needed to recognize demographic and geographic differences, temporal fluctuations – partly to aid in detecting sudden increases in their occurrence (due perhaps to the presence of novel teratogens or zealous observers) – and to evaluate the efficacy of efforts to prevent them. An older and still important objective of such efforts is to gauge the dimensions of malformations in pre- and perinatal deaths. For all these purposes it is obviously necessary to conduct ongoing surveillance of malformation occurrence, another subject discussed below.

Meanwhile, we continue the topic begun above of further conditions and circumstances that can cause malformation estimates to vary; including flawed records, unreliable data sources, and human failings.

Ascertainment

Beyond vagaries of definition the discovered frequencies of major congenital malformations have varied, sometimes widely, because of differences among studies in the procedures by which they have gone about establishing them, i.e. in the elements of what is called ascertainment.

The term ascertainment as used in epidemiology or medical genetics refers to the selection of individuals or families for inclusion in a study; and biases, as encountered in these disciplines, are practices that lead to faulty estimation of the role of etiological factors in the pathogenesis of a disease. By contrast, in teratology ascertainment refers to the means – the how, where, and when – of obtaining the information used to estimate malformation frequency, and biases to the factors that lead to faulty estimation of malformation frequency (Sackett 1979).

As noted cursorily above various sources of information are available to studies surveying malformation frequency: public health records, i.e. birth and death certificates, etc., of a particular geographic area during a specified time span; hospital or domiciliary births (the latter becoming ever rarer) during a stated period of time; malformation registry data, as well as various others, overlapping and not, that have also been called upon at times (see e.g. McKeown and Record 1969).

Regardless of the source of data all studies differ in numerous ways, in manner, method, and circumstance, that to one degree or another affect the estimate arrived at. The most elementary cause of distortion of frequency is of course selection, often unknowingly, of subjects with negative or positive risk potential.

Underestimating Frequency

Limiting estimation of malformation frequency to liveborn infants – although its primary and principal source – leads to its systematic underestimation, since the total load and array of malformations cannot be appreciated when abnormalities in spontaneous abortions and perinatal mortalities are excluded (Sentrakul and Potter 1966, Harlap et al. 1980, Poland et al. 1981, Shepard et al. 1988, Kalter 1991, Shiota 1993).

Establishing with certainty the malformation burden in spontaneous abortuses is beset with difficulties – among others, of examinability, recognition, diagnosis, and ascertainment – making it understandable why such abortion studies have not been plentiful. It is obvious, risking tautology, that the elimination of some fraction of malformed abortuses reduces the frequency of malformations seen at birth, perhaps significantly, and often modifies its spectrum as well. Selective abortion of prenatally diagnosed malformed fetuses has an impact on malformation birth frequency as well, but probably not greatly. Such elimination is of particular interest in special studies, like that of folic acid and malformations, gone into in detail below.

Malformations in stillbirths and neonatal deaths are not so easily overlooked. As over time some of these deaths have become less common, at least in parts of the world, their malformed fraction accordingly increases inversely (Kalter 1991). The literature is befouled when it does not explicitly state whether such instances are included in the tally of malformations or not.

There is a further underestimation when the relatively small fraction of congenital malformations is disregarded, mostly less serious except for cardiovascular ones, that are not expressed or found for some months or even years after birth. But, to reiterate, by far the main focus of the study of congenital malformations and the estimate of their frequency is the neonatal period.

Overestimating Frequency

A powerful biasing force leading to exaggerated estimates of the frequency of some abnormalities is the undue attention that may be given to what have been called 'interesting' cases. Examples are few however: clubfoot (Ehrat 1948, Woolf and Turner 1969, Kaminski et al. 1981); pilonidal cyst, usually a trivial abnormality, which "appeared in epidemic proportions in December 1940 and January 1941 ... [when] examination of the records disclosed the workings of a pediatric intern with an undue fondness for this diagnosis" (Stevenson et al. 1950); absent umbilical artery, in newborn children of diabetic women (Neave 1967), which a pathology study from the same hospital found rarely in autopsies of infants of diabetic births (Driscoll et al. 1960).

A perhaps common and only slowly recognized cause of an increased frequency are changed diagnostic methods and criteria. For example, a jump in the frequency of ventricular septal defect, first thought to be due to surged diagnosis, was later found to have been due to the detection of small septal defects overlooked by less intense diagnostic methods (Laursen 1980, Newman 1985, Spooner et al. 1988, Martin et al. 1989, Fixler et al. 1989, Anon 1994, Meberg et al. 1994).

Biological Factors

Certain malformations are more frequent or less frequent in some racial or ethnic groups than in others, a trait that can be useful in etiological investigation. Examples are an increased frequency of a mild type of polydactyly in blacks, of orofacial clefts in Japanese and some North American Indians, and a lower frequency of neural tube defects in blacks and Ashkenazi Jews (Tanaka 1963, Altemus and Ferguson 1965, Naggan and MacMahon 1967, Lowry and Trimble 1977, Christianson et al. 1981).

But in all the annals of human teratology, among the most frequent and noteworthy of congenital malformations are those that have been called neural tube defects, i.e. the major malformations anencephalus and spina bifida, resulting from failure of the neural tube to form in early embryogenesis. They are remarkable in several respects: they vary in frequency over 50-fold by race, ethnicity, and geography; their frequency has decreased in recent times; females are invariably more often affected, sometimes far more often, than males (MacMahon and Yen 1971, Elwood and Elwood 1980, Little and Elwood 1991, Murphy et al. 1996).

To these features two further ingredients recently entered the mix, to result in modifying their neonatal occurrence: the innovation of prenatal diagnosis, permitting elective elimination of affected feuses, and maternal consumption of folic acid, said to prevent the development of some proportion of them. Many unanswered questions swirl around the latter allegation (Eurocat Working Group 1991, Limb and Holmes 1994, Oakley et al. 1994, Cragan et al. 1995).

See later chapters for further consideration of these and other matters pertaining to malformation frequency and its estimation.

Pioneering Studies

We go now to the search for the causes of human congenital malformations, the principal objective of teratologic investigation in the present era. The interests and professional backgrounds of those engaged in this quest have been varied – clinical, genetic, epidemiologic, experimental, ecologic; a variegation that has been a strength, but also a detriment in not constituting a unified purpose.

Experimental teratology, in a picturesque sense, had its start in the ancient world, with the occasional production, although inadvertently, of embryonic abnormalities in chicken eggs improperly handled during artificial incubation. (See Walter Landauer's 1967 great history detailing this story; and also the account of Landauer's personal history, by Clark and Pierro 1994, to get a full flavor of his life and work.)

Only in the nineteenth century did this science take off, as embryogenesis and its mishaps came into focus; with its interests then largely confined to avian and amphibian forms (see e.g. Dareste 1877). Attention turned in the first years of the just ended century to external forces affecting prenatal mammalian development. It was the newly discovered roentgen rays and the revelation of their prenatal harmfulness that precipitated the interest in the environment that characterized the new era.

X-Irradiation

Animal Studies

The revelation that the barrier supposedly shielding mammalian embryos from pernicious surroundings could be penetrated came about through the discovery of x-rays in 1895 by Wilhelm Conrad Röntgen; made evident by the fact that this new exploratory tool was soon found capable of disrupting prenatal development in mammals (Kalter 1968). This tool was to have numerous applications.

Earlier investigators however were unable to realize its full potential, because they failed to time the onset of pregnancy and were not able to measure dosage accurately; hence organ susceptibility could only be loosely related to dose and

H. Kalter, *Teratology in the Twentieth Century Plus Ten*,
DOI 10.1007/978-90-481-8820-8_2, © Springer Science+Business Media B.V. 2010

prenatal stage. Ever more precise understanding of these relations came in time. Numerous details of these studies were summarized by Russell (1954), Wilson 1954 and Kalter (1968).

From these and later studies three major principles emerged: the specificities of the outcome are governed by the type and quantity of the injurious agent, the timing of its application, i.e. the stage of prenatal development exposed, and the genetic constitution of the subjects concerned – maternal and fetal; with understanding of the interrelation of dose and stage becoming ever more refined with time. All of these equally pertain with suitable qualifications to teratological happenings generally.

These relations have been defined only vaguely for human beings, opportunities for discovering them, thankfully, having been few. The contrast between experimental animals and humans in this respect is amply illustrated by the detailed knowledge of them in the one (Russell 1954, Hicks and D'Amato 1966, Kalter 1968) and their paucity in the other (Dekaban 1968).

A contrast again are the teratological consequences of irradiation in animals and humans. In animals many different malformations are induced, each with a generally clear prenatal age specificity: in the earliest sensitive period exencephaly (corresponding to anencephalus in humans) and other abnormalities of the central nervous system and sometimes cardiovascular malformations, and at later times eye and various other defects (Hicks and D'Amato 1966). In human beings irradiation causes a much smaller assortment of abnormalities, as will be discussed below, for reasons that are not clear, but may be related to the more protracted human embryofetal development.

As for the means, the mechanism, by which irradiation causes teratogenesis, that too is not clear. A wide-ranging review of the question (Kalter 1968) looked at numerous earlier lines of evidence and opinions: was the target the pregnant animal or the embryos themselves? in the embryo was the damage due to mitotic damage, somatic mutation, cell death, tissue repair, etc.? All of these to one extent or another may be involved, and yet none be the vital process. Perhaps the right question has not been, and now never will be, asked.

Human Pelvic Irradiation

An Irradiation-Caused Abnormality

Within a year of its discovery x-irradiation was put to medical use, pregnant and nonpregnant women being treated for various gynecological ailments then thought to be amenable to this novel instrument. But then, beginning in 1920 and throughout the decade, there came reports of abnormalities in children of irradiated pregnant women, especially microcephaly, i.e. small head circumference (Goldstein and Murphy 1929), which, by the way, corroborated earlier experimental findings (described in Kalter 1968).

In Philadelphia, Douglas P. Murphy started an inquiry into the subject, having become interested through his experiments with rats on the effects of ovarian irradiation on fertility. By searching the medical literature and through questionnaires sent to many gynecologists and radiologists in the US he learned of 625 women who had been exposed to therapeutic pelvic radium therapy or roentgen irradiation, a sufficient number for a close examination of the data (Murphy 1928).

An objective analysis required that the ill health of children of women irradiated before as well as during pregnancy be considered. Conducting a forerunner epidemiological study, Murphy balanced the possible roles of various elements, maternal age, health before and during pregnancy, and previous reproductive history; and found that only postconception irradiation had possible serious untoward outcomes – abortion, stillbirth, infant death, and congenital abnormality (Murphy 1929, Goldstein and Murphy 1929).

The last was most obvious with respect to a particular abnormality, microcephaly. Of the 402 children born to women irradiated before conception seven were malformed, six with various defects and one with microcephaly, a frequency of 0.25%; while of 74 exposed in utero 25 were malformed, eight apparently nonspecifically and 17 with microcephaly, a frequency of 23%, a difference that needed no statistical test to demonstrate its significance (Murphy 1929).

Thus was discovered the first environmentally induced congenital malformation in human beings.

It remained to learn what period during prenatal life was most sensitive to the irradiation, and the answer was that about 70% of the microcephalic children had been exposed before the 5th month of pregnancy (Goldstein and Murphy 1929). Questions remained: exactly when during this interval was the damage done, and what was the effect of time of exposure on severity of the condition. The answers awaited a future calamity.

Another important finding was that in only one-third of the children was the microcephaly present at birth, and in the others became apparent over a wide postnatal period. Meaning in essence that damage sustained prenatally may be latent and expressed overtly after birth.

Eye Abnormalities

Irradiation had another effect. Women with medical conditions for which at the time elective abortion was indicated were irradiated with large doses of x-rays in the 2nd month of pregnancy, and the pregnancies interrupted some time later. Rosettes were present in the retinas of all the abortuses, and in the oldest specimen the embryonic optic cleft persisted and the iris and ciliary processes were rudimentary. Although the eyes were of size appropriate for fetal age it was commented that "the possibility of microphthalmia, had the embryo grown to full term, cannot be excluded" (Goldstein and Wexler 1931).

The rosette formation also duplicated earlier animal findings (Hippel 1907) and presaged ophthalmic defects produced by irradiation in mice years later (Rugh and Wolff 1955).

Dose, Time, and Effects

The relation of outcome to radiation dose and gestation age exposed was examined in later studies. The first, by drawing developmental parallels between human and animal gestational intervals, pointed to a partial correspondence with experimental findings (Russell 1954). The other, more successful, based on a small number of irradiated women, found that exposure at particular stages of gestation caused specific types of outcomes, death but not malformation by exposure in week 2–3, i.e. prior to implantation, while in weeks 4–11 eye defects and microcephaly resulted; dosage information however was imprecise (Dekaban 1968).

Based on this information, on Murphy and Goldstein's findings, and on the effects of the calamity to be discussed below, the teratogenic effects of irradiation in humans appeared to be limited, as noted above, and to have as the major target the central nervous system.

Dose Matters

What is the lowest amount of irradiation having detrimental human embryonic effects? This is an important question, for diagnosis and treatment, as well as theoretical considerations. The consensus is that 5 rad or less of x-rays poses no threat to structural development (Mole 1979). However, for more subtle injury, such as cellular brain damage, responsible perhaps for severe mental retardation, while earlier there appeared to be no clearly detectable threshold (Otake and Schull 1984), this opinion was modified later, for which see below. The important topic of no-effect levels or thresholds in teratogenesis will be discussed elsewhere in this work.

Murphy's Contribution

Douglas P. Murphy's pioneering inquiries, and their remarkable revelations, deserve comment. His interest in the children of irradiated women stemmed from experiments with ovarian-irradiated rats, with their emphasis on offspring ill-health (Goldstein 1928, Murphy 1928). These studies were never published, but it seems prompted a review of numerous experimental studies, which confirmed his suspicion of the harmful effects of maternal therapeutic irradiation (Murphy 1928). Which in turn were supported by the two dozen or so scattered articles that had appeared in the first 8–9 years of the decade noting the unhealthy children of such women, notably with microcephaly.

Remarkably, not for some years after Murphy announced this danger did it get the attention it deserved, sensitization to teratological events not yet having been awakened in the medical world. It was Murphy nevertheless who was the first to draw attention to the prenatal harmfulness of maternal irradiation, and the first to discover a human environmental teratogen. Murphy died in 1971 at 78 years of age, and is remembered as a busy physician, obstetrician, teacher, gardener, pipe smoker, but strangely not as a ground-breaker in the field of human teratology (Fields 1972).

Atomic Radiation

The Hiroshima and Nagasaki Bombs

Needless to say as the harmfulness of therapeutic pelvic irradiation of pregnant women became better known the practice declined. But such danger did not disappear, and another, a far more deadly source of irradiation, emerged as a cause of fetal defect: the atomic bombs dropped on Hiroshima and Nagasaki in August 1945. Study of these and other effects of radiation in atomic-bomb survivors were undertaken by the Atomic Bomb Casualty Commission, established in 1946 by US presidential directive.

In surviving children of pregnant women exposed to the bomb in Hiroshima only one malformation type, microcephaly, reduced mean head circumference, was significantly increased in frequency (Plummer 1952). The all-important variables, dose and timing, were soon looked into.

The defect occurred in seven of the 11 children of women who were within 1200 m of the hypocenter (the point directly beneath the bomb when it exploded) at 11–17 weeks of gestation, almost all mentally retarded at about age 4 years. A study in Nagasaki similarly found that mothers relatively close to the hypocenter had children with significantly reduced head size (Yamazaki et al. 1954). A search for skeletal abnormalities found none (Sutow and West 1955).

A more precise account of the Hiroshima findings noted microcephaly in almost 20% of surviving children whose mothers were at less than 1200–2200 m from the hypocenter, about two-thirds at 7–25 weeks of gestation, over half mentally retarded (Miller 1956). Overall growth and development were reduced, but head size was disproportionately affected (Blot 1975), and retarded physical development was not reversed by later growth (Wood et al. 1967).

In sum, apparently the only teratogenic effect of the fetal exposure was microcephaly, predominant sensitivity to which was during weeks 7–15 of gestation, and whose frequency and severity increased with dose, i.e. closeness to the explosion (Miller and Blot 1972).

Microcephaly and Mental Retardation

More complicated was the situation with regard to mental retardation per se (Miller 1956). Of ten children exposed in early pregnancy, at 6 weeks or less, two were microcephalic – but not mentally retarded (the sparse number perhaps due to undetected prenatal elimination of most irradiated this early, or to the nature of the original plan of patient selection).

The results seemed to contradict a venerable and widely held belief, that "...children with a head circumference below minus two standard deviations from the mean ... are, probably with few exceptions, mentally subnormal" (O'Connell et al. 1965). There seem to be more than a few exceptions to this generalization.

For example, looking into the relation of microcephaly and intelligence Dorman (1991) commented that "... microcephaly reflects a pathological change in brain structure, usually occurring in early fetal life, with an effect upon intelligence dependent upon the extent and type of underlying pathology...[but] reduction in brain size without such pathology ... does not affect intelligence."

Additional considerations regarding brain structure clarified the question. The minimum dose producing an effect at less than 18 weeks of gestation in Hiroshima was 10–19 rad, but in Nagasaki there was no consistent effect below 150 rad, the discrepancy perhaps due to the difference in radiation quality of the two bombs (Miller 1968, Miller and Blot 1972). Also, for the first time, the subject of the cellular basis of head-size reduction was brought up, but only in passing.

Dekaban (1968), alluding to this matter, noted that "microcephaly reflects abnormal smallness of the brain." (As an aside, since the latter is secondary to reduced brain size, Warkany 1971, p. 237, advocated the more accurate name micrencephaly. Miller 1999 avoided the semantics entirely by suggesting the term "small head circumference.") Animal studies found that radiation destroys brain cells, which in early embryos inhibits expansion of the skull and causes small head size; and with increasing cell death, mental retardation in humans. This scheme was later realized to be an oversimplification of the histopathological events, but without seriously modifying the understanding of the fundamental processes.

A closer analysis explained that while microcephaly resulted from exposure at all stages up to 25 weeks of pregnancy, especially 8–15 weeks, exposure before 8 weeks did not cause mental retardation, the reason being the difference in brain cell composition and possibly in cell behavior or replenishment at various times (Otake and Schull 1984).

The story seems to have come to a rest here, except for a statistical examination of the data which suggested the presence of a threshold in the low-dose area, i.e. a dose below which mental retardation did not occur, of about 0.57 Gy, much larger than was originally calculated (Otake et al. 1991, 1996, Miller 1993, 1999, Schull and Otake 1999).

In essence then ionizing irradiation, whether of medical or military origin, displayed its target specificity by almost exclusively causing microcephaly. And its effects were time- and dose-limited, the former since the abnormality occurred virtually entirely through exposure in the 6th–11th weeks of pregnancy; and the latter, as calculated by the relation of offspring head circumference to maternal distance from the Hiroshima atomic-bomb hypocenter (Miller and Blot 1972, Miller and Mulvihill 1976).

In conclusion, as was the case with the therapeutic studies, the only proven congenital abnormality produced by atomic bomb irradiation was microcephaly, whose degree of occurrence was proportional to the dose received.

Parenthetically it should be mentioned that no statistically demonstrable increase of major congenital malformations, in toto or of any specific type, was found among the later-conceived children of atomic-bomb survivors in Hiroshima and Nagasaki, i.e. due to mutations induced in parental gonads (Neel and Schull 1991).

Radiation of innumerable types and exotic sources, hardly worth enumerating, have been suspected or accused of being teratogenic in more recent years, but none proven to be a human teratogen.

The most important outcome of the discovery of the teratogenic potential of ionizing irradiation was in showing that human embryos are not shielded from deformative environmental forces, man-made or otherwise. But the full meaning of this vulnerability was not immediately grasped or even heeded, since the harmful effects of irradiation were considered unphysiological and irrelevant for humans.

The next human environmental teratogen discovered could hardly be so accused.

Rubella

The belief that human embryos were invulnerable to ordinary sources of damage, as distinct from the alien danger of x-irradiation, was changed forever when a momentous discovery was made implicating a mild and familiar environmental circumstance, one whose potential harmfulness had never previously even in the least been suspected. A discovery, remarkably, that coincided with the time of publication of the earliest experimental teratological findings of human relevance.

The Discovery

Early in the year 1941, at the Royal Alexandra Hospital for Children in Sydney, something extraordinary was seen, "an unusual number of cases of congenital cataract . . . what might almost be regarded as a mild epidemic" (Gregg 1941). Having made this observation, the writer, Norman McAlister Gregg, by communicating with fellow ophthalmologists, learned that the condition had also been seen in several other, widely separated, regions of Australia, with the suggestion of a common etiology.

Gregg's description of the cataract pointed to a defect specific and characteristic in form, different in his opinion morphologically from previously depicted congenital lenticular opacities. The process causing the cataracts, because they seemed to involve all but the outermost layers of the lens, he considered to have begun early in embryonic life. In addition they were very often accompanied by microphthalmia, which may have been due to inhibition of development of the eye by the cataract, as suggested by their frequent same-sidedness.

Other malformations were also found, deafness and microcephaly in a small number of children, and congenital defects of the heart in "an extremely high percentage of the babies," identified in several autopsies as a widely patent ductus arteriosus with foramen ovale combined with interventricular septal defect (Swan et al. 1943).

The German Measles Epidemic

Gregg soon deduced, and maternal histories diligently probed by him confirmed, that it was no coincidence that the pregnancies bearing the affected babies dated to a time of maximum intensity of an unusually severe epidemic of so-called German measles or rubella that had swept Australia in 1940. His keen observations and astute reasoning had led to the discovery of the first "natural" human teratogen (Gregg 1941). Greeted at first with mild skepticism (Anon 1941), and a notation in the Lancet that he said "resembled a verdict of the Scottish courts, 'not proven'" (Gregg 1944), the association between the syndrome and rubella was before long confirmed and widely accepted.

Teratological Principles

Gregg had not only made a profound discovery, but also revealed a fundamental teratological principle, only provisionally disclosed by Murphy's findings, which could clearly be seen to apply to humans, as radiation studies had shown it did to animals. His observation that "in the vast majority of the cases infection occurred either in the 1st or 2nd month of pregnancy. . .," coupled with finding that later infection may cause various abnormalities but not malformations, was the earliest statement of the cardinal precept that it is principally if not only during the organogenetic period of prenatal development that malformations may be induced.

The prenatal effects of the rubella virus also exemplified a second major tenet of teratology, that the malformative effects of a teratogenic agent are determined by its properties in conjunction with the age of the embryo at the time of exposure to it. It is this combination that, acting on a sensitive genotype, produces the particular abnormality or set of abnormalities that characterizes each teratogen; a principle that prepared the way for later applications in clinical and experimental teratology.

Timing and Malformation Pattern

Maternal rubella was associated with a set of defects – eye, heart, microcephaly, and deafness – but their timing was unclear. To understand this pattern knowledge was required of the time of onset of the infection. Where this was known the record showed that the disease occurred before the 5th month of pregnancy, and in 85% of these within the first 3 months (Gregg et al. 1945).

With respect to the individual abnormalities, again where timing was known, 91% of the eye defects occurred before the end of the 2nd month and 60% of the instances of deafness and heart defects, alone or together, after the 2nd month, though diagnosis of deafness may have been incomplete. A later prospective study of older children exposed in the first 16 weeks of gestation to rubella confirmed that the extent of hearing loss had been underestimated; and also noted that despite this handicap deaf children were normal in social adjustment and intelligence (Sheridan 1964).

Many children also had some degree of microcephaly, but whether it accompanied the frequent growth retardation is uncertain. Reduced birthweight however was later found to be transitory, as microcephaly may also have been (Cooper 1975). Finally, in older children head size was correlated with stature but poorly with intellect, the children thus possibly being generally small rather than microcephalic (Macfarlane et al. 1975).

In sum, while there was overlap in the time of induction of the different abnormalities, it was concluded that there appeared to be "a definite relationship ... between the time of onset of the maternal infection during pregnancy and the type of defect...." (Gregg et al. 1945).

Additional information about the relation between the time of infection and malformation, gathered through various means, showed with few exceptions that cataracts occurred after infection at various times within the first 8 weeks of pregnancy and heart malformations and deafness within the first 12 weeks (Swan et al. 1943, Swan and Tostevin 1946). Which means that the relation between the time of infection and malformation type was wide and overlapping, and only in part coincident.

A British study years later found a largely similar pattern (Munro et al. 1987). In malformed children with confirmed maternal infection and confident assignment of gestational stage infected, eye and heart defects were induced during weeks 3–12 of gestation, and deafness almost always during weeks 3–16. Thus, while susceptibility to the infection appeared to be confined to the first 12–16 weeks of gestation, within this period induction of particular malformations, or even of specific types, was not narrowly time restricted.

There are several possible reasons for these fuzzy temporal relations: poor knowledge of pregnancy onset, of the exact time of maternal infection during pregnancy, of when the virus reaches the embryo, etc. But probably the main reason, which did not become evident until the 1960s, when the rubella virus was isolated, is that the syndrome is associated with a chronic fetal viral infection. It then became clear that the teratological consequences were due to viral tissue invasion during extended sensitive stages of individual organ development and hence little delimited by morphogenetic constraints, within the broad overall sensitive period. Thus for rubella the syndrome was more the product of the agent than of the time of its action, more so perhaps than for most noninfectious teratogens.

For a compendiously full summary, in small font, of the pathological consequences to the offspring of maternal infection with this disease, one can hardly do better than consult Warkany's modestly entitled encyclopedic tome of 1971.

The 1964 Epidemic

The extent of the damage this disease can inflict became better able to be gauged following an epidemic of rubella in Sweden in 1951 (Lundström 1962), and even more so after the epidemic in the US in 1964, whose devastating effects can be judged from studies of the resulting 20,000–30,000 children with rubella-associated malformations (Cooper 1975). These included various transitory effects, such as low birthweight, hepatosplenomegaly, etc., and permanent manifestations, the classical malformations present at birth, if sometimes only covertly, like cataract, deafness, and heart abnormalities, and others not clinically apparent till some years afterward. It was not always clear as well whether some transient and later-appearing conditions were prenatal or postnatal in origin.

An incidental note in Lundström's (1962) detailed monograph underscored what has been a predominant finding with regard to environmental forces, namely that even when they can derail early embryonic development they are usually incapable of affecting the rate of spontaneous abortion.

Time Versus Agent: The 'Critical' Period

This is an opportune place to interject a reflection on the question of the relative importance of agent versus time in determining teratological outcome; and especially on the concept of the 'critical period,' topics which were of concern at times in the past when philosophical and theoretical matters were of more interest than they generally are today.

Swan and Tostevin (1946), in replying to critics who suggested that the association between rubella and congenital defects may have been fortuitous, noted the elements that clearly refuted this challenge: time limitation and defect specificity. To strengthen their argument, without looking into the implication of his statement it would seem, they quoted Stockard (1921) who, based on his experimental work with the common minnow *Fundulus heteroclitus*, was perhaps the first to maintain the principle that "the type of abnormality is determined by the particular developmental 'moment' at which the noxa acts." Put explicitly, this means that the specific abnormality that results from disruption of normal developmental processes is determined solely by when – the critical moment – during development the disruption occurs, regardless of what the injurious agent is. Whether the latter is so was the crux of the question.

The Debate

A discussion – perhaps better characterized as a debate – that took place in 1953 gives the flavor of the differences of opinion swirling about this question in those days (see discussion following the article by Warkany 1954 for the entire exchange). It began with the pathologist Peter Gruenwald declaring "I am glad to hear that the idea of the critical period being the only determining factor is losing ground. Particularly in the medical literature on rubella, there is considerable misuse of this concept when people claim that one thing happens during the 2nd month, and another after the third.... I think it should be made clear that the specific influence of a given agent is in many instances more important than the exact time with which it acts, particularly because [this is an important point] some of these agents apparently act for a long period of time."

The chairman of the meeting, the experimental embryologist Paul Weiss, answered. "At the same time is it not a truism that if we choose to designate the period during which a tissue can be affected as the 'critical' period, then that tissue can be affected only during the critical period? The question, therefore, is merely whether this 'critical' period is the same for all agents ... Apparently this is not the case, but there are still some critical periods." To which Gruenwald noted that biologically it must be assumed that "both factors act together; in some instances the critical period is more important and in some instances the agent is more important..." Weiss, though, demurred: "There is no point to the question as to which of the two is more important. If you take either one away, you get nothing."

That seemed to put a halt to the discussion; until, some time later, Liane Russell, a mammalian geneticist, reintroducing it, said "I should like to emphasize that this phrase [critical period] is used by us merely to designate an end result, i.e. the developmental stage at which [an agent] has to be applied to produce a given change. The fact that different patterns of sensitivity may be demonstrated by the use of different teratogens is quite compatible with the concept of 'critical periods' as used by us."

This had gone on for some time, it seems, when Josef Warkany, a physician as well as an experimentalist, emphasizing a pragmatic view, broke in. "I should like to state here why some of us are so concerned about the indiscriminate use of the term 'critical period.' We are concerned about the ... misinterpretation of the idea ... as used at the present time in the medical literature. There you find statements which indicate not only that there are critical periods for the origin of certain malformations, but it is also asserted that one can ascertain from a malformation the time at which the injurious agent acted." He continued...

> This is not just a point of theory. Let us assume that the symptoms of a clinical syndrome can be attributed to arrest of development of several structures in the eighth week of fetal life. Does that permit the conclusion that the arrest of development is due to environmental interference at the time the fetus was eight weeks old? ... There are statements to that effect in the recent medical literature. The morphology of congenital anomalies does not permit us to draw definite conclusions as to their date of origin.
>
> There are practical considerations involved. If the simplified idea of 'critical periods' should reach the lay public we must expect serious consequences. If, for instance, a woman who had an automobile accident during the seventh week of pregnancy, gives birth to a Mongol child [the name given to Down children in a less sensitive time], she could sue the taxi company. [Warkany had in mind a specific case, in which this very charge was made (Ingalls 1947).] [We can state] that a malformation must have been caused before a certain time but one cannot say how early before the 'critical period' the injurious agent acted...

He explained further, at a later time (Warkany 1971, p. 49), by saying that the usual concept of the critical period is fallacious, since what is for all practical purposes the same malformation can result from insults at different times during the sensitive period, i.e. before the latest time a given malformation can be produced, designated the 'termination period.' This is the concept fully accepted today. Questions of agent-specificity and time-specificity with especial emphasis on the experimental point of view were explored in detail by Wilson (1957).

As things will do, the question of critical periods has lately resurfaced, in the form of an article devoted to its reexamination; this time, however, delimited to what was called: 'specified' periods, apparently thought to be more 'scientific based,' but whose rationale and heuristic usefulness were beyond my discernment (Czeizel et al. 2008a).

Gestational Age and Frequency

To return to rubella, it was learned early that malformation type was loosely determined by time of maternal infection, infection in the first 2–3 months of pregnancy leading to eye and heart defects, in the 2nd trimester to deafness, microcephaly, and mental retardation (Swan et al. 1943); and later fetal and neonatal exposure to immunological and various tissue injuries (Monif 1970, p. 106, South and Sever 1985). The validity of these findings, suspected as being exaggerated because of being established retrospectively (Bass 1952), was later vindicated by a prospective study (Miller et al. 1982).

Time of infection also determined abnormality frequency, the earlier generally the greater. But another risk component, dose – contrary to its usual importance in nonviral teratogenesisis – appears not to be involved: malformation type, severity, or frequency was not related to the intensity of the maternal disease (Anderson 1950).

A more exact delineation of the congenital risk was made following an outbreak of rubella in 1978 in England and Wales (Miller et al. 1982). Heart malformation and cataract were present only in infants infected in the first 10 weeks and deafness in all 16 weeks, frequency decreasing with time of infection in the first trimester. The figures are less certain than they appear to be, based as they were on a small number, owing to many pregnancies having been terminated.

These temporal relations were supported by a fuller evaluation of a large number of affected children, with decreasing frequency and wide periods of susceptibility again found, heart, eye, and central nervous system malformations after exposure in weeks 3–12 and deafness weeks 3–17, with none affected after week 18 (Munro et al. 1987).

The very early finding that in neonates cardiac defects far exceeded ocular ones (e.g. Swan and Tostevin 1946) was also confirmed later. The frequent deafness due to rubella, it must be remembered, is a later manifestation, not usually recognized until the children are 1–2 years old (Gregg et al. 1945, Sheridan 1964).

An Old Disease

In adults rubella is a mild disease, causing fever, a rash, and achy joints. The name means red, from rothelm, and was called German measles from the time it was first differentiated from rubeola in Germany in early 1800s.

It is also an old disease, despite having first been clinically recognized only quite recently (Griffith 1887, Forbes 1969, Cooper 1985). Epidemics of rubella no doubt recurred periodically throughout history, as was detailed by Griffith (1887) for Europe, and Cooper (1975) for New York City. And untold thousands of pregnant women in long past ages contracted this infection, yet it was only the day before yesterday that the wartime conditions in Australia made the discovery possible that rubella is teratogenic. One writer noted that "nowhere in the literature prior to 1941 have I been able to find any trace of evidence that rubella in the course of pregnancy is a source of danger to the fetus" (Wesselhoeft 1947); though it was occasionally postulated, looking backwards, that outbreaks of births of deaf children had been associated with rubella epidemics (Lancaster 1954). The prenatal damage this otherwise trivial disease is capable of doing was forcefully demonstrated by the aforementioned severe 1964 rubella epidemic in the US, which caused harm to an estimated 30,000 infants, the last such event before mass immunization programs were instituted (Cooper 1968).

A Teratogen Disappears

Rubella is the second example, after irradiation, but the far more important one, of the virtual disappearance of a cause of human teratogenicity following its recognition. The isolation of the rubella virus and development of the vaccine that it permitted, followed by mass vaccination, all but wiped out the congenital rubella syndrome (South and Sever 1985, Anon 1989, Tookey and Peckham 1999); not forgetting that certain immunization patterns still pose a risk of return of the syndrome (Panagiotopoulos et al. 1999).

Apparently not much of one however. A study in Toronto found no evidence of the congenital rubella syndrome, nor of effects on birthweight and developmental milestones, in children of women exposed to the vaccine soon before and soon after the periconception period (Bar-Oz et al. 2005). And the same was true in vaccination studies in Iran and Brazil (Hamkar et al. 2006, Minussi et al. 2008).

The transcendent importance of this accomplishment is due – in distinction to the other human teratogens discovered in the twentieth century – to the conquest of a natural and ubiquitous accompaniment of human existence, not something newly introduced by man.

It took Norman Gregg's rare qualities, and the coincidence of the spread of the infection by wartime conditions of urban crowding, to bring this previously little regarded disease to prominence (Dods 1966, Burgess 1991). Gregg was 50 years old when he made this discovery, the "most important contribution ever made to medicine in Australia" (Burnet 1985), made furthermore while he was engaged in a busy private and hospital medical career, a gigantic onetime achievement that won him many honors during the very active 25 remaining years of his life.

Later Studies

Now that it seems rubella as a teratogen is no longer a threat, and it is to be hoped never again will be, studies have turned to looking at its cellular and molecular basis, with of course a variety of theories, concerning which it is premature to venture into (Lee and Bowden 2000, Atreya et al. 2004, Cao et al. 2008), so we won't.

Other Infectious Diseases

Influenza

It is understandable that the remarkable discovery of the teratogenic effects of the rubella virus prompted much speculation about and investigation into the possible prenatal harmfulness of other human infectious diseases (e.g. Brown 1966, Kurent and Sever 1977). Very early suspected, and easy to study because of its common recurrence and great frequency, was influenza.

Soon reported, in fact, was an association with central nervous system malformations, particularly anencephalus (Coffey and Jessop 1955, 1959, Saxén et al. 1990), but the association was weak and further evidence contradictory (Dudgeon 1976, Saxén et al. 1990, Lynberg et al. 1994, Arvin and Maldonado 1995). The current impression is that the influenza virus or viruses are weak teratogens at best, which is consistent with the rarity of their causing intrauterine infection, and the fact that their infrequent, supposedly prenatal, effects form no recognizable pattern (Arvin and Maldonado 1995).

Déjà vu again. Now from northern China comes a retrospective inquiry finding a high risk for neural tube defects (NTD; this entity described fully below) associated with maternal fever and influenza, especially accompanied by the use of antipyretics (Li et al. 2007). Many questions unanswered.

Cytomegalovirus

In contrast are two infectious diseases, cytomegalovirus and toxoplasmosis, which are usually transmitted placentally and are without doubt responsible for abnormality patterns (Alford et al. 1983). Cytomegalovirus is the commonest human intrauterine infection, affecting about 1% of all live births in the US. Ten percent of affected infants show congenital signs of the disease, many with micrencephaly and microcephaly, but no other malformation, probably because first trimester transmission is rare; the great majority however develop severe mental and physical handicaps by early childhood (Stagno 1995).

Toxoplasmosis

Congenital toxoplasmosis, due to transplacental infection with the protozoan parasite *Toxoplasma gondii*, is also fairly common, affecting 1–3/1000 live births in the US. Only about 10% of infected newborns have congenital signs of the condition, especially hydrocephalus, chorioretinitis, and cerebral calcification; but by early childhood a large majority develop severe physical and mental abnormalities. None of these effects can be called true malformations however, which conforms with the fact that maternal infection before week 10 of pregnancy rarely leads to severely damaged infants (Alford et al. 1983, Remington et al. 1995).

Varicella-zoster Virus

The varicella-zoster virus has been associated with fetal damage, though infrequently. The typical signs of the congenital varicella syndrome are skin scarring, limb hypoplasia, and eye and brain damage – none considered malformations – all probably largely attributable to viral invasion and damage of neural ganglia and the spinal cord (Alkalay et al. 1987, Higa et al. 1987, Gershon 1995).

Though the virus has been suspected of such consequences from the mid-1940s very few affected pregnancies have been identified with certainty, many women being immune to the virus, having had chickenpox as children. Recent estimates of the risk have found it scanty.

Problems Regarding Intrauterine Infection

The intrauterine infections differ from one another in the specificities of the timing of the damage as well as in the types of damage they cause. Rubella interferes with developmental processes mostly during the earliest weeks of pregnancy. Varicella-zoster leads to abnormalities predominantly in the second trimester. Cytomegalovirus and toxoplasmosis cause abnormalities in fetal and postnatal life, from ongoing inflammatory and necrotizing tissue damage. In addition, rubella, and perhaps all fetal viral infections result in chronic states, with various neurological and other repercussions, which may persist for long periods after birth (South and Sever 1985). What accounts for these the differences it seems has not been addressed.

Another question that has been left in limbo is how to classify the abnormalities caused by intrauterine infections. Warkany (1947) glimpsed this difficulty when he wrote that the hydrocephalus, chorioretinitis, and so on due to toxoplasmosis "are not the result of arrested development in the period of organogenesis, but rather the outcome of a prenatal disease which affects the fetus in the growth period. The absence of major developmental defects indicates that the infective agent invades the fetus late in prenatal life."

This is equally true of the defects caused by rubella infection and cytomegalovirus later than the first trimester. But the difficulty of classification even extends to the cardiovascular malformations associated with rubella infection in the first trimester, since they are traceable to viral invasive processes causing not conventional malformations but damage of vessel walls (Korones 1986). See also other thoughts, e.g. those considering that rubella affects the heart along one or more of several pathways (Rosenberg 1987). But even damage associated with varicella-zoster, which though originating in the late organogenetic period appears to stem not from disruption of developmental processes but from effects of viral invasion of fetal sites. This distinction also applies to the defects associated with varicella infection named above, different in developmental process and usually in appearance from conventional congenital malformations (Alkalay et al. 1987).

Intrauterine Infection in Animals

In the years before rubella became a preventable disease it seemed of practical value to study the possible prenatal consequences of the virus in animals. Afterward its practical benefits evaporated, which may explain why the few tentative findings were not followed up and have remained inconclusive.

Rubella studies have been made with macaque monkeys, baboons, hamsters, rats, mice, rabbits, and ferrets. The findings were varied and inconsistent. Transplacental passage of virus seldom occurred, but when it did infection often led to fetal growth retardation, abortion, and neonatal death. Some young fetal monkeys had histological changes in ear, eye, and skin, and rats had lenticular opacities and atrophic interventricular septa, but none of these outcomes was independently confirmed. In other words, the teratogenicity of the rubella virus in non-human animals is unproven (Oxford and Sutton 1968, Elizan et al. 1969, Elizan and Fabiyi 1970, Cotlier 1972).

Similar efforts were made with numerous other viruses, human and animal. Prenatal infection with cytomegalovirus was attempted only in mice, with no conclusive evidence that the virus crossed the placenta and caused fetal maldevelopment. Studies in mice with influenza A virus were similarly unsuccessful, as was true of other viruses, which though inducing congenital infection in some animals were not teratogenic (Elizan and Fabiyi 1970). A late finding regarding maternal influenza infection in mice concluded that whatever its effects on fetal brain development may be are likely to be indirect, since viral RNAs were not detected in fetal brains from infected mothers (Shi et al. 2005).

Rubella thus remains unique among the viral infections, in its ability to cause multiple malformations in human embryos, in the high frequency of the abnormalities it causes, in its high infectivity leading to epidemics, and finally for inciting research that led to a successful vaccine program and the eventual virtual disappearance of a human teratogen.

Pioneering Experimental Studies

The discovery early in the twentieth century that x-irradiation caused abnormal prenatal development in laboratory mammals received little attention. And even when it was found that irradiation could cause malformations in humans, the enormous implication – that the supposed invulnerability of human embryos to environmental forces was fallacious – was little appreciated.

But studies in a remote corner of the US changed all that. The newest hazard, different in kind from rubella, was a quantitative one, and originated not from the outside but was part of the organism, maternal and fetal, itself.

Vitamin Deficiency

Hale and Vitamin A Deficiency

It is discoveries in out-of-the-way central Texas that were the beginning of experimental mammalian teratology. These are the discoveries by one Fred Hale in the early 1930s at the Texas Agricultural Experiment Station in College Station, which, in time-honored scientific tradition, were entirely fortuitous (Hale 1933).

Hale had received degrees in animal husbandry, specializing in swine nutrition, and was concerned with the effects of vitamin A on his favorite subjects, when he came upon an area of missing knowledge. He had noted that while "numerous publications have appeared relative to the effect of vitamin A from birth to maturity literature is lacking concerning the relation of maternal vitamin-A deficiency to embryonic development." He decided to fill this gap by making pregnant female pigs severely vitamin A deficient by dietary means, and as he later said, "unexpectedly" found that their offspring had various congenital malformations.

A brief note reported his preliminary findings. A registered Duroc-Jersey gilt (for the non-agriculturist, a young sow), fed a vitamin A deficient ration for several months before breeding, on March 29, 1932 farrowed 11 pigs, "all of which were born without eye balls," as was determined macroscopically (Hale 1933).

He realized that because of the diehard convictions of the age he would need to give good reasons why the condition could not be hereditary; but since all the

H. Kalter, *Teratology in the Twentieth Century Plus Ten*,
DOI 10.1007/978-90-481-8820-8_3,

offspring were stillborn or died within a few days proving this by breeding them was foreclosed. This first report was soon followed by others of more extensive and better controlled experiments (Hale 1935a, 1937), which reproduced the ophthalmic defects, and noted that other types of malformations also occurred, cleft lip, cleft palate, malformed hind legs, accessory ears, cryptorchidism, and ectopic ovaries and kidneys.

[Parenthetically, for the sake of historical completeness it should be noted that a communication, made over 10 years previously, was actually the first report of malformations produced in mammals by environmental means (Zilva et al. 1921). A pregnant pig fed a diet highly deficient in what was then called the fat-soluble factor – now known to be vitamins A and D – had a litter of eight stillborn or soon dying offspring with malformed hindlimbs, "represented by thin tail-like appendages." The authors were hesitant to ascribe the defects to the diet, but said that "further experiments were in progress," of which however no record has been found. Hale (1935b) himself also quoted observations, by Hughes and co-workers, from the Kansas Technical Bulletin 23, of "eye lesions developed in pig fetuses apparently because of the insufficient amount of vitamin A received by the mother during the first part of the gestation period."]

Various important details emerged in the course of his work, all to be confirmed by others in future studies. To cause malformations pregnant animals had to be severely vitamin depleted, but just short of jeopardizing their viability or fertility; deprivation had to continue into the stage of embryonic organ formation; various degrees of deprivation caused defects of a variety of severities, e.g. more severe, anophthalmia, less severe, microphthalmia.

By this work Fred Hale (who, incidentally, did not have a doctorate degree) demonstrated conclusively for the first time that an absence or insufficiency of an environmental factor – a nutritional element – of great and vital human necessity, could cause serious malformations in mammalian embryos. And he clinched the proposition with breeding and nutrition experiments that ruled out the possibility of genetic factors being responsible in any way for them.

Discovery Greeted Skeptically

Even so, the nutritional origin of the eye defects was greeted skeptically, as shown by an exchange of views following the delivery of his paper at a meeting of the Association for Research in Ophthalmology in Atlantic City on June 11, 1935 – in which he held his own very well (Hale 1935a). A skepticism, nevertheless, that was not fully squashed till years later when his studies with pigs were decisively confirmed (Palludan 1961). Pediatricians, on the other hand, a less obtuse breed than their ophthalmologic colleagues, soon received an objective summary of his findings, supporting the nutritional origin of the abnormalities (Anon 1934).

Hale realized the significance of his work, but cautioned against overinterpreting it: "It must not be forgotten, of course, that the nutritional conditions which brought about our litters of blind pigs are extreme and exaggerated. It would be almost impossible for an expectant mother to be as nearly depleted of vitamin A as were the animals in our experiments." But, he added, "on the other hand, vitamin A deficiency is by no means uncommon in human diet. . . ." An excellently balanced approach.

After these pioneering contributions, Hale disappeared from the teratology scene, as if to say his work needed no embellishment. He was born and spent almost his entire professional life in Texas, had a prominent career at the Texas A&M University, published extensively on swine nutrition, but, living to the age of 79 years, never again published on teratology (Anon 1981).

Warkany and Riboflavin Deficiency

Just a few years later, unaware at first of Hale's work, another investigator, Josef Warkany, opened a chapter in teratologic history that on the contrary he was to continue to be part of for over 50 years. In fact, though, he too blundered into teratology, coming upon his discovery by chance.

He was a pediatrician who never had a private practice, making his whole career in children's hospitals. He was born in 1902 in Vienna, Austria, attended its prestigious medical school, and when he found that the direction he wished to take, into pediatric research, was closed for him there, he came in the early 1930s to the US. At the Children's Hospital in Cincinnati, where he was made to feel welcome, at first he took on various jobs, which more and more turned toward developmental problems (Warkany 1988a).

As a youth, during the First World War, while on summer hikes into the mountains of lower Austria to attempt to supplement the family larder, he had seen people with cretinism, a severely debilitating congenital physical and mental condition afflicting persons born to mothers with endemic goiter. These experiences left him determined to study the disease by attempting to reproduce it in laboratory animals (Warkany 1971, p. 121).

He followed through on this ambition, when the opportunity arose, by feeding female rats a goitrogenic diet from immaturity through pregnancy, preventing its rachitogenic effects by periodic vitamin D supplementation. As expected, they became goitrous (Nelson and Warkany 1938), but unexpected was the effect produced in the offspring, not cretinism but skeletal and other malformations of a specific pattern (Warkany and Nelson 1940, 1941).

Searching for the Cause

A systematic search was then made to identify the basis of the outcome. It was not absence of iodine, as the maternal goiter suggested, since iodized salt added to the diet prevented the goiter but not the malformations. Focusing on the rachitogenic properties of the diet, its Ca:P ratios were modified, but with little benefit. Similarly, casein, manganese (whose deficiency in fowl had been found to cause micromelia, one of the defects produced in the rats), and numerous other ingredients, added to the diet one by one, were also ineffective (Warkany et al. 1942).

Adding a concentration of dried liver to the diet, in trying to modify its goitrogenic effects, did the trick. Females fed the mixture had offspring free of malformations (Warkany and Nelson 1941, Warkany et al. 1942). Years of trial and error were then spent seeing which of the numerous components of liver was the preventive one.

Attention turned to the recently available B vitamins; and of the different combinations of five of them added to the original diet, thiamin, niacin, pantothenate, pyridoxine, riboflavin, normal young were only had by females fed a combination containing riboflavin.

The crucial experiment was performed when a purified diet supplemented with the other four vitamins resulted in abnormal young, but the same diet supplemented with riboflavin proved preventive (Warkany and Schraffenberger 1943, 1944a). Ever cautious, Warkany refused to accept these findings until they had been confirmed in several independent laboratories (see references in Kalter and Warkany 1959).

The All-Important Details

These few lines summarize what were years of tedious work, each step of the way built on the previous ones (reconstructed by this writer, the experimenters themselves describing them only meagerly). The first task, duplicating Hale, was to find the delicate balance in young female rats that would achieve a degree of nutritional deficiency and not overly inhibit growth and sexual maturation; to disrupt normal embryonic development but not to the extent of killing all offspring. Merely to aspire to this purpose was to defy the prevailing belief that maternal stores of nutrients shielded embryos from external dangers.

In rebuttal of this opinion they found that "between the two there exists a narrow range in which maternal nutritional deficiency may result in arrest of the embryos' development without causing death. In this case congenitally deformed offspring may be the result" (Warkany and Schraffenberger 1944a).

Many difficulties had to be overcome. Depleted female rats did not always breed successfully. In some instances all offspring perished early in pregnancy. In others offspring were all normal. Only in a relatively few litters were there some malformed offspring. Offspring in first or second litters were not always abnormal, so females were bred repeatedly; and in successive pregnancies, as they became progressively more nutritionally depleted, malformations appeared and then increased in frequency.

This meant, of course, that females had to be allowed to give birth, and thus that newborn offspring had to be rescued, day or night, to prevent their mothers from mutilating or devouring the deformed ones. It cannot be proven, but there may have been another reason, at least at first, for breeding the females over and over again: scarcity of money and animals. Warkany once remarked that back in Austria, when as a young investigator he had asked his 'chief' for a rabbit to continue a study, the latter had replied, ‘but you already have one’! If this had left a shadow of a memory, it was a lucky one, since the frugality led to the discovery.

Later studies, built on these foundations, made many important discoveries, e.g. regarding the so-called sensitive developmental period: riboflavin fed to depleted females prevented malformations only when added to the maternal diet before the 14th day of gestation, an example of what was called the ‘termination period’ (Warkany 1971, p. 49), discussed above; a finding that led to pioneering studies of nutritional requirement for normal bone histogenesis (Warkany and Nelson 1942, Warkany and Schraffenberger 1943).

Cautions and Critics

Like Hale, Warkany urged caution in extending his results to maternal nutritional deficiencies in humans, at first only in a tentative note (Warkany 1944), but later more explicitly and inclusively, when he said “the fact that procedures have been found which permit the experimental production of malformations in mammals should not imply that similar factors determine malformations in man” (Warkany 1947).

Warkany, like Hale, also met skepticism as to the environmental basis of the abnormalities. Long-held beliefs of older generations in the environmental invulnerability of human and other mammalian embryos, that systemic, symmetrical, and familial malformations could only be explained by the action of abnormal genes (Baur et al. 1921, 1931), were not to be willingly relinquished.

Still, dogmas gave way as the older generation passed on – as unfortunately often seems to be required for outdated beliefs to vanish – and these studies served to solidify the growing acceptance that nutritional elements, components necessary for human postnatal health, were also vital for normal prenatal development, that their insufficiency could drastically impair prenatal development. These discoveries thus opened an entirely new chapter in human thinking and in medical concerns, establishing the field of embryonic or prenatal pathology, signaling that for human embryos the environment was not always benign, and that no doubt there were other disturbances to its normality and well-being to be wary of.

Early Experiments

New Needs and New Ideas

The decrease during the twentieth century in the rate of infant death from infectious disease made deaths associated with congenital malformations ever more conspicuous. This growing presence, catching the attention of investigators in various medical and biological disciplines, began to shape attitudes to new needs of pediatric investigation. And in the belief that understanding the fundamentals of the phenomena of malformations might help in the prevention of early deaths, work got underway in several parts of the world to accomplish this goal, and thus began the flowering of the field of experimental teratology.

The initial work concentrated on basic matters such as the effects of dose and timing of experimental intervention, as well as attempts to understand certain puzzling matters, such as inconsistency and variability of outcome. For example, why, even at uniform dosage and timing of the disruptive factor, the teratogen, the frequency of malformations varied, not all offspring were malformed, and abnormal ones had different types, severities, and combinations of defects.

The studies of this period, in the late 1940s and 1950s, laid the groundwork for the expanding field of experimental mammalian teratology (for detailed summaries of many aspects of which see Kalter and Warkany 1959 and Wilson 1959).

And from these experiments the basic concepts were derived which define teratologic susceptibility to environmental agents to this day. Namely, that susceptibility is determined by the genotype of the embryo and pregnant animal, by the embryonic stage exposed to the teratogen, and by the dose and properties of the teratogen, the responses being growth impairment, malformation, and death, their degree increasing with the dose of the teratogen.

Vitamin A and Diaphragmatic Hernia

The influence of the first of these principles, that of genotype, the genetic constitution of pregnant females and their embryos, was vividly demonstrated by a study that inadvertently discovered the interworking of teratogen and heredity and became a prototype of the phenomenon – and another example of serendipity.

H. Kalter, *Teratology in the Twentieth Century Plus Ten*,
DOI 10.1007/978-90-481-8820-8_4, © Springer Science+Business Media B.V. 2010

Dorothy H. Andersen was a pediatric pathologist with an early and abiding interest in cystic fibrosis of the pancreas (Andersen 1938, 1958), a serious autosomal recessive disease of infancy, common in whites, even today over half a century later not fully understood (Doull 2001). In performing autopsies on infants and children dying of the condition she had consistently noted destruction of the pancreas; and several lines of evidence and several hypothetical considerations led her to believe that vitamin A deficiency during pregnancy might be connected to this phenomenon.

To test this possibility she conducted an experiment. Female rats of a stock colony that had been maintained by random breeding were fed from a young age and throughout pregnancy diets containing the lowest level of vitamin A compatible with fertility and offspring viability. She was no doubt disappointed when no offspring had pancreatic changes suggestive of fibrocystic disease. But, unexpectedly, many were found to have something else, a congenital malformation, diaphragmatic hernia of a severe extent (Andersen 1941, 1949).

This was not so unusual since females fed a nutritionally adequate diet also had offspring with the abnormality, but only rarely. The remarkable thing here is that a spontaneous developmental abnormality, ordinarily seldom occurring, had been greatly increased in frequency by an environmental situation. [Which raises the by-the-way question of whether such augmentation might not be uncommon, with predisposition to developmental irregularity, ordinarily covert or negligibly expressed, made manifest by an imposed condition.]

That a predisposition was needed for the intensified expression, i.e. raised frequency, of diaphragmatic hernia, was shown when in experiments with a different breed, Long-Evans rats, in which the defect did not occur spontaneously, the dietary deficiency was virtually without this consequence. Evidence of genetic involvement was strengthened when induction of the defect by the vitamin deficiency was enhanced by selecting stock-colony females for breeding that had borne offspring with high frequencies of the abnormality. In the stock colony about 3% of control and 18% of experimental offspring had the defect, and selection increased these frequencies to 8% and 34% respectively.

As Andersen aptly commented, "this series of experiments provides one more example of failure to attain the initial objective of an experiment with an incidental discovery of greater interest than the one originally sought." She also warned that "it would be rash to infer that the deductions made from these experiments are directly applicable to the diaphragmatic hernia of man," echoing the caution voiced by others cited about overenthusiastic extrapolation of such experimental findings to human conditions. Yet, as the future was to exhibit, experimental revelations may sometimes have unexpected human relevance.

How little Andersen's "discovery of greater interest" was regarded by her pathology colleagues was indicated by a biographical sketch of her that did not mention this discovery at all (Collins 1995). Described as "her hand containing the everpresent cigarette," she died just short of 62 of lung cancer.

This chance discovery was the first example in mammalian teratology of a genetic tendency interacting with and reinforced by an environmental stimulus.

Coming when it did, it was of great influence and inspired the search for other such interactions, which were soon to be discovered. Similar phenomena in human beings, it may be mentioned, would be difficult to detect. An example may be deafness due to maternal rubella infection interacting with a genetic predisposition to the condition (Anderson et al. 1970).

The Power of Genetics

Further testimony of the power of the genetic component in teratogenesis was provided by differences between Andersen's results and those of Warkany and coworkers, both using vitamin A deficiency. In the former the only malformation described was diaphragmatic hernia, while in the latter many other types of abnormality occurred as well.

After his pioneering experiments with riboflavin deficiency Warkany turned to another problem, the repeated suggestion that congenital blindness in animals may be caused by maternal nutritional deficiency. Reversing Cannon's (1940) earlier failure, he confirmed Hale's findings that eye defects could be induced by deficiency of vitamin A (Warkany and Schraffenberger 1946); abnormalities that were soon expanded to include a wide variety of malformations, not only of the eye, but also of the cardiovascular, urogenital, and respiratory systems, as well as diaphragmatic hernia, such as to comprise a specific syndrome (Warkany et al. 1948, Wilson et al. 1953).

The difference between Andersen's and Warkany's findings were without doubt due to attributes inherent in the stocks of animals they used. Warkany and coworkers, using Albino Farms and Sprague-Dawley rats, induced the defects named above along with a high frequency of diaphragmatic hernia, but saw no defects in several hundred controls; while Andersen's 'home-grown' rat stock, with its small spontaneous percentage of diaphragmatic hernia, was amazingly resistant to induction of any other defect. It is impossible to imagine these disparities as being due to anything but the differences in the genetic makeup of the stocks of animals.

I digress for a moment to comment on the papers of Warkany and Wilson cited above: one is awed by the meticulous description, still unsurpassed, of the morphology and pathogenesis of the experimentally induced urogenital and cardiovascular malformations, undoubtedly mainly the work of Wilson.

James G. Wilson had graduated Ph.D. from Yale University in 1942, and was in the Department of Anatomy in Rochester, New York, when beginning in 1947 he spent several muggy Cincinnati pre-air-conditioner summers working with Warkany, joint endeavors that continued to benefit from his considerable embryological knowledge and observational skills when he joined the anatomy department at the medical school in Cincinnati in 1950. From those years he went on to a lengthy creative and productive career as an experimenter, teacher, mentor, writer, and editor in teratology (see especially his *Environment and Birth Defects* of 1973, and editorship with F.C. Fraser of the four-volume *Handbook of Teratology* of 1977).

Basis of Diaphragmatic Hernia

A study of the embryological basis of the diaphragmatic hernia found that it consisted of failure of the diaphragm to undergo the normal process of prenatal closure, the location of the gap, or discontinuity as anatomists term, varied but was most often right-sided. The complex normal formation of the diaphragm was described by Wilson et al. (1953).

Andersen offered two hypotheses for the pathogenesis of the defect. The first, less likely, is that it was due to the closure being prevented by the forced protrusion of abdominal organs. The second, more reasonable, she ascribed to delayed growth of the diaphragm, which aggravated by the dietary deficiency increased the occurrence of the defect [somewhat analogous as we shall see to the delayed development of the palate leading to its remaining open, ending as a cleft palate].

To end on a bit of a negative note it is necessary to consider the possibility that Andersen's major finding – that the vitamin deficiency potentiated a spontaneous abnormality – contained a flaw. The malformation in the colony controls occurred in the offspring of females that had received a regular supplement of vitamin A. Intriguing to ponder, is whether the low frequency of the defect means that the control females were still somewhat vitamin A deficient, and that a larger supplement would have abolished the defect entirely. Even determinations that showed that the liver content of vitamin A was many times greater in control females than in those not supplemented do not entirely satisfy the question whether a threshold of vitamin requirement for normal diaphragm development had not been reached.

The Early Investigators

Early in the story of modern teratology its practitioners came from a diversity of backgrounds, Hale animal husbandry, Warkany and Cohlan pediatrics, Andersen and Thiersch pathology, Wilson and Giroud anatomy and embryology, Lois Murphy chemotherapy, Fraser genetics, Marjorie Nelson nutrition. The diversity of professional backgrounds was to expand as individuals from many fields – mammalian and clinical genetics, developmental biology, numerous areas of medicine, etc., joined in the pursuit of the causes, development, and prevention of congenital malformations, each from his specialized outlook.

This variety and the interactions they created gave teratology its vigor; but has been its weakness also, with no one area predominating to provide a concentration of effort; and also with the field sometimes perceived as the refuge of persons enthusiastic but only temporarily engaged. Another outcome of this diversity is that no medical school and few research institutions have had departments emphasizing the goals of teratology, which has meant that few have been the facilities for training students in the field.

In the few years following the first studies there burst forth a small renaissance of experimental teratological studies, in which numerous instrumentalities were used to disturb and learn from disturbing prenatal morphological development (see Kalter and Warkany 1959). It is important to note however, while the relations between prenatal growth retardation, malformation, and fetal death were being clarified (Wilson 1973, Kalter 1980), that not every agent and procedure succeeded in causing abnormalities, that many substances applied to pregnant animals produced no teratological changes (Kalter and Warkany 1959, Kalter 1968).

Why particular agents were selected for study was not always stated or obvious and is now hardly to be discovered. Sometimes the choice had a reasoned basis, and led indeed to malformations, but the hypothesis it was based on was not always borne out. Two examples follow.

Trypan Blue

Trypan blue, a vital stain first synthesized by Paul Erlich in 1904, is so-called because it can kill trypanosomes, the parasites than cause sleeping sickness. It is commonly used to distinguish viable from nonviable cells, the former excluding the dye, the latter absorbing it.

Scientists at the University of the Witwatersrand, motivated by the prevalence of adverse perinatal outcomes in South Africa due to the malnutrition common at the time, hypothesized that chronic maternal malnutrition might cause abnormal fetal development by "steady flooding of the circulation with particles derived from abnormal metabolism." Finding that trypan blue caused alterations in the blood picture and plasma proteins of rats that simulated certain effects of malnutrition in humans, they administered this dye to pregnant animals, and found that it caused numerous malformations, among them those of the central nervous system (Gillman et al. 1948).

Joseph Gillman and his colleagues were anatomists, which explains their careful and detailed descriptions of the numerous induced malformations. They also showed imagination, in discussing the significance of their studies, by attempting to draw parallels between the possible mechanism of teratogenic action of trypan blue and that of the rubella virus.

It is in the attempts over the years to understand how teratogenic agents cause congenital malformations – their 'mechanism of action,' as one group called it, "that is, where, when, and how these agents affect the embryo" (Wilson et al. 1959) – that perhaps the greatest failure of experimental teratology lies. The very first of these efforts, those of Gillman and his colleagues to get to the root of trypan blue's teratogenicity, was followed by the disappointment that has beset most such efforts at plumbing the basis of fetal maldevelopment due to various teratogens since then, recounted below.

Trypan Blue's Teratogenic 'Mechanism'

As learned above, viable cells exclude trypan blue, hence it is not surprising that the dye did not reach embryonic tissues from the maternal circulation. Because of this exclusion the explanation for its teratogenicity offered by Gillman and his colleagues – disturbance of many maternal metabolic processes with prenatal maldevelopment as its consequence – seems hard to accept, but was plausible to them. And of course, no disturbed maternal locus was identified. But the studies went on.

They turned to the embryos when it was found that malformations occurred only when the dye was administered during a fairly limited prenatal interval (Gillman et al. 1951, Wilson et al. 1959), which with other evidence pointed to a direct action of the teratogen upon the embryo. What this action consisted of however was unclear.

Numerous theories addressing this question followed in the next 50 years, mainly invoking disturbed transfer mechanism between mother and embryo (Beck and Lloyd 1966) and altered fluid balance in and around the embryo (Rogers et al. 1985). While these ideas may have taken the trail a step further, they were inadequate to the task of resolving the nature of the proximate step, the one that leads directly to the induction of specific malformations. The problem, in fact, seems to have been too discouraging to invite continued pursuit, and no efforts daring enough to do so have been reported since the mid-1980s. A most difficult and puzzling episode in experimental teratology.

Hypoxia

Also chosen to be studied for what seemed to be a logical reason was hypoxia, i.e. reduced oxygen concentration. The logic happened to be fallacious, but the procedure once more turned out to be teratogenic. The hypothesis that "anoxia may be an important mechanism" was suggested by associations between chronic maternal disorders, e.g. gestational hemorrhage, and mongolism, as Down syndrome was then called, and other abnormalities (Ingalls 1947, 1953). These abnormalities, it was postulated, all had in common that the mothers had experienced lack of oxygen during pregnancy, as a result of rheumatic heart disease, vaginal bleeding, etc.

The theory led to experiments in which pregnant mice were exposed to severe degrees of hypoxia, which did indeed induce certain abnormalities, especially of the ribs and vertebrae, when applied at specific times during gestation (Ingalls et al. 1950). But to conclude from such experiments that abnormalities in children having some resemblance to the induced ones were due to their mothers' experiencing a supposed anoxic episode was far out.

The invalidity of the theory was made plain by Ingalls' attempt to apply it to Down syndrome. At that time the ingredient most distinctly associated with the syndrome was advanced maternal age (Penrose 1954), which according to Ingalls "gives proof of a maternal factor . . . in causation of the condition."

A few years later, before he became aware of the discovery that Down syndrome had a chromosomal etiology (Lejeune et al. 1959), Warkany (1960) reviewed the history of the futile search for its causation; among the 39 theories he enumerated, all soon to be relegated to the dustbin of history, were "diseases and accidents in early pregnancy."

A Little Break: Down Syndrome

I take a small break to talk about what has come to be called Down syndrome. It was originally known as the "Mongolian type of idiocy" by its "discoverer," the London physician John Langdon H. Down in a short paper published in 1866. There in a few words he described the condition and conjectured about its causation: "Had the nurse dosed the child with opium? Has the little one met with an accident?" But no, "They are always congenital idiots and never result from accidents after uterine life."

But it is to be wondered at that, this state in which so marked are the typical features "that when placed side by side, it is difficult to believe that [they] are not children of the same parents," was not described in medical writings long before the last half of the nineteenth century (Volpe 1986).

Down syndrome children are still being born, and because of demographic realities, perhaps have even increased in frequency. Knowledge of the cytogenetic basis of the condition has not been enough to change that. As was written years ago, "[I]n medicine the riddle of a disease is not solved until we learn how to prevent it." (Warkany 1975).

Cortisone Studies and By-Products

Returning to the question of why particular agents were chosen for experimentation – sometimes the choice was made for no better reason than 'some stuff' happened to be around and sounded promising. The following is one such example, of how an involved chapter in experimental teratology got its start. It was told by Fraser (1990) in an autobiographical sketch, that while he was preparing to explore Ingalls' hypoxia findings, as he told the story, " . . . a plastic surgeon, Dr. Hamilton ('Happy') Baxter, who was making his cleft lip and cleft palate patients available to me for genetic studies, happened to get hold of some cortisone, at that time a newly discovered wonder drug that nobody knew much about except that it was good for arthritis. He thought that since cortisone was a steroid, and the embryonic organizer was a steroid, maybe treatment of pregnant mice with cortisone would cause NTD [neural tube defects]. We didn't think much of the argument . . . but what did we have to lose?" It didn't cause NTD, but it was teratogenic (Baxter and Fraser 1950); and a long-running ball was thereby set rolling.

Fraser's aim was to use hypoxia – being a mammalian geneticist, as well as a newly minted M.D. – to counterbalance the opinion that malformations were mostly due to environmental factors. He would demonstrate the contrary by using for his experimental subjects strains, as they were called, of highly inbred mice, each with a characteristic genetic constitution. These, the supposition went, with their individual physiological and developmental proclivities would yield individual patterns of malformations; and thus demonstrate that heredity was a major actor in experimental teratology.

Administering donated cortisone to pregnant mice, not different types of abnormalities, and not NTD, but a single malformation, cleft palate, was induced. But vindicating that genetics had a part in defining prenatal sensitivity to environmental forces the frequency of this abnormality varied greatly among the different strains used, as even a preliminary report noted (Fraser and Fainstat 1951).

Relevance to Humans

Before continuing, it should be asked of what relevance to people is the finding that cortisone is teratogenic in mice. If the question is taken to mean, did this discovery foretell human teratogenic susceptiblity to corticosteroids, the answer is a guarded no, guarded because that's how scientists express themselves.

Over 40 years had to elapse before this long-lingering question was resolved. Of the possible prenatal effects of the frequently used corticosteroid drugs in humans, in ever-cautious scientific parlance, it was concluded that such medications present so low a teratogenic potential as to be undetectable (Fraser and Sajoo 1995, Czeizel and Rockenbauer 1997). Later reports claimed a supposed increase of cleft lip (Park-Wyllie et al. 2000, Pradat et al. 2003), mistakenly equating that defect with the cleft palate induced by corticosteroids in animals.

But, to return to the matter of relevance, if the question asked above is taken to mean, of what relevance was the discovery in furthering teratology research, the answer is different. Probing metabolic aspects and especially the embryological basis of the differences in strain sensitivity to the teratogen (Walker and Fraser 1956, 1957) led to new ways of thinking about developmental processes (Fraser 1965, 1976), which proved to be useful in understanding human problems (Fraser 1996); as will be seen later on.

Induced and Spontaneous Malformations

One of the mouse strains used, the A/J strain, in addition to being highly susceptible to the induction of cleft palate by cortisone, displays a seemingly related defect spontaneously, cleft lip; which raised the question of whether this was a coincidence or not. Were induced malformations, as Andersen had found for diaphragmatic hernia, simply the outcome of intensified expression of spontaneous ones? Sometimes yes, but as time showed, not always.

Other investigators using inbred mouse strains to examine genetic influences on malformation induction made other discoveries. For example, Ingalls et al. (1953) found that the pattern of the relation between spontaneous defects and those induced by hypoxia varied in almost every way possible. Thus, one malformation, umbilical hernia, occurring in a small percentage of controls of two of the strains used, was greatly augmented in frequency by hypoxia in one of the strains but not the other; cranioschisis and cleft lip and palate, also with small spontaneous occurrences in some strains, were not increased in any; microphthalmia, seen in no controls, was induced by the treatment; and last, rib and vertebral abnormalities, occurring in a small spontaneous frequency in all the strains, were markedly increased in all but one of them. Obviously, no generalization could easily emerge from such dissimilarities.

Studies directed at learning which of the malformation doublets were identical in development and morphology and which not might have explained these disparities. But these are matters that were never explored, and the endeavor remained at this unfinished level, despite the promise the topic seemed to hold of clarifying certain developmental interrelations. But many other interests existed at this early stage of experimental teratology, and the few hands plowing that field had other concerns to turn to.

Genetics and Individual Responses

Another question left unanswered was the genetic basis of individual differences in teratologic response. Might not a genetic study of a single abnormality, such as cleft palate induced by cortisone, be more productive than one of the multiple abnormalities caused by hypoxia?

Early investigations of this question, exploiting the properties of inbred strains (Kalter 1978), revealed some of the analytical complexities, when the differences between strains in sensitivity to cortisone teratogenesis were seen to be controlled by complex genetic arrangements together with fetal physiological components (Kalter 1954, 1957). Studies of this question continued intensively for many years, growing ever more convoluted, with the final word probably still not written (Biddle and Fraser 1977, Vekemans 1982). Another frustrating end it would seem of the search for mechanism.

The Multifactorial/Threshold Concept

A by-product of the cortisone-cleft palate research was the formulation of the multifactorial/threshold concept. It evolved from studies by Bruce E. Walker, an early doctoral student of Fraser's, which, enabled by his deft skills, opened an important chapter in experimental teratology.

The embryology of the palate must first be described. Its formation begins with the appearance of the palatal processes, rudimentary projections hanging down from the cranial base and bordering the lateral margins of the tongue. At a certain stage in early fetal life the processes turn inward and come to lie above the tongue, where they form shelves, which grow, meet, and fuse to constitute the definitive palatal structure (Walker and Fraser 1956). How the transposition of the processes from vertical to horizontal occurs is not clear, but it probably involves coordinated developmental events in the shelves, tongue, mandible, and head, features mentioned again below.

Walker discovered that cortisone, in essence, delayed the movement and perhaps the growth of the palatal shelves, preventing them from meeting, resulting in a cleft (Walker and Fraser 1957). But of equal and perhaps even greater ultimate significance was his finding that during normal fetal life mouse strains differed in the timing of shelf elevation; and that the more laggard the normal process in a strain the more susceptible it was to the induction of cleft palate (Walker and Fraser 1956).

It was from the effort to integrate these two sets of observations and from conjectures about the process of palate formation, that the so-called multifactorial/threshold concept or model emerged. The idea of a threshold had been suggested at different times to explain aspects of human and animal development (Wright 1934, Grüneberg 1952, Carter 1961). But a grand summary of years of work made clear that Fraser's (1980a) formulation was novel, and furthermore pointed to a different way of addressing questions of embryotoxicity. Aside from that, the concept has found other applications. For example, the prediction that "in conditions appearing more often in one sex than the other, the sex ratio should change as the frequency changes" (Fraser 1971), has been vindicated by being found to apply to novel situations (Fraser 1998); and well might provide an opportunity to explore its implications regarding the sex difference in NTD frequency.

A New Concept of Embryotoxicity

It had been found, as noted, that among mouse strains the time of palate closure varied from strain to strain, being normally distributed relative to fetal developmental maturity. That is, strains had different but overlapping distributions, each with its own mean time of normal closure. By happy chance, the three strains used in the cortisone studies possessed individually distinct patterns of palate development and, hence, of closure times: in C57BL fetuses, with the most accelerated development, closure took place about 10–12 h before it did in A/J fetuses, the slowest, with the DBA fetuses intermediate (Walker and Fraser 1956).

The bottom line: if closure was not achieved by the time a critical developmental stage – the threshold – was reached cleft palate resulted (Trasler and Fraser 1977). Consequently, when the normal distribution was shifted, as it was by the delay caused by cortisone treatment, some fetuses failed to reach the threshold, and defects resulted, whose frequency was determined by the dynamics of normal palatal development.

A quantitative attribute, time of closure, was thereby transformed by the threshold into a qualitative one, a malformation (Fraser 1980a, 1996). It was thus shown that in addition to continuous and discontinuous variations in biology, there can exist quasi-continuous variations, which appear to be "discontinuous, but behave ... as if they reflected an underlying continuous distribution of a variable" (Fraser 1965).

But most startling of all, analysis of these relations revealed that the amount of delay caused by cortisone was the same in all the strains studied! Responsiveness – teratologic susceptibility in this case – it thus became apparent, may depend largely on inherent variabilities in rate of developmental processes. Other elements possibly also contributing to variable responsiveness – movement of tongue, jaw, etc. – are as yet hypothetical. It was thus shown that there exist developmental bases of variable adverse embryonic reactions to environmental forces that are independent of metabolic or pharmacological actions. The important question, of course, is how prevalent this maldevelopmental mechanism may be, a consideration that like so many others has been bypassed and left in limbo.

A Variety of Experiments

During the 1950s chemical substances of many sorts were found to have teratogenic properties (Kalter and Warkany 1959), and the reasons for their having been chosen for experimentation was that they often reflected the specialties of their discoverers. Lois Murphy, e.g., confronted with problems of treating cancer patients, examined cancer chemotherapeutic agents of the day (Murphy 1960); and Marjorie M. Nelson, a nutritional scientist, studied acute vitamin deficiency, including that of folic acid (Nelson 1960), discussed further below.

Very different were studies of large doses of vitamin A, which were found to cause central nervous system and other malformations in rats (Cohlan 1953, 1954). It was administered by oral intubation (nonoral routes were later found to be ineffective) over long periods of gestation, which accounted for the many types of malformations induced. These experiments, quickly confirmed and expanded (Giroud and Martinet 1955, 1956a, Kalter and Warkany 1961), led to many studies in later years. They also led to the numerous experiments with vitamin A congeners, the retinoic acids, discussed below, and from them to the recognition of the basic role of retinoids in normal embryonic development.

Vitamin Antimetabolites

A different class of chemicals, antimetabolites, were soon found to be powerful teratogens; and indeed the first discovered instance of the causation of malformations in human embryos by a man-made chemical was by such a substance. That story will continue below; first we go to experimental studies with these chemicals.

Studies of the role of nutritional elements in mammalian embryonic development accelerated in the last years of the 1940s, greatly aided by the discovery of structural analogues of vitamins. These agents, antagonists or antimetabolites, interfere with vitamin function, and when given to pregnant animals rapidly induce severe nutrient deficiency; eliminating the need for lengthy prepregnancy depletion.

One such chemical, galactoflavin, a riboflavin antimetabolite, allowed a greatly diversified investigation of the effects of vitamin deficiency, allowing quantitative studies relating dosage and teratogenicity, and enabling study of the effects of acute deficiency during selected periods of pregnancy (Baird et al. 1955, Nelson et al. 1956, Kalter and Warkany 1957).

Using vitamin antimetabolites created new opportunities for inducing and studying many aspects of a wide range and variety of malformations (Kalter and Warkany 1959); and fortuitously led to an avenue of human studies.

Folic Acid

It was studies of one such antimetabolite, that of pteroyglutamic acid, or folic acid as it is most commonly called, that led to interest of this vitamin in abnormal human development, an interest that has extended into the twenty-first century. In the first such experiment the antimetabolite, when administered to pregnant mice and rats at the time of uterine implantation or earlier, caused early prenatal death, but not malformation; suggesting an all-or-none action (Nelson and Evans 1949, Thiersch and Philips 1950).

This dire outcome led Thiersch to suggest this as a means of interrupting human pregnancy. Thus he gave 12 women (with illnesses for which at the time therapeutic abortion was indicated) 6–12 mg aminopterin (4-amino folic acid, a powerful antagonist) for several days during the first trimester. And indeed rapid fetal death resulted, with spontaneous delivery of the aborted conceptuses. But several of them had various malformations – hydrocephalus, cleft lip and palate, and myelomeningocele (Thiersch 1952). And in a second series of women one receiving aminopterin beginning at what was estimated to be the 17th day of pregnancy had a prematurely delivered child with anencephalus (Thiersch 1956, 1960).

As a result, for the first time, unwittingly, human embryos had been made malformed by a chemical substance, i.e. not by a deprivation, but by a positive force.

Folic Acid Antimetabolite Human Use

Despite this outcome aminopterin and an even more potent folic acid antimetabolite, methotrexate, have been used as abortifacients, with high doses taken before 8 weeks of gestation (Lloyd et al. 1999). But they too caused various malformations, especially a typical syndrome of skeletal defects (Warkany 1978a, Lloyd et al. 1999, Schardein 2000).

It is to be noted however that these folic acid antimetabolites did not cause central nervous system defects, neither anencephalus, spina bifida, or encephalocele and myelomeningocele (McElhatton 2000). Recent use of methotrexate as an abortifacient avoided possible teratogenic outcome by administering low-risk doses or early abortion of the failures (Davis et al. 1999, Borgatta et al. 2001).

It is of interest, in the light of later events, to note a comment in an early Ciba Foundation Symposium on Congenital Malformations (Wolstenholme and O'Connor 1960). Noting that folic acid deficiency had been found in 1–10% of pregnant women in Britain, Walker (1960) asked is "it likely that a maternal deficiency of an order such that it would only appear at midpregnancy, and only then on careful examination, would have any foetal deficiency effects early on?"

This inquiry, judiciously worded as it was, proved a foretelling of decades of pondering and study of the folic acid question, discussed fully below.

New Challenges

Infant Mortality and Malformation

Halfway through twentieth century medicine was ready for a reappraisal of its responsibilities toward children. This was made urgent by the changed spectrum of the causes of death in early childhood brought into focus by the decreased infant mortality rate. Deaths during the first year of life in the US had decreased over the preceding 50 years from 150 to about 29 per 1000 live births (US Bureau of the Census 1960), which had been achieved especially by advances in the control of infectious disease, the major cause of illness and death of the very young in that era.

As infant deaths diminished, those from congenital malformations, the most unyielding of the still prevailing causes of such deaths, grew ever more frequent, a trend that continues to the present day. The impact of these statistics is most telling if the denominator is deaths under 28 days, since most deaths related to congenital malformations occur during these earliest weeks. As US numbers show, in 1910 5.0% of deaths in the first 4 weeks were due to congenital malformations, in 1959–61, 12.3% (Anon 1965), and in 2004 20.2% (Mathews and MacDorman 2008); making them, dating from 50 or more years ago, the most frequent cause of early deaths in the US, and in other advanced countries of the world as well (Anon 1967).

Professionals in medicine and biological studies, governmental officials, and lay leaders, recognizing that these facts presented new challenges, sought new means of pursuing their solution. Bringing together individuals with common interests directed to this task was a beginning.

Teratology Conferences

The dominant role of malformations in early childhood death, coupled with the discoveries of the vulnerability of mammalian embryos to serious disturbance by environmental means, roused interest in abnormal prenatal development generally and in experimental teratology in particular. To discuss these matters several conferences were convened to chart ways of addressing them.

H. Kalter, *Teratology in the Twentieth Century Plus Ten*,
DOI 10.1007/978-90-481-8820-8_5, © Springer Science+Business Media B.V. 2010

The first was held in Cincinnati in January 1956. It was attended by individuals invited as much to consider the feasibility at that time of setting up a teratology society to promote the subject of prenatal abnormal development (in the end it was thought premature to do so) as to present their experimental and clinical work. Others followed – at the Jackson Laboratory in Bar Harbor, Maine in August 1956, in Bethesda at the National Institutes of Health in April 1957, Portland, Oregon in March 1959, Palm Beach in April 1959, and the Sloan-Kettering Institute in New York City in April 1960 (Warkany 1957, Mintz 1957, Anon 1959, Rivers 1959, Rivers et al. 1959).

Signs of the attention teratology was starting to get were signaled by the support several of the conferences received from various interested organizations, the Association for the Aid of Crippled Children, the Human Embryology and Development Study Section of the National Institutes of Health, newly formed in 1955, and from the National Foundation.

The sponsorship of the 1959 Palm Beach conference by the National Foundation should be explained. This organization, founded in 1937 as the National Foundation for Infantile Paralysis, had recently modified its name to the National Foundation, to announce that – vaccination for poliomyelitis having had such excellent results – the foundation wished to turn some of its attention elsewhere and "was broadening its program to include congenital malformations" (Rivers et al. 1959).

The foundation's change of focus notwithstanding, its scientific advisory committee did not recommend congenital malformations as one of its main new areas of attention. Something memorable took place at the time nevertheless. It was between presentations, while walking on the beach, not "deterred by ... misquided remarks" of the committee, believing that the right time had come, Warkany, Wilson, and Fraser revived the idea of the creation of an organization of teratologists (Warkany 1988a), and out of this seaside stroll sprang the Teratology Society. So it can be considered that was the National Foundation's most important contribution to teratology – and greatest missed opportunity.

The Teratology Society

This idea was brought to fruition the following year, at the 1960 conference in New York, when following the presentation of scientific papers, a session was convoked to organize the proposed society, its formation declared, and the name Teratology Society chosen. Future meeting places were decided upon and the 62 individuals present declared charter members of the fledgling society. In a letter sent to prospective new members the object of the society was announced to be "to stimulate scientific interest in, and promote the exchange of ideas and information about problems of abnormal development and malformations." (The initial annual dues was set at US$3.00; in 2009 it was $126.00).

Interest in abnormal development was not limited to North America; almost concurrently there was formed a sister society, the Anomalies Research Association of Japan, renamed the Japanese Teratology Society, and not too many years later the European Teratology Society, the Australian Teratology Society, the Midwest Teratology Association, and the Neurobehavioral Teratology Society came into existence. The story of the early history of 'organized teratology,' as they termed it, was told by Wilson and Warkany (1985).

Thalidomide

By a terrible coincidence, just the year after the formation of the Teratology Society, a shattering happening put congenital malformations and teratology on the map. The event that shook scientific circles, government officials, the pharmaceutical industry, and the public worldwide, was the revelation that a seemingly harmless therapeutic substance had caused thousands of children in many countries of the world to be born with severe limb deformities.

The brief initial notices of these occurrences, appearing in the December 16, 1961 issue of the widely read Lancet, and the next week in the Medical Journal of Australia, sounded quite frantic. They stated that in recent months in Australia "multiple severe abnormalities" had been seen in babies whose mothers had taken a medication, thalidomide, during pregnancy – and the author asked if similar abnormalities had been seen by others in babies exposed to this drug (McBride 1961a, b).

The response soon came, in the January 12, 1962 issue of Lancet. Yes, in the then West Germany, children with variable though specific malformations had also been seen whose mothers had taken this drug early in pregnancy (Lenz 1962). The author stated as well that after his earlier-voiced suspicion regarding the drug (Lenz 1961a, b) he had been informed of other such occurrences in Belgium, England, and Sweden. All together he calculated that perhaps over 3000 'thalidomide babies' had been born in Germany alone since 1957, the year of its first mass marketing in that country – the large number undoubtedly the result of the drug's reputed efficacy as a sedative, and as an antinauseant and antiemetic for use during early pregnancy, and of its being considered so safe that it was available without prescription, as it also was in Great Britain when put on sale there in 1958.

These horrible events, despite some lingering doubts of its culpability, caused the drug to be quickly removed from the German and English markets and elsewhere soon thereafter. A tentative final estimate put the total worldwide number of births with such malformations definitely attributable to thalidomide at nearly 6000, a number which did not include the probably considerable number of abortions and fetal and neonatal deaths associated with the drug (Lenz 1988).

A calculation with respect to the frequency of the limb defects in Germany at the time of the epidemic made a shocking point (Neubert and Neubert 1997). Namely, that despite being 150–300 times greater than their ordinary, i.e. background, frequency, they might have been barely noticed against the total level of congenital

H. Kalter, *Teratology in the Twentieth Century Plus Ten*,
DOI 10.1007/978-90-481-8820-8_6, © Springer Science+Business Media B.V. 2010

malformations, except for their extraordinary rarity and severity, which enabled the epidemic to be detected even as soon as it was.

The Thalidomide Syndrome

The malformation syndrome, a highly specific complex, was described in great detail (Lenz and Knapp 1962, Knapp and Lenz 1962, Taussig 1962, Weicker 1963, Anon 1964, Swinyard 1969, Warkany 1971, Smithells and Newman 1992, Miller and Strömland 1999). Most children had variable degrees of reduction deformities of the limbs, usually the upper limbs, with the most severe consisting of amelia and phocomelia (with digits emerging from the shoulder), as well as numerous other types of malformations, skeletal and nonskeletal – but, remarkably, no certain instance of major malformations of the central nervous system.

With few exceptions none of the individual malformations was unique to thalidomide; rather it was their frequent concurrence that labeled them as comprising a distinctive syndrome. For all that, however, it was the severity, conspicuousness, and extreme rarity of the landmark sign of the syndrome, the phocomelia, that allowed the revelation – delayed, nevertheless, as will be noted below – that a drug-induced epidemic was in progress.

There have been numerous defect-by-defect descriptions of the syndrome, but few qualitative accounts of them. In a passing remark Lenz (1985) mentioned that the "severe malformations had attained an incidence of more than 1 in 1000 births (8 cases among 6420 births from September 1960 to August 1961 as compared to about 1 in 200,000 between 1930 and 1958)." He himself had seen 52 babies probably malformed by thalidomide and had received reports of 115 others; and estimated that 2000–3000 such babies were born in West Germany in 1959–61 (Lenz 1962). Smithells (1962), in what he called the "first population study," published over 30 cases of ectromelia and 12 of other defects, 11 of the former of which were stillborn or soon died, giving a perinatal mortality rate of 37%.

An account of malformations in children born in England and Wales between 1/1/1960 and 8/31/1962, collected through a Ministry of Health survey (Anon 1964), suggested that there were 200–250 surviving children with limb defects and another 50 with other defects, or using other assumptions 430 and 150 respectively; with the total loss due to abortion and stillbirth not calculable. The report included a detailed listing of the defects in 894 babies, about 81% of whom were alive when the survey was completed, 64% with limb deformities only, 7% nonlimb deformities only, and 29% limb and nonlimb deformities. The most common nonlimb defects were those of the ear, eye, and heart, alone or together with limb defects.

The defects in more than 400 children seen at the University of Bonn Department of Pediatrics in 1959–62, the subject of an address at an international congress (Weicker 1965), were later summarized in English (Warkany 1971). By early 1964 about 45% of the children had died of congenital heart disease, intestinal atresias, kidney malformations, etc. In about 60% of the survivors only arms were affected, in a few only legs, and in 10% both arms and legs, with a great many variations in site and degree. In 20% ears and arms were affected, in about half ears only; and in 5% eyes and arms were affected; again of a great variety of degress and types.

A largely similar account of malformations in 148 children exposed early in pregnancy seen personally by Smithells and Newman (1992) named, as the most frequently occurring, after limb defects, those of ears, eyes, and nerve supply to the face, eye muscles, and lacrimal gland, and internal defects most commonly affected the heart, kidneys, urinary, alimentary, and genital tracts – none as already mentioned unique to thalidomide; the 40% early mortality rate much eliminating the internal ones.

These and additional diversities and multiplicities of defects are of significance with regard to concepts regarding mechanism of action of thalidomide, discussed below.

The Revelation

The remarkable fact must be recalled that two of the first three human teratogens discovered, the rubella virus and the drug thalidomide, were announced to the world by Australian physicians, Gregg and McBride. Next it is amazing that McBride, on the basis of evidence that in hindsight seems to have been confusing and flimsy, should have got it right. When one reads of the prolonged and convoluted efforts to unravel the problem of the etiology of the epidemic of bizarre limb defects emerging in Germany (Weicker 1963, Sievers 1964), one must be impressed at the perspicacity of the obstetrician McBride, getting it right, almost overnight. But how true is this?

In McBride's letter to the Lancet (1961a; the second, he later contended – since a slightly earlier one making the same assertion had been rejected by this journal; an allegation denied by its editor – Munro 1979), he gave no hint of how he came to his conclusion, merely stating that these abnormalities (he mentioned only "abnormally short" limbs) occurred in "babies delivered of women who were given thalidomide...." He stated that the hospital records showed that the pregnant women, while in the hospital, received no other medication (Knightley et al. 1977).

Nevertheless the women may have taken other sedatives or other antiemetics, may have been exposed to viral infections, x-rays, etc., which would have complicated the picture. Finally, regardless of all other considerations, the small number of women the conclusion was based on would hardly support a definitive judgment. According to a private communication from McBride (Taussig 1962) it appears that he rested his deduction on six children, three seen in the spring and three more in the autumn of 1961, whose mothers had taken thalidomide. But it would appear that the conjecture was based on the first three only, and supported by two others born later in the year (Knightley et al. 1977).

A further insight comes from Lipson (1992) who noted that McBride was conducting a trial of thalidomide for morning sickness in Sydney, when in May/June 1961 he saw three affected babies. But he hesitated to implicate the drug because the "next 23 babies in the trial" were normal; only a further two abnormal ones in

September convinced him. McBride (1961a) himself did not mention this sequence, and never further explained how he came to his dramatic conclusion, no word of the reasoning, the intuitive leap. But these facts obviously were the basis of his stating that the defects had a frequency of almost 20%.

The co-discoverer, the other rank amateur, the pediatrician Lenz, on the other hand, recounted in great detail the inquiries over months in 1959–61 which, mounting to a crescendo, finally compelled him to announce that the agent responsible for the outbreak of a type of limb defects otherwise seen only extremely rarely was thalidomide. And he always gave recognition to the others who were close on the trail, but whom circumstances delayed reaching that conclusion (Wiedemann 1961, 1964, Wegerle 1962, Weicker et al. 1962).

Even after his announcement, he sent a letter to Chemie Grünenthal, the manufacturer of the suspected agent, asking them to withdraw it from the market, "I am, of course, perfectly aware, that these data are not sufficient proof of a causative connection between the drug treatment and the malformations" (Lenz 1985). But as he stated at a pediatric meeting in 1961, "from the scientific point of view it seems premature to discuss it. But as a human being and as a citizen, I cannot remain silent about my observations."

Thalidomide: Safety and Sales

Soon after its synthesis in 1954 thalidomide was studied in adult laboratory animals of several species for various kinds of toxicities, and very high doses found to have no apparent adverse effects. Clinical trials conducted in Germany and the US in the same period noted the drug to be an effective sedative, with overdosage apparently impossible since huge amounts merely led to prolonged sleep without harmful aftereffects (Mellin and Katzenstein 1962); results that confirmed its low acute toxicity in animals. The drug, found safe, was marketed in Germany in 1956, at first for the treatment of influenza under the name Grippex, and with the discovery of its supposed sedative properties was next promoted to combat sleeplessness, in Germany in 1957 as Contergan, in England in 1958 as Distavil, in Canada the same year as Kevadon, etc.

But only later, when the animal and clinical studies were closely examined, a harsh critique of them emerged. Years after he had spoken out Lenz (1979) wrote, "thalidomide was released for sale before carefully controlled animal and clinical experiments had been done." And further commented, "the papers published . . . on animal experiments and . . . on clinical experiences with thalidomide have so little value that . . . they should not have been accepted for print" (Lenz 1988).

The inadequate methods for testing drug safety in that earlier era must bear some blame for the disaster. But ultimately, behind these failings, the éminence grise was the venality of Chemie Grünenthal, the pharmaceutical company marketing the drug, and its licensees, abetted by sometimes naive medical people and sycophantic journal editors who forsook their responsibilities.

Toxicity in Adults

In 1959 and 1960 increasingly numerous reports noted that some adults taking thalidomide for several months developed serious peripheral polyneuritis; and that when they stopped taking the drug the symptoms improved. Only afterward did the parallel become obvious between this sequence (Mellin and Katzenstein 1962) and the appearance of the malformation syndrome some months after the drug was first marketed and its disappearance within months of its withdrawal (see figure in Lenz 1965a,b). But while the connection between the drug and the neuritis was accepted, bewilderingly its teratogenic connection was continually denied by the manufacturer.

Thalidomide in the USA

Thalidomide under one name or another was sold, and children with the distinctive malformations due to it were born, in about 30 countries of the world. But it was never sold and virtually no such children were born in the United States. How this came about is a story worth remembering.

An application to permit the distribution of thalidomide in the US was submitted to the Food and Drug Administration (FDA) on September 12, 1960, by the American licensee of the parent drug firm, William S. Merrell Company. It was assigned for review to a medical officer, Frances Oldham Kelsey, who had become an FDA employee just a few weeks earlier. This newcomer, after a short orientation program, was assigned this particular application since it was thought best for her to start with one that was "simple and straightforward" (Kelsey 1988).

During her reviews of the application Kelsey found that it had many inadequacies, among others lack of full details of clinical and animal studies, incomplete chronic toxicity data, and disregard of side effects; which Kelsey pointed out in a communication to Merrell on November 10, 1960, the last day before the application would otherwise have become effective. Another side effect, peripheral neuritis associated with prolonged use of the drug by adults, would have been of especial concern, had she known of it. Although the applicant was aware of this fact it was not mentioned in the application, and this withheld bit of information came to her attention only after the application was withdrawn.

Prior to 1962 the FDA had statutory control only over drug safety, negative decisions regarding which had to be made within 60 days of receipt of a drug application to halt automatic approval and release of the product for marketing (Lasagna 1989). To delay this release Kelsey used the tactic of repeatedly rejecting the application, which then being resubmitted instituted a new 60-day review period. But Kelsey was spared further action when a few days before a new deadline, because of recent reports from Germany of the drug's suspected teratogenic effects, the application was withdrawn. A recounting in great detail of these sorry events is found in McFadyen (1976) and Green (1996).

Kelsey, a Canadian, at 46 years of age, with undergraduate and graduate degrees from McGill University in Montreal, with a Ph.D. in pharmacology and an M.D. degree from the University of Chicago, after academic and private-practice experience, obviously was not to be easily cowed. She remained adamant and defied the applicant's attempts to force the release of the drug. It is thus to Kelsey's having been dissatisfied with the application, and to her strength in withstanding pressure, that credit and gratitude have justly gone for preventing in the US a catastrophe of unimaginable proportions.

Just how huge this might have been is intimated by the disclosure that some 1200 US physicians, most of whom had not been notified of the drug's suspected teratogenicity, received samples of thalidomide for distribution among their patients. Extraordinary efforts by FDA personnel to inform physicians and to retrieve unused drug samples limited the damage to ten cases of injured children from this source and some seven from pills obtained abroad (Kelsey 1988).

Dose- and Time-Response Relations

Although suspicion fell early on thalidomide as the cause of the epidemic of limb malformations, it took an arduous epidemiological investigation to establish the connection firmly (Weicker 1963, 1969). Various systems were affected, but abnormalities of the limbs, as noted above, were by far the most common. A most meticulous time specificity was discovered as well, exposure only during a 2-week period in pregnancy being teratogenic, with different parts affected at different times during this interval (Nowack 1965, Kreipe 1967).

This was learned only by analyzing information gleaned from hospital records, prescription information, and interviews with physicians and mothers, which enabled the construction of a timetable charting the relation between the time during pregnancy thalidomide was taken and the type of malformation that resulted; as had been done in past years for irradiation and rubella. From this it was discovered, first, that the period of sensitivity was extraordinarily brief, malformations being produced only when the drug was taken during the roughly 2-week period of pregnancy from the 35th to 50th day after the first day of the last menstrual period (i.e. postconception days 21–36); medication taken entirely before or after this interval being without teratologic consequence (Lenz 1965b). (Attested to by a Japanese experience which found that of 113 pregnant women taking thalidomide in 1959–61 only those few exposed during the critical period had malformed babies – Kajii et al. 1972.)

Within this window, days 35–38 were associated with anotia or gross abnormalities of the external ear, 41–44 amelia of the arm, 42–47 three-finger phocomelia, 44–48 absence of the leg, 47–50 triphalangy of the thumb (Lenz 1964, 1990), a relationship conforming in a roughly cephalocaudal gradient with the period of drug ingestion. Though fairly reliable overall, the accuracy of these timings may have been weakened by faulty maternal memory and prenatal developmental variability (see Neubert and Neubert 1997).

It was claimed that there was no evidence of an embryo escaping being malformed during this sensitive interval (Lenz and Passarge 1965). If true this would be quite extraordinary; and in fact seven instances were identified of mothers most likely taking thalidomide during this interval and having normal children (see references in Khera 1984). An estimate of mothers taking the drug during that period and having offspring with the typical malformations arrived at about 20% (Shepard 1992).

Also marking thalidomide as extraordinary is the small amount of the drug that was teratogenic. Though not always accurately recorded, available documents showed that typical malformations resulted from the ingestion of as little as 25 mg three times a day or 100 mg/day for 3 days, equivalent on a maternal weight basis to the astonishingly small dosage of about 1 mg/kg. This is 10–50 times smaller than doses found to be teratogenic in species of susceptible animals.

The thalidomide syndrome challenges the theory that the particular abnormality produced by a teratogen is solely determined by the part vulnerable to disturbance at the time of embryonic exposure, rather than by the properties of the teratogen (what Wilson 1957 called time specificity in contrast to agent specificity); though in all probability both factors enter the equation. Although thalidomide caused various malformations, the hallmark of its teratogenicity, the most remarkable example of its specificity, was phocomelia. This ultimate degree of limb reduction deformity has occurred extremely rarely in the annals of experimental teratology, sharing this rarity with other unique defects, namely diaphragmatic hernia (Andersen 1941), atresia of the esophagus (Kalter and Warkany 1957), and cyclopia (Kalter 1992).

Thalidomide's Teratogenic Mechanism

Thalidomide made it imperative to learn how this chemical causes malformations, and especially the particular ones it causes. The question of the basis of teratogenesis – its mechanism – "that is, where, when, and how these agents affect the embryo" (Wilson et al. 1959), especially the "how" – a matter in the past merely of academic curiosity – had, as we have seen, been previously raised with respect to trypan-blue teratogenesis. But the thalidomide episode, bringing a new dimension to the quest, emphasized the urgent need for understanding how teratogens work to cause such phenomena, with the ostensible goal of avoiding future such disasters.

In the search for suitable experimental subjects for such mechanistic investigations the animals usually favored for teratologic study, mice and rats, failed the task, being barely sensitive to thalidomide, at best responding with only relatively low frequencies of nonspecific malformations (e.g. Scott et al. 1977). A wide-ranging search then ensued among almost the entire domestic and feral menagerie for a susceptible species, which included hamsters, pigs, cats, dogs, ferrets, armadillos, chickens, rabbits, and nonhuman primates, a search that found them all failing the requirement of reproducing the crucial malformations, but for the last two.

Limb defects were regularly induced in rabbits, but they were not analogous to the characteristic human ones (Felisati 1962, Kemper 1962, Somers 1962, Fabro and

Smith 1965, Sawin et al. 1965), although it was sometimes claimed otherwise (Sterz et al. 1987). Nonhuman primate species, especially macaque and baboon, however, did have the typical limb reduction malformations, varying from amelia to forms of phocomelia, with dosage and timing comparable to those in humans (Lenz 1968, Hendrickx and Binkerd 1990). Hence they should have been the species used for such studies – as later they were (Ema et al. 2009) – but practical considerations prohibited their routine use. Consequently, though hardly ideal, rabbits became the subject of such investigation. But even so, with no satisfactory revelation of the how and the why.

Teratogenesis Mechanism Continued

Rarely induced malformations, implying a unique connection between cause and effect, offer exceptional opportunities for investigating the means by which they are elicited. The question of how thalidomide, a seemingly ordinary chemical compound was able to have this extraordinary teratogenic property has engaged many minds. Innumerable theories have been propounded and studies directed at this question in the last 30 years and more.

Whether the failures were because of experimental inadequacy or, as suggested by Fabro (1981) and others, poor understanding of the biochemistry of normal embryogenesis, or just beyond present understanding, it will be interesting to see whether studies of the molecular or other advanced varieties, when by such means it is attempted to tackle this matter, will be more successful.

For now, it is not news that mechanism studies have been virtually disappointingly fruitless. Such riddles concerning other teratogens have been equally challenging, and students directing their efforts thereunto, even while charting apparent early steps on the way to the outcome, have failed to deliver the goods. Just what 'mechanism' means is not easy to pin down. One class of malformations clearly falls into the category, that of those due to dysynchronization, such as delay of the palatal shelves to become elevated, or the neural folds to meet, in both cases preventing timely fusion, leading to cleft palate and spinal bifida respectively. Here the question is the reason for the delay, perhaps ultimately to be traced to a form of growth retardation. But the limb malformations caused by thalidomide are due not to this pattern but to failure of organization, i.e. to disorganization. And tracking that back to its ultimate basis is a problem of a different order of complexity.

A multitude of studies in the past 50 years have brought much imaginativeness to this challenge. The list of the conjectured theories is long and varied – hardly necessary to document here. Provocative especially has been one among these, proposals beginning at least some time ago (D'Amato et al. 1994) alleging inhibition of the formation of blood vessel growth in the developing fetal limb bud, a view lately gaining momentum (Sauer et al. 2000, Ng et al. 2003, Wells et al. 2005, Tamilarasan et al. 2006, Therapontos et al. 2009). But even its proponents said that "the mechanisms underpinning the teratogenic effects of thalidomide are unclear" (Therapontos et al. 2009).

What is also unclear is how relevant studies with embryonic chickens are to the question; and whether the one analogue found to have an effect on embryonic limb development by this study is identical to the human teratogenic molecule (which incidentally to my knowledge has not been definitively identified among the many metabolic products thalidomide is transformed into – Schumacher et al. 1965 – nor the reason for the teratologic resistance of rodents to it – Nau 1986). May it not thus be somewhat presumptuous to assert that "… the trigger of thalidomide-induced teratogenesis has been uncovered …" (Vargesson 2009)? Not to forget that not only limb defects were induced.

Postscript: Was Anyone to Blame?

Was anyone to blame for the thalidomide tragedy? Could something have been done to prevent it? Contrary opinions have been voiced. Sir Ernest Chain, a Nobel laureate, said in a newspaper interview (Henningham 1975) "[O]f course it was a tragedy … it could not have been foreseen … no one was to blame … even if you do all these things [drug testing] there is still a risk factor … you cannot achieve safety in any case – safety is an illusion." He was right. Were thalidomide subjected to the currently mandated teratogenicity testing procedures, in which pregnant rodents are subjected to the largest tolerated doses of agents, with little doubt it would pass as acceptably safe with flying colors.

Another opinion argued that someone must have been to blame. Why? Because it was asserted to have been well known in the prethalidomide era that embryos could be deformed by drugs, in fact that "there was already evidence of [thalidomide's] teratogenic potential" (Dally 1998). This individual was wrong. But that did not stop her from asking, why "had all this evidence been ignored."

Of course that chemicals can cause congenital malformations in laboratory mammals was known then. But that is not the same as saying, and as we saw above, that all chemicals have that capability. Though a large number of chemicals have been found to be teratogenic in animals (Schardein 1983, 2000), very few were found to be teratogenic in human beings when a thoroughgoing overview of the question was made almost 30 years ago (Kalter and Warkany 1983), and 30 years later that is still the case.

But Chain made a further statement, which *can* be strongly challenged, namely that "the firm that manufactured thalidomide could not be condemned." On the contrary, if negligence or even deliberate fault were to be held responsible for the tragedy, the blame should be laid first and foremost at the door of the manufacturer and marketer of thalidomide, Chemie Grünenthal; not because the drug was not tested by the means available at the time for potential reproductive harm, but because when instances of peripheral neuropathies and later of malformed children began to be reported to them by physicians, sometimes through their own sales representatives, the company ignored them and denied that the drug was at fault, until they could no longer dismiss the truth.

The Future?

Thalidomide is an unusual chemical. In addition to its teratogenic properties and its long-known sedative and hypnotic effects in adults, it has been discovered to have a host of apparently beneficial applications. The list of its current or potential therapeutic uses is long (Patrias et al. 1997, Miller and Strömland 1999, Neiger 2000, Powell 1999, Lafille 2006, Kaur et al. 2009). But in putting thalidomide to these desirable uses, its tragic past cannot be forgotten, and this dark page from pharmaceutical history must guide the future.

Detailed protocols must be formulated to enable the drug to be administered to those who would benefit from it, while providing quarantees that women of reproductive age are shielded from its inadvertent exposure. Thus the FDA, in approving thalidomide for treatment of a form of leprosy, has gone to great lengths to restrict its distribution, setting in place a network of safeguards (Nightingale 1998, Uhl et al. 2006).

Even so, set loose in the world, regardless of the precautions, the harm the drug is capable of doing may reemerge – toxicity in adults and malformation in embryos; the latter already seen in parts of the world where stringent controls of the sort promulgated by the FDA were absent (Gorman 1994, Marwick 1998, Ances 2002). One takes comfort in the hope that with wide controls the danger will recede; and in the expectation that such eventualities would be quickly alerted to (Yang et al. 1997).

This chapter cannot be ended without pointing out the contrast between the riveting attention given thalidomide and the almost forgotten rubella (see Webster 1998). The reason cannot be the number of victims, because the virus damaged tens of thousands of children and the drug comparatively few. Yet it is the latter that remains a dread in the public mind.

This is not hard to understand. The harm caused by the last rubella epidemic in the US, 50 years or more ago, has been overshadowed by the poignant malformations caused by thalidomide. Thus one tragedy eclipses another.

Testing for Teratogenicity

The thalidomide episode demanded a thorough rethinking of how drugs were to be tested to safeguard the unborn. In time new and expanded ways of determining whether pharmaceutical and other chemicals posed teratogenic risks were instituted.

Proposals for Drug Testing

Procedures for testing drugs for potential reproductive toxicity had existed in the years before thalidomide, but were judged inadequate for the specific purpose of detecting possible embryonic damage (Wilson 1979). This dreadful happening opened eyes to the urgent need of setting up standardized procedures for testing the potential teratogenicity of drugs. Clearly commercial self interest was also in play.

In August 1962 the Pharmaceutical Manufacturers Association, as it was called at the time, the lobby for the US pharmaceutical industry, established a Commission on Drug Safety, to suggest ways of remedying the shortcomings of the then existing animal tests for teratogenesis. The commission formed a Subcommittee on Teratology for the purpose of proposing meaningful testing procedures (Anon 1962). And in its turn the subcommittee convened a conference on prenatal effects of drugs with this aim; in effect kicking the can down the road.

In the end the only contribution of any consequence that emerged from all of this was the proposal that workshops conducted by experimental teratologists be held to acquaint pharmaceutical personnel, toxicologists, and others with experimental methodology (Anon 1963b). The first workshop, setting the pattern for those in ensuing years, consisted of lectures and laboratory demonstrations for participants and observers from drug companies, governmental regulatory agencies, and university medical science departments (Wilson and Warkany 1965). An interesting question to ponder is whether the rarity of medication-induced maldevelopment in subsequent years was partly to be attributed to the take-home lessons of these lectures.

A more tangible outcome of the thalidomide scare however was the promulgation in 1963 of new investigational drug regulations, making premarketing reproductive and teratology drug testing mandatory (Kelsey 1982). What can be called the

H. Kalter, *Teratology in the Twentieth Century Plus Ten*,
DOI 10.1007/978-90-481-8820-8_7, © Springer Science+Business Media B.V. 2010

revolution in the science and politics of testing pharmaceutical products for potential prenatal harmfulness was described by Sullivan and Barlow (1985) and Schardein (1988).

The Procedure

A superabundance of writings and official documents expatiating upon the subject of how to test for the teratogenic potential of drugs followed upon this mandate. The first requirement, especially emphasized, was that test animals should resemble as closely as possible human anatomical, physiological, and pharmacological features and conditions; at the same time realizing the impossibility of expecting an ideal fit (Anon 1973).

For an extensive description and discussion of the complexities of the drug testing process see Wilson (1973). A detailed outline of the recommendations for conducting the teratology test appeared in a Health and Welfare Canada document (Anon 1973, pp. 135–74). Of the hundreds of pharmaceutical chemicals that have been tested since the thalidomide scare, how many were rejected at one stage or another of the process, there is no way of knowing. What is known is that a miniscule number have posed teratogenic risks to humans. These will be discussed below. So by that judgment the program was a success.

The Dose-Response Curve

May I digress for a few lines to consider another matter, one perhaps less pragmatic than theoretical (Daston 1993). That is the nature of the dose-effect relation in teratogenesis; which brings us to a discussion of the basis of teratogenic action, and in order to sharpen the focus to what distinguishes it from that of mutagenesis.

Teratogens and Mutagens

Mutagens and teratogens and carcinogens too are sometimes considered to be intertwined (Kalter 1971). Most mutagens are held to be carcinogens, many carcinogens to be teratogens, and "most. . .mutagens might be expected to be teratogens" (Ferguson and Ford 1997). But does this imagined intertwining also mean that their primary targets are identical or even similar?

One of these conjunctions was suggested by experiments in which mutagenic chemicals caused retarded prenatal development, but not malformations, when administered to pregnant mice in early embryonic stages (Generoso et al. 1991, Rutledge et al. 1992, Polifka et al. 1996). The effect was not associated with chromosomal aberrations and not thought to be due to gene mutations but to altered programming of gene expression, and so was interpreted as due to nonconventional, perhaps 'epigenetic,' mechanisms. The term epigenetic, once used by Waddington

(1957) in referring to canalized embryonic development, was left undefined. [It is not amiss to mention a conventional teratology study of long ago, which found that chemical treatment in the preimplanation period caused prenatal death and retarded development but not malformation (Nelson and Evans 1949).] There is no doubt much still to be learned; but thoughts on the subject of preimplanation teratology (Rutledge 1997) did little to clarify the question.

Other attempts to link mutagens and teratogens, even to the point of exaggerating the role of mutant genes and chromosomal aberrations in the causation of human congenital malformations, equally missed the mark (Ferguson and Ford 1997, Bishop et al. 1997). The latter authors' unattributed assertion that "transmitted gene mutations and chromosome aberrations account for as much as 25% of all human birth defects," is an exaggeration; the best estimate of the frequency of serious congenital malformations caused by single mutant genes and major chromosomal abnormalities is 7.5% and 6.0% respectively (Kalter and Warkany 1983). Whatever the figure, it is irrelevant to the argument. Even though Bishop et al. failed to say it themselves, they were referring to abnormalities caused by mutant genes present in ova before fertilization, and not to the actions of mutagenic agents upon already fertilized ova, an entirely separate kettle of fish.

But the fundamental distinction, the characteristic that tells mutagenesis apart from teratogenesis, is the shape of the dose-response relation, which is the palpable expression of underlying mechanism, the latter a risky but still useful term. It is currently accepted that the response to mutagens, as to carcinogens, in the absence of adaptive responses, is linearly related to dose such that there is no dose that is without an effect. Thus even after years of study the shape of the dose-response curve at low levels of radiation and chemical carcinogens has not been drawn with absolute certainty.

This so-called no-threshold concept, as it pertains to carcinogenesis, was taken to its ultimate degree when it was given governmental imprimatur, almost set in concrete, by the much-decried Delaney Clause. This was wording included in the 1958 Food Additives Amendment to the Federal Food, Drug, and Cosmetics Act of 1954, added at the instigation of a congressman of that name, which established that no amount of a chemical found to cause cancer in animals would be allowed in processed foods.

The theoretical basis of this concept is that mutagens and carcinogens produce their consequences by hitting single targets, molecules in DNA without interaction with other entities or the outer environmental world, and thus that no matter how small the dose of the agent applied an effect, mutated genes or cancers, occurs.

In teratogenesis, on the contrary, the shape of the dose-response curve is not straight, but sigmoid, rising quite steeply from the no-effect dose and flattening at doses associated with maternal toxicity and total fetal death. This of course is a generalization, describing as it does the paradigm. What accounts for this unique pattern? Instead of the single unmediated, invariant target theorized of mutagenesis, teratogens deal with embryos, dynamic multicellular entities with reparative capacities and hereditary inputs, which in interacting with its enveloping membranes and the maternal body, create unimaginably versatile physiological, pharmacological, and metabolic efficiencies that continually divert and modify the effect of impinging agents. A totality that combines to produce unpredictability, and adds to the complexities of the drug tester's task.

The whole fabric, therefore, that attempted to interrelate in some metaphysical manner, these diverse processes, falls apart; and as it happened was never to be woven again.

Teratological Detours

The thalidomide tempest left a scar on the public and scientific consciousness, and the anxiety and mistrust stemming from it readily engendered suspicion of other potential causes of fetal defect; suspicion, often unjustified, that came to be pointed in a number of directions. Only with time were many of these teratological scapegoats exculpated. But one most difficult to exonerate was a common pharmaceutical product.

Bendectin

Many women suffer from so-called morning sickness, hyperemesis gravidarum, severe and debilitating nausea and vomiting in the early months of pregnancy; so many so that it was postulated to have evolutionary significance as a fetoprotective mechanism (Hook 1976). The alleviation of this common illness is big business, for which antihistamines and antiemetics have been the usual prescribed medications (Mazzotta et al. 2000, Lacroix et al. 2000, Quinlan and Hill 2003). (Incidentally, it was thalidomide's reputed effectiveness as an antinauseant that was one reason for its popularity.)

The Product

The most widely used antiemetic medication at one time was a product called Bendectin in the US and Canada, Debendox in the UK, and by other names elsewhere. With a much longer life span, so to speak, than the ill-fated thalidomide, from the time it was first marketed in 1956 until it was withdrawn in 1983, it was taken in the US alone by over 33 million women (Biggs et al. 1986).

It was not a single compound, but originally consisted of three ingredients, doxylamine succinate, an antihistamine with antiemetic properties, dicyclomine hydrochloride, an antispasmodic, and pyridoxine hydrochloride, i.e. vitamin B6, also with antinauseant properties; its formulation was changed in 1976, and dicyclomine omitted because of its believed inefficacy.

H. Kalter, *Teratology in the Twentieth Century Plus Ten*,
DOI 10.1007/978-90-481-8820-8_8, © Springer Science+Business Media B.V. 2010

Alleged Teratogenicity

Only years after Bendectin became available did claims appear that it was teratogenic, based on isolated occurrences of malformed children of women taking the product (Anon 1977, Paterson 1969, 1977, Donnai and Harris 1978, Frith 1978, Fleming et al. 1981). The abnormalities the product was associated with were a hodgepodge, ranging from minor to severe, including heart defects, orofacial clefts, digital defects, limb abnormalities, genital tract anomalies, diaphragmatic hernia, prune belly syndrome, lung hypoplasia, cloacal agenesis, spina bifida, anencephalus, etc.

Animal studies were not helpful in substantiating or clarifying these findings, since doses of Bendectin far greater than the therapeutic amounts consumed by pregnant women, when administered to rats, rabbits, and nonhuman primates in early stages of pregnancy, produced no consistent evidence of teratogenicity (see Tyl et al. 1988).

Early surveys of congenital malformations in children of pregnant women taking the drug found no statistically significantly different overall frequency than in those not taking it (Bunde and Bowles 1963, Bunde and Leyland 1965). These studies may have been primitive epidemiologically, but more refined ones, soon conducted, reached similar conclusions, regarding malformations in general or specific ones; thus the judgment was that if there was a risk it was small and clouded by various uncertainties (Yerushalmy and Milkovitch 1965, Henderson 1977, Shapiro et al. 1977, Rothman et al. 1979).

[The page turns. A recent case-control study, casting a different light on the subject, concluded that Bendectin seemed to protect against the occurrence of congenital heart defects (Boneva et al. 1999). And now medication for nausea and vomiting of pregnancy was said to be associated with increased IQ in 3–7 year old children (Nulman et al. 2009) – can you beat that!]

In the course of time the teratogenicity of Bendectin was examined by a total of 27 cohort and case-control studies, whose findings were subjected to intense analysis (see references to them all in MacMahon 1981, Holmes 1983, Einarson et al. 1988, McKeigue et al. 1994). These universally agreed, as one writer said, in scientificese, that while "the possibility that Bendectin – or anything else – is causally associated with some form of malformation can never be totally excluded [nevertheless] a substantial body of evidence now indicates that if Bendectin is teratogenic at all in humans it can be so only rarely" (MacMahon 1981).

Similarly, two meta-analyses concluded that "Bendectin is not related to teratogenic outcomes in humans" (Einarson et al. 1988), and that "thirty years of Bendectin data published through 1993 ... [indicated] neither an increase nor a decrease in birth defect risk following first-trimester use of Bendectin" (McKeigue et al. 1994). Concurring in these judgments were reviews by official bodies – an FDA panel, a UK committee, a Canadian advisory committee, and those of other countries (Kolata 1980a, note 26 in Sanders 1993, Ornstein et al. 1995).

Yet none of this body of expert opinion convinced parents of malformed children who had taken Bendectin during pregnancy of the innocence of the product. Especially, and no doubt with memories of the horrible damage fetuses had suffered from the apparently innocuous thalidomide still vivid in the public mind, the danger posed by this new thalidomide aroused anger and anxiety anew. As a commentator wrote, "the safety of Bendectin ... has become the subject of an emotional and intense debate among parents, lawyers, and medical scientists;" and, that, as a lawyer expressed it, scientific considerations will not deter law suits (Kolata 1980a).

Legal Action

Thus began protracted litigation, involving hundreds of lawsuits against the manufacturer of Bendectin, the then Richardson-Merrell Company – whose unsavory history in the thalidomide episode (see Green 1996) was not in its favor, and seemed to support sensational allegations of concealment by the company of knowledge of Bendectin's fetal harmfulness (Dowie and Marshall 1980).

The problem, formidable enough, of making the biological facts and concepts at the bottom of the question of human teratogenesis understood by the public, juries, and the legal profession, was in addition burdened by the conflicting testimony of expert witnesses and its attendant psychological aspects (Skolnick 1990, Brannigan et al. 1992, Sanders 1993, Green 1996, Marshall 1999, Weinstein 1999).

Merrell, however, a decade before the Supreme Court decision regarding admissibility of scientific evidence in effect absolved the company of liability (Gold et al. 1993, Sanders 1993, p. 64 et seq.), because of the prohibitive costs of insurance, declining sales of the product, legal fees, and the large amounts of money that were expected to be paid to settle the lawsuits against it, felt it had no choice but to voluntarily cease manufacturing Bendectin (Skolnick 1990).

Pregnant women in the US were thus deprived of an efficacious medication with a well-established record of fetal safety, and in its place were left with medicines that were less effective and less well vindicated teratologically (Jewell and Young 2000). Paradoxically, in the US doxylamine succinate, the antihistamine in Bendectin, is present in several commonly used nonprescription products (Kalter and Warkany 1983), and in other countries, among them Canada and Great Britain, ingredients in Bendectin continue to be medically available, albeit under different names (Koren et al. 1998).

The second, and no doubt even longer-lasting and more profound consequence of the Bendectin episode, is the incidental beneficial effects it has had on the legal and philosophical climate, even culture, regarding liability litigation, having had to force a generation of legal scholars and practitioners to become educated about the difficulties of dealing with scientific imponderabilities (Brannigan et al. 1992).

Blighted Potatoes

I'll get to the blighted potato story in a moment, but first: to the search for the cause of NTD. The search for discrete causes of congenital malformations has often led into exotic byways. This is especially true of the search for specific entities that may cause malformations of the central nervous system, especially NTD – efforts that have often given rise to numerous imaginative claims, so many in fact and so trivial most of them that I merely enumerate them here without the bother of citation. By category, first genetic causation: HLA immunogenetic factors, methylenetetrahydofolate reductase gene, X-chromosome inactivation, maternal Rh phenotype; X-linked defect in the selective abortion mechanism, ovulation induction, in vitro fertilization, fetus–fetus interaction, maternal XO/XX mosaicism, X inactivation, amniotic band syndrome;

next, maternal disease, practice, etc.: the common cold, influenza, maternal occupation, hyperthermia, cigarette smoking, alcohol, tea drinking, consumption of cured meats and canned peas;

much miscellanea: drinking-water hardness, trace element composition, chlorination, geomagnetism, cosmic radiation, hurricane, iodine deficiency, supplemental iodine, zinc and copper deficiency, maternal selenium status, intergenerational factors, rainfall;

and maternal nutritional status: nutritional pattern, various foodstuffs, methionine deficiency, cyanide in cassava, goitrogens in brassicae, blighted potatoes, vitamin deficiency, including most prominently deficiency of folic acid, whose history is told below.

The reader may find lists of many more environmental substances and drugs and drug classes that were once suspected of being human teratogens elsewhere (Kalter and Warkany 1983). The record of human folly is endless.

Were Potatoes the Answer?

Among this exotica, one supposition that was most intensely dwelt on was blighted potatoes. Its story is fascinating. First, based on a variety of strands it was considered plausible that NTD had a dietary explanation; and that a, if not the, responsible factor was a substance present in potatoes (Renwick 1972). Next, based on the correlation of variables regarding NTD distribution, such as season, region, and time, with patterns of the social-class gradient of NTD prevalence and potato consumption, etc. – the diverse and detailed exposition of which the original paper must be read to be appreciated – it was hypothesized that the malformations were due to fungal-infestated potatoes. And thus that "95% of [NTD] occurrences . . . could be prevented" by potato avoidance (Renwick 1973a).

Replies were rapidly forthcoming. Epidemiological data contradicted the hypothesis (Emanuel 1972, Emanuel and Sever 1973); examination of the relation between consumption of blighted potato and NTD prevalence in many worldwide regions failed to uphold it (Morrow 1972, Baird 1973, Carter 1973, Clarke et al. 1973, Cruz-Coke 1973, Elwood 1973, Elwood and MacKenzie 1973, Field and Kerr 1973, Kinlen and Hewitt 1973, MacMahon et al. 1973, Masterson et al. 1973, Roberts et al. 1973, Smith et al. 1973, Spiers et al. 1973).

This plethora of contrary findings was argued away as due to limited sample size, inadequate analysis, secondary interference with the blight-NTD correlation, as well as a "rash supposition" regarding variable relation of new maternal intake and slow release of the teratogen (Renwick 1973b, Renwick et al. 1974).

Animal Studies

Animal studies it was hoped would clarify the question. Blighted potatoes or glycoalkaloids extracted from potatoes administered to pregnant rats and rabbits had no teratogenic effect (Poswillo et al. 1972a, Swinyard and Chaube 1973). Better success it was thought would be had with marmosets (*Callithrix jacchus*), a primate species, for no better reason than that they were susceptible to the teratogenic effects of thalidomide (Poswillo et al. 1972b). But in a preliminary study blighted potato concentrate fed to pregnant marmosets at fetal stages caused only a cranial osseous defect without affecting neural tissue (Poswillo et al. 1972a). But even this outcome was not confirmed in studies with other primate species (Poswillo et al. 1973, Allen et al. 1977).

Avoidance Trials

The decisive way of dealing with the theory was to conduct an avoidance trial, as Renwick had proposed. One scholar was so bold as to say that "in view of the extensive, considerable, and consistent data on the recurrence risks of [NTD] it is difficult to see why a trial should be controlled. It is only necessary for a few hundred women to fail to produce a second affected child on a potato-free diet, or for a few women to have this misfortune while dieting, to clarify this issue. . . ." (Edwards 1973).

Trials indeed were decisive. A woman with a previous NTD child avoided potatoes throughout pregnancy and had a NTD fetus (Lorber et al. 1973). One case proved nothing. In a larger trial as well women who avoided potatoes had NTD children (Lorber 1974). A controlled trial was conclusive. Women with previous NTD children avoided potatoes as soon as they decided on another pregnancy, while closely matched controls did not. The NTD recurrence rate in the former was not significantly different than in the latter (Nevin and Merrett 1975).

The last words on the subject gave the theory its death blow (Anon 1975, Leck 1977), however murkily put: "blighted potatoes have come to appear much less to cause NTD as a result of correlative studies of groups in which no significant differences in prevalence were found between children who were embryos during periods of high and low potato blight." Why so hesitant?

Female Sex Hormones

And then there were female sex hormones. Certain classes of which, the progestins and progestogen–estrogen combinations, were once taken for contraception, pregnancy testing, supportive therapy for threatened or recurrent abortion, etc. Their use for some of these purposes declined after the FDA warned of potential harmfulness, but others continued and were suspected of teratogenic potential.

Genital Defects

Synthetic female sex hormones had often been linked with congenital malformations, genital and nongenital. There was little disagreement about the former – that progestins administered in early pregnancy sometimes masculinized female fetuses (Schardein 1980). The changes however were often transient or easily correctable surgically.

Nongenital Defects

With regard to nongenital defects the story was more complicated. A heterogeneous assortment of malformations was said to be associated with these drugs when used as oral contraceptives or for pregnancy testing, namely cardiovascular malformations, neural tube and other central nervous system defects, limb malformations, certain malformation clusters, as well as malformations generally (Wiseman and Dodds-Smith 1984).

Years of suspicion were kicked off with a report that seemed to implicate hormonal pregnancy testing with myelomeningocele and hydrocephalus, defects of very different pathogenesis, unconvincing also because the testing was made late in pregnancy (Gal et al. 1967). More clearly positive were findings noted in a case-control study conducted by a UK Committee on Safety of Medicines, namely nonspecific associations with some malformations, including oral clefts, limb reduction malformations, and perhaps NTD; supported later by a larger series, but with little consistency among the several malformation types (Greenberg et al. 1975, 1977). An overview listed a potpourri of malformations seen in exposed pregnancies, but concluded that the association was "still in need of confirmation" (Shapiro and Slone 1979).

Defect Nonspecificity

It is this patternless picture, as was the case with Bendectin, that was the major argument against any association with sex hormones. This was made abundantly clear by a vast review of the sex hormone-congenital malformation literature as it stood at the time, a review that was decisive in turning the tide (Schardein 1980). One by one the reports regarding the individual malformations attributed to these compounds were discussed, and the conclusion was reached that the available data were "not convincing." In agreement was an analysis of the epidemiological data that concluded similarly (Wilson and Brent 1981).

Most later examinations of the question worldwide concurred in finding no association between sex hormones administered during pregnancy and individual nongenital malformations and malformations generally. Evaluations also failed to confirm earlier findings of association with a number of disparate defects (see citations in Kalter 2003). All these and other such accusations were dismissed by perhaps the last word on the subject, the FDA publication of wording for drug packaging inserts which removed warnings for nongenital malformations for all progestational agents (Brent 2000).

Diethylstilbestrol

Fetal Wastage

The long ongoing and still continuing story of diethylstilbestrol (DES), the first orally active nonsteroidal estrogen, got started soon after its synthesis (Dodds et al. 1938). It was not long before it was administered to women with previous reproductive difficulties – on the supposition that progesterone deficiency was the basis of spontaneous abortion and that DES could reverse the deficiency – with apparent success in reducing pregnancy loss (Smith and Smith 1946).

Disagreement ensued when further evaluation found it to be valueless in preventing abortion (Dieckmann et al. 1953, Giusti et al. 1995); findings that were criticized in turn as being based on inappropriate design and analysis (Horne 1985). Any attempt to reconcile these discrepant views is impossible today in the face of contemporary patient treatment standards. In any case DES is currently believed to be of no use in preventing spontaneous abortion (Harremoës and Gee 2002).

DES Usage

Doubts of its efficacy in this respect notwithstanding, the use of DES continued, encouraged by some positive pregnancy outcomes; hence for many years it was administered to women with high-risk pregnancies. For example, at the height of its popularity it was received by 5–7% of all pregnant women in two clinics (Lanier et al. 1973, Heinonen 1973), and from 1948 to 1971 perhaps 4 million women in the US alone were prescribed DES during pregnancy (Mittendorf 1995).

In time its usage tapered off. A report from certain hospitals noted that in 1959–65 less than a half percent of pregnant women received DES; and the number of liveborn female offspring (those to be worried about as will be seen below) exposed prenatally to DES in the US in 1960–70 had subsided to about 10,000–16,000 per year (Heinonen 1973).

The Revelation

The appearance of a rare type of a vaginal cancer – clear cell adenocarcinoma – in a small number of women prenatally exposed to DES brought its use to an abrupt halt (Herbst and Scully 1970, Herbst et al. 1971). Pointing to this cluster having a common etiology – maternal ingestion of DES – were its otherwise extreme rarity in premenopausal women, its occurrence in a brief period, and its uniformity of histological features. Thus being the first example of transplacental carcinogenesis in humans. Confirmation soon came, and clarification was urgently demanded (Greenwald et al. 1971, Langmuir 1971, Noller and Fish 1974).

Dosage and Timing

Considered teratologically the outcome was unusual, primarily because it was not expressed until years after birth. Large amounts were administered, totals of 450–4875 mg according to one source (Greenwald and Nasca 1974), and 135–18,200 mg another (Herbst et al. 1977), in contrast with the dose used in studies on monkey species of about 1 mg/day throughout the fetal period (Walker 1984).

The sensitive period was the first trimester, since no cancer resulted from exposure after the 17th week of gestation, and with sensitivity decreasing with later onset of treatment. It was not possible to separate the effects of timing and dose (Herbst et al. 1979), but neither seemed of greater importance (Poskanzer and Herbst 1977).

The Registry

Many additional cases of the cancer became known in the years following the discovery of the first patients in 1970, through a registry and other sources in the US and elsewhere (Herbst et al. 1972b). In the first registry report there were 34 affected women, all of whom were exposed to DES in utero (Herbst et al. 1972a). In the second report the fraction exposed dropped to 74% (Herbst et al. 1972b); which diminished gradually with each successive addition to the number accessioned, till recently it was 59% (Mittendorf 1995). Still a problem at the outset of the present century (Treffers et al. 2001), a January 2008 update mentioned two-thirds of about 760 female babies with possible structural changes in the vagina, etc. were associated with prenatal DES exposure; but no third generation effects as yet (CCARegistry@babies.bsd.uchicago.edu).

Identification of a large number of affected women made it possible to estimate the risk of exposed female offspring developing the cancer. Currently it is 1 per 1000 through age 27 for DES-exposed women compared to 1 per 10,000 through age 34 for all women, i.e. a rapid falloff (Melnick et al. 1987). There has been a steady decrease in the number of cases accessioned annually, from its peak in the 1970s to 1–2 case per year later (Mittendorf 1995), which probably means that the epidemic has all but receded. Meanwhile warnings of later generation risk have been expressed (Treffers et al. 2001, Veurink et al. 2005), but with little soundness.

The Reevaluation

Two accounting systems were set up to chart the occurrence of the cancer, the Registry for Research on Hormonal Transplacental Carcinogenesis, to identify women with it, and the National Cooperative Diethylstilbestrol Adenosis Project, to determine how often it occurred in women exposed to DES (Labarthe et al. 1979). In essence it was found that many instances in young women were not associated with maternal hormone treatment during pregnancy; and none of the DES-exposed women had this type of cancer (Robboy et al. 1984).

The concerted approach revealed that this cancer was not as rare in this age group as was once thought; and that the cancer and DES were weakly associated if at all. [The medical encyclopedia offered to consumers by PubMed said that "(w)omen whose mothers took diethylstilbestrol (DES) during the first trimester of pregnancy are at increased risk for developing clear cell adenocarcinoma." But of course the government is a slow learner.]

Critique

Looking back, flaws in the original case-control studies had made the claimed relation between maternal DES medication and vaginal cancer in offspring doubtful (Mantel 1985, McFarlane et al. 1986, Horowitz et al. 1988). The critics admitted the difficulty of investigating a rare outcome expressed years after the supposed initiating agent. But at the same time they pointed out a fault in the choice of the controls. Since DES, they argued, was prescribed to women who had difficulties in pregnancy, the women chosen as controls should also have had the same difficulties, to avoid what they termed a susceptibility bias. Mainly for this reason, but also because of serious problems of ascertainment, they believed that the existing data were too weak for the causal role of DES to be regarded as firmly established.

This criticism was rebutted by a complex, perhaps questionable, analysis, which concluded that it was unlikely that maternal history of bleeding during the index pregnancy (the only feature considered) was related to vaginal clear cell adenocarcinoma in daughters (Sharp and Cole 1990).

Nevertheless doubts remained. The registry found that genital tract cancer occurred in about 60% of DES-exposed women accessioned, a paltry few hundred of the millions supposedly exposed. But even this may have been an overestimate, relying as it did on what was undoubtedly unrepresentative volunteered information. Furthermore the cancer also occurred in more nonexposed younger women, less motivated to present themselves to the Registry, than were previously believed to exist (Kaminski and Maier 1983, Horowitz et al. 1985). Finally, erroneous recall by mothers of DES ingestion (Tilley et al. 1985) may also have led to faulty estimation of the association.

Congenital Defects

Prenatal exposure to DES, it was alleged, led not only to cancer, but also to certain congenital anomalies in the female genital tract, such as epithelial changes (Herbst et al. 1972a, Kaufman et al. 1977). Contrary to the supposedly rare, late-appearing cancer these were quite common and already present at birth (Kurman and Scully 1974, Johnson et al. 1979). As was true of the cancer their susceptible period was the first trimester – the time of vaginal development – and the incidence was higher the earlier in pregnancy DES treatment was started, as well as being dose related (Sonek et al. 1976, Johnson et al. 1979).

These uterine defects raised questions about the reproductive capability of affected exposed women, and others as well about effects on males exposed utero to DES. Concerning the latter early gestation exposure was also alleged to be correlated with a disparate variety of male urogenital abnormalities, as gleaned from mailed questionnaires and collaborative follow-ups, both of suspect credibility (Palmer et al. 2009).

Summary and Conclusion

Final remarks about this medical episode can be brief. Fifty years after the fad began of treating pregnant women with a nonsteroidal estrogen for the purpose of preventing pregnancy complications – never proven efficacious – the suspected harmful effects on especially female offspring exposed prenatally to the substance were found to be negligible, transitory, unproven, or nonexistent; and the one possible actual effect – a seldom occurring vaginal cancer – has disappeared from history with the aging of the supposedly affected exposed cohort.

The lesson to be learned from this is an old one: hesitance and skepticism in the face of attractive novelties; and doubt of the validity of registry-collected data, attested by other such efforts discussed elsewhere in this work.

A final word, addressed to the concern with estrogenic embryonic imprinting. Following the lead of studies finding that the rodent brain is permanently altered by prenatal exposure to sex hormones, a study was made to learn whether the human brain is similarly vulnerable (Wilcox et al. 1992). The subjects were surviving adult sons and daughters of women who had received high doses of DES or a placebo in a clinical trial in 1950–2. College entrance examination scores were obtained, in the perhaps dubious assumption that they were a measure of cognitive function. Among daughters there was no difference in test scores, while exposed sons had marginally higher scores, "probably due to chance." Draw conclusion for yourself.

Surveillance of Congenital Malformations

The thalidomide episode had ramifications that continue to shape regulatory and scientific activities till today. Causing a not wholly irrational alarm, it aroused public pressure to demand meaningful changes in the philosophy and practices of the FDA. Thereby called into being was not only the refurbished field of reproductive toxicology, but also the unprecedented task of devising means of foreseeing and forestalling outbreaks of malformation epidemics.

Having addressed the preliminary requirement – preempting fetal harm through premarketing drug testing – the next responsibility was to institute ways of curtailing if not preventing such harm; which required knowledge of the baseline of malformation occurrence.

Surveillance and Monitoring

This was to be accomplished by detecting unusual clusters, frequencies, or types of malformations by close and continuous monitoring of newborns; for which, as noted, knowledge of the level and array of defects ordinarily occurring, the so-called background, was the basic necessity. Attempts to estimate this had begun decades before thalidomide was dreamt of. But the findings of the early efforts, based as they were on inconsistent and unreliable means of obtaining information, not surprisingly differed considerably; and relying on such shaky foundations to recognize unusual happenings would have been precarious.

Consequently new procedures were needed to establish dependable baselines and for rapid detection of untoward events, due to newly introduced prenatal hazards. This was to be accomplished by creating means for constant and direct notification of congenital malformations from various sources to linked central facilities, which would then promptly analyze the information and communicate suspected irregular events to responsible authorities.

The earliest such monitoring system was the Swedish system, commencing in 1964 and in continuous operation since then. Soon after its initiation the periodic reporting of severe malformations to a register became compulsory (Källén and Winberg 1968), with its accomplishments and limitations periodically and candidly

H. Kalter, *Teratology in the Twentieth Century Plus Ten*,
DOI 10.1007/978-90-481-8820-8_9, © Springer Science+Business Media B.V. 2010

discussed up to the present (Källén et al. 1984, Källén 2008). Similar systems were established over the years in several regions (Miller 1971a, Flynt 1974, Miller and Lowry 1978, Edmonds et al. 1981, Czeizel et al. 1983). But all were not coordinated, and their differences in method, scope, and purpose may well have diminished their efficient functioning.

Two broad-based registries were set up with the aim of overcoming such limitations. The International Clearinghouse for Birth Defect Monitoring Systems, established in 1974, is a nongovernmental organization representing in 2009 more than 44 malformation monitoring programs worldwide. It helps local congenital malformation registries exchange routine information about prevalence of congenital malformations, provides collaborative epidemiologic research consultation, etc.

The second, the European Concerted Action on Congenital Abnormalities and Multiple Births, EUROCAT, established in 1979, consists of a network of 37 population-based registries in 19 countries across the continent. Its aims have been to pool congenital anomaly data with standardized definition, diagnosis, and terminology, monitor the occurrence of malformations in the participating registries, identify temporal or spatial clusters, act as a data base for research into suspected causes of malformation, provide material to allow collaborative studies on rarely occurring defects, etc. (Dolk 2005).

Monitoring's Limited Abilities

It is now, in 2010, 50 years since the thalidomide disaster. In that interval there has been no affirmed major teratological episode. What should get the credit for this blessing? Premarketing drug screening, monitoring systems, good luck? (Tuchmann-Duplessis 1972)

During this time, had there been the threat of another disaster or anything approaching it, would monitoring systems have averted it? Monitoring systems, it must be admitted, have only a limited ability to detect changes in the background level of congenital malformations, and thus to forewarn of new teratogens.

There are many reasons for this. Although congenital malformations of consequence are fairly common, in epidemiological terms; nevertheless detecting an increase in their frequency requires that a large population be surveyed; and that a particular malformation whose increase is suspected be distinct from the run of the mill types. Other hindrances are low rate of exposure to new teratogens, limited teratogenic potential of most teratogens, and incomplete reporting of malformations (Khoury and Holtzman 1987).

It may be hazardous therefore to place great reliance on surveillance and monitoring, as they now stand. It is fervently to be hoped that thalidomide was unique in its harmful properties, and that another such epidemic will never again happen. But 'never again' cannot be relied on. Animal testing may have been, and may continue to be, useful in keeping possibly prenatally harmful medicines from the marketplace; ever more astute practitioners may nip teratogens in the bud; and innovative surveillance techniques and improved communication will transcend Leck's (1983) doubts and even marginably questionable occurrences be reliably identified. Thus in essence we put our trust in the hands of fate.

Epidemiology of Congenital Malformations

Before malformation surveillance there was malformation epidemiology. Its original concerns, as of epidemiology generally, were "the distribution and determinants of health-related states or events in defined populations" (Last 1988); which in the context of malformations refers to determining whether and how specific abnormalities may be associated with time, place, condition, and circumstance, for the implicit purpose of clarifying etiology.

The pioneering investigations of congenital malformations of this sort were made in Birmingham, England after the Second World War, and dealt with variables associated with anencephalus and spina bifida, which later came to be designated in toto as neural tube defects or NTD (Record and McKeown 1949, 1950a, b, 1951). These, and other malformations, were found to vary in frequency as they were associated in complex ways with numerous biological, social, and miscellaneous variables – maternal age, birth order, spontaneous abortion rate, social class, season, temporal cyclicity, geography, ethnicity, race, genetic and nutritional factors, etc. (Anon 1969, Laurence et al. 1967, 1968, Elwood and Elwood 1980, Knox and Lancashire 1991, Little and Elwood 1991).

It was studies of this kind that were influential in leading to similar studies of other so-called common congenital malformations, usually defined as having frequencies of one per 1000 births and greater, particularly cardiovascular malformations and orofacial clefting (Leck 1976, Rothman and Fyler 1976, Ferencz et al. 1987, Bear 1988, Robert et al. 1996).

Such epidemiological studies were enabled by the establishment of birth registers of malformations and development of the case-control method; the credit for both of which as well goes to Thomas McKeown and Reginald Record, whose innovative work created the field of teratoepidemiology (Leck 1996). McKeown, who became chair of social medicine at the University of Birmingham at age 32, was "one of the world's most original and distinguished epidemiologists ... one of the more profound thinkers in modern medicine" (Anon 1988), as his later writings testify (McKeown 1976, 1979, 1988). The descriptive studies of McKeown and Record laid the foundation for extensions of their technique into several areas of applications of epidemiology – correlative, analytical, interventional.

H. Kalter, *Teratology in the Twentieth Century Plus Ten*,

DOI 10.1007/978-90-481-8820-8_10,

Epidemiology of Malformation Communities

Many challenges have faced the study of malformation etiology. An especial problem is presented by communities of malformations, varied in form but apparently related in that they involve the same organ or part. One way of dealing with this problem is to consider morphologically similar abnormalities a unit, in order to relate risk elements to more statistically manageable defect combinations. The purpose is to form a grouping, by lumping as it is called, that minimizes regarding the defects individually whose differences would obscure etiology. The goal thus is to achieve analytical practicability without loss of biological sense.

This approach depends on more or less arbitrary descriptive systems or classifications, which themselves can vary greatly. Limb malformations, for instance, present a particularly difficult problem. Various schemes for assembling them into a unit have been formulated (see Frias et al. 1977), but few if any have given insight into their etiology and pathogenesis.

Two recent epidemiological studies, each of a specific subset of limb malformations, serve as examples. In one, preaxial defects – most heterogeneous in themselves – were analyzed, but careful attempts at logical grouping were not sufficient to draw more than minimal conclusions (Robert et al. 1997). Reduction defects are another large family of limb malformations which have also been variably classified. These are known to be etiologically diverse, with many known or suspected modes of causation – genic, chromosomal, maternal infection, external teratogen – but few clear risk factors for which have been identified (Calzolari et al. 1990).

Familial Studies

Families of children with congenital malformations may be thought of as examples in micro-epidemiology, with such families the 'study population.' Its purpose is to discover how often family members of different degrees of relatedness to affected individuals have the same malformation, with theoretical and pragmatic goals in mind. One of which is to discover the pattern of inheritance of the malformation; and another to counsel families as to the likelihood of malformation recurrence. Studies of clefts of the lip and palate illustrate this type of study.

Clefts of the Lip and Palate

Facial clefts are prominent in fact and fiction. Robert Malthus – he whose work when read by Darwin sparked the latter's theories of survival of the fittest – was born the sixth child of seven in 1766 in England. His only distinguishing feature would seem to be that he was born with a cleft lip and palate, although this did not seem to impair his speech or social life.

Before Malthus there was Shakespeare who in King Lear (iii,4) drawing upon the superstition of the age invoked the mischievous elf " ... the foul fiend

Flibbertigibbet; he begins at curfew, and walks till the first cock; he gives the web and the pin, squints the eye, and makes the hare-lip; mildews the white wheat, and hurts the poor creature of earth." And many others.

Clefts of the lip and palate, being in the class of commonest congenital malformations, have a frequency in live births of about one per 1000 in Caucasoid, twice that in Mongoloid, and half that in Negroid populations. The vast majority of them are isolated or nonsyndromic, i.e. are not associated with other malformations or syndromes of malformations. Nevertheless, environmental features may contribute to their risk, including epidemiological variables frequently involved in multifactorial teratogenesis, namely geography, race, but not temporality or parentage. And also, as is often true of multifactoriality, its neonatal frequency is diminished by spontaneous abortion of abnormal fetuses (Iizuka 1973, Bear 1988).

These defects are composed of two developmental, etiological, and epidemiological forms: cleft lip with or without cleft palate and cleft palate without cleft lip. This fundamental distinction was discovered by Fogh-Anderson (1942), upon noting that siblings and other family members of persons with cleft lip have an increased frequency of cleft lip, but not of cleft palate, and vice versa; patterns often confirmed since (Fraser 1980b).

That they have a multifactorial etiology with a superimposed developmental threshold is indicated by the risk of recurrence being increased as the number of affected relatives is increased and in relatives of probands of the less-often affected sex (Curtis et al. 1961, Carter 1977, Fraser 1980b) and by the male excess (Fraser and Calnan 1961).

A balloon-puncturing word: the concept of multifactorial or polygenic inheritance, although explaining familial patterns of most common congenital malformations, as Fraser (1980a) has written merely invokes a "simplifying assumption to reduce the complexity of real life to manageable terms," similar in intent to the procedure of lumping mentioned above. In other words, it is an expedient for managing perplexity, which perhaps will some day be superseded by deeper understanding. It is to be hoped that when it eventuates it will allow more accurate family counseling.

Neural Tube Defects

The mammalian central nervous system originates early in embryonic life as a flat plate which normally becomes a tube that fuses in the dorsal midline to form the spinal cord. When this process fails a family of serious malformations ensues, the most frequent being anencephalus and spina bifida; often regarded as related because of their shared epidemiological features. These are by far the most conspicuous, serious, and intriguing of the neural tube defects. For the long history of these malformations and many other details see Kalter (2009).

I note here the following enigma. The most glaring of the NTD is anencephalus; which despite its inevitable early lethality and casting out generation after generation – along with whatever genetic factors may play a part in the fact of its occurrence – recurs generation after generation without cease. And if this were not enough, the extraordinary variation in its frequency – racial, ethnic, temporal, geographic, socioeconomic, plus the as-yet unexplained facts of its being far more common in girls than boys and its declining temporal prevalence – have absorbed the interest of epidemiologists, pathologists, embryologists, teratologists, geneticists in this phenomenon, for decades (Saint-Hilaire 1832–1837, Penrose 1957, Lemire et al. 1978, Michie 1991).

As mentioned above, in depth epidemiological studies of anencephalus and spina bifida began only after the Second World War, with studies in Birmingham (Record and McKeown 1949, 1950a,b, 1951). The numerous biological and social variables affecting their frequency were also discussed above. But it was environment – because many of these associated factors indicated a strong environmental basis, and perhaps also because its ingredients often are or seem most controllable – to which greatest attention was turned. The long list of discrete environmental factors that have been considered in the etiology of NTD (see list of accused factors given above) is testament to the frustration of its students. But it was to one in particular, deficiency of folic acid, to which much attention turned. This is discussed at length below.

Human Disease as Teratogen

In a few instances human disease itself may be teratogenic (Scriver et al. 1995). One such disturbance is the fairly common insulin-dependent diabetes mellitus, while another is the comparatively rare phenylketonuria.

Phenylketonuria

Discovery and Basis

Phenylketonuria (PKU) is another of those medical disorders whose cure had unforeseen health consequences. It consists of a recessively inherited inability to metabolize the essential amino acid phenylalanine, whose consequent accumulation in the body, including the brain, almost invariably leads to severe mental retardation – or as is said in these politically sensitive days, intellectual impairment. The condition doubtlessly existed before its discovery in 1934 (Følling 1934, 1994), but a set of special circumstances was necessary for that to happen (see Centerwall and Centerwall 2000). The discovery by Følling was recently nicely retold (Williams et al. 2008).

Only years later was it recognized that the mental retardation may be prevented by a diet limiting intake of phenylalanine begun early in infancy (Bickel et al. 1953). Early intervention however awaited the devising of a program of early newborn screening (Guthrie and Susi 1963); which becoming mandatory in time, together with the development of a phenylalanine-restricted diet, reduced the damage to the vanishing point.

Because phenylalanine is present in many foods – meat, fish, eggs, cheese, etc., even mother's milk – only by a restricted diet can it be avoided. But the usual semisynthetic low phenylalanine diet is unpalatable, and children as they age often refuse it, with IQ decline usually following (Schuett et al. 1985).

The diet nevertheless has often allowed phenylketonuric women to attain normal intelligence, to lead normal lives, and to become pregnant, which had seldom happened before.

H. Kalter, *Teratology in the Twentieth Century Plus Ten*,
DOI 10.1007/978-90-481-8820-8_11,

Pregnancy Outcome

PKU is the only known human metabolic disorder that is teratogenic (Levy and Ghavami 1996). Intimations of unfavorable outcome of the occasional pregnancies of phenylketonuric women had been appearing for some years before therapy of the condition was invented. This took the form of mental retardation in their non-phenylketonuric children (Dent 1956, Mabry et al. 1963), as well as indications of increased spontaneous abortion, congenital malformation, and fetal growth retardation, stemming from the fetotoxic effects of phenylalanine of maternal origin (Fisch et al. 1996, Stevenson and Huntley 1967). As for the abortion rate, when ascertainment biases were discounted and enlarged data allowed earlier estimates to be revised, it was found to be within the expected range (Lenke and Levy 1980, Platt et al. 1992).

Congenital Malformations

About malformations and other defects there was no doubt. These occurred frequently and consisted of persistent microcephaly and various cardiovascular malformations, especially tetralogy of Fallot and coarctation of the aorta, but others as well, ventricular septal defect, patent ductus arteriosus, etc. (Lenke and Levy 1980, Levy et al. 1982, 2001, Koch et al. 2000a, Platt et al. 2000, Rouse et al. 1990, 2000, Smith et al. 1990). Other frequent outcomes were fetal growth retardation and anomalous facial features – epicanthal folds, long philtrum, upturned nose, micrognathia, etc. (Bovier-Lapierre et al. 1974, Lipson et al. 1984, Platt et al. 1992, Levy and Ghavami 1996, Rouse et al. 1997).

The close resemblance of the abnormal facies to those occurring in the fetal alcohol syndrome (which see below) has been remarked upon and extensively discussed (Lipson et al. 1984, Levy and Ghavami 1996).

Grave effects were later confirmed. By the end of the twentieth century several hundred pregnant women with PKU seen in a program begun in 1984 in several countries had given birth to liveborn offspring, of whom 7.5% had cardiovascular malformations (about 10 times the background level); one-third with microcephaly and later developing mental and psychomotor impairment (Rouse et al. 2000, Rouse and Azen 2004).

Dose and Time Matters

Previous opportunities to examine the relation in humans of teratologic outcome to dose and time variables – occasioned by rubella, radiation, and thalidomide – yielded vague, sparse, or limited information. PKU on the contrary yielded unambiguously clear relations between maternal first-trimester blood phenylalanine concentration and frequency of microcephaly and cardiovascular malformations in untreated pregnancies (Lenke and Levy 1980, Levy and Waisbren 1983, Rouse et al. 1997, 2000, Platt et al. 2000).

[As an aside, it may be recalled that microcephaly is a nonspecific malformation of varied and heterogeneous etiology – chromosomal, genic, and environmental – the latter including irradiation, rubella, organic mercury, toxoplasmosis, cytomegalovirus, and others, as recounted below.]

Microcephaly, usually the most frequent harmful outcome of PKU pregnancy, was associated with relatively high maternal blood concentration, head circumference being normal in the presence of concentrations of less than 600 μM/L phenylalanine achieved by 8–10 weeks of pregnancy and maintained through pregnancy (Drogari et al. 1987, Platt et al. 1992). Cardiovascular malformations also occurred rarely at less than 600 μM (Lenke and Levy 1980, Platt et al. 1992, Rouse et al. 2000); the strongest predictor of these defects being increased concentration at 4–8 weeks of gestation (Rouse et al. 2000).

An international prospective study found cardiovascular malformations in 14.5% of offspring of women with a basal phenylalanine level greater than 900 mol/L by the 8th week of gestation, which thus may be the threshold for such malformations; no such defect occurring after metabolic control was achieved (120–360 mol/L) by 8 weeks of pregnancy (Koch et al. 2000b, Levy et al. 2001).

All in all, optimal outcomes resulted from 120–360 μM/L maternal blood phenylalanine achieved by 8–10 weeks of gestation and maintained throughout pregnancy (Koch et al. 2003).

PKU Varieties

In addition to the so-called classical form of PKU, in which individuals on an unrestricted diet have blood phenylalanine concentrations of 1200 μM or greater, there are two other forms of the disease, at lower concentrations: atypical PKU with intermediate (600–1199 μM) and mild hyperphenylalaninemia, with low concentrations (240–599 μM) (Koch et al. 1994).

In one study whether the atypical form was teratogenic could not be made clear since the outcomes of pregnancies of women with the classical and atypical forms were not considered separately (Rouse et al. 2000). But when looked at alone, mild hyperphenylalaninemia, though it caused no serious fetal consequences, was associated with slight head-size reduction, along with reduced birth weight (Levy et al. 1994, 2003).

Maternal PKU Therapy

The degree of offspring protection afforded is related to timing of maternal dietary treatment. Therapy begun before conception was most successful in preventing the abnormalities, and the earlier therapy was begun after conception the lower the malformation frequency (Levy 1987, 1996). There also was an apparent inverse relation between phenylalanine concentration at conception and birthweight and head circumference, perhaps indicating absence of a threshhold. It seemed therefore that irreversible damage occurred very early in pregnancy and that only strict therapy from before conception might prevent fetal damage (Drogari et al. 1987).

This chilling outlook was contradicted by a study finding normal length and weight and only slightly reduced head circumference in offspring of women achieving metabolic control in the first 10 weeks of pregnancy, but much worse head dimensions with poor control afterward (Koch et al. 1994). Although treatment at any time during pregnancy appeared to reduce the risk of cognitive impairment and developmental delay, the best chance of offspring being free of the defects and with optimal cognitive development was associated with concentrations of less than 600 μM attained by 8–10 weeks of pregnancy and maintained throughout pregnancy (Platt et al. 2000, Waisbren et al. 2000).

PKU Frequency

What began as a medical advance, promising to correct a serious hereditary disease and eliminate a major cause of mental retardation, led to creating many medical and ethical difficulties (Hall 2000, Paul 2000). As inherited diseases are reckoned, PKU is considered common, with estimates of its frequency in live births varying from 1 in 10,000–25,000 in most European areas (Bickel et al. 1981, Scriver 2001) and 1 in 13,500–19,000 in the US (Hellekson 2001).

In ordinary terms, though, PKU is rare. Taking an intermediate estimate in the USA of 1 in 15,000 births, the number of girls [the focus of interest here] with the PKU phenotype born yearly in the US, with about 2 million annual female births, is about 130, and the total number from 1966, the year widespread neonatal PKU testing was mandated, to the end of the century, if the numbers follow, about 4600. This approximately matches an earlier estimate that put the number of women with PKU by the 1980s at 2000 (Drogari et al. 1987), and a bit later at as many as 3000 (Levy et al. 1994). Another estimate, of the number of women of reproductive age with classical PKU residing at the time in the six New England states, gave it as about 183 (Waisbren et al. 1988). The latest reports, 12 or so years later, based on pregnancies followed since 1984, the year the collaborative PKU study began in the US, Canada, and Germany, noted a mere 260 PKU women (Koch et al. 2000a, Platt et al. 2000, Rouse et al. 2000, Waisbren et al. 2000).

One asks, are these all of them, and if so how representative are those included in the latest reports? Is it possible that not all women with PKU have been seen? One study described the difficulty of complete ascertainment, which underscored the likelihood that cases had been missed (Waisbren et al. 1988).

Population Malformation Load

Has the newly won fertility of women with PKU increased the population load of microcephaly and cardiovascular malformations? Whether this has happened and its possible extent depend on the completeness of detection of PKU and especially on how well phenylalanine is maintained at normal levels during pregnancy. Since undiagnosed maternal PKU (Hanley et al. 1999) and PKU inadequately controlled during pregnancy (Waisbren et al. 2000) may not be uncommon, the following projections may be fairly correct.

In the worst possible case, in the US with an overall birth prevalence of microcephaly of 3.4/10000 total births (Anon 1993) and of cardiovascular malformations of 8–10/1000 live births (Hoffman 1990, Meberg et al. 1994), in the aforementioned 35-year period there were about 48,000 background occurrences of microcephaly and about 1.12×10^6 of cardiovascular malformations. Using as an approximation the above-noted 32% frequency of microcephaly and 7.5% of cardiovascular malformations in PKU offspring yielded 2240 and 525 additional occurrences of these malformations respectively. Therefore over this period 16% of all infants with microcephaly and 0.02% of all infants with cardiovascular malformations in the US were born to PKU women, the latter difficult if not impossible to identify, the former obviously constituting a conspicuous medical and social problem.

It is surprising, then, and difficult to understand, why at the same time that many young children of PKU women inadequately controlled during pregnancy were significantly impaired cognitively, an association of this impairment with congenital and persistent microcephaly was not mentioned (Waisbren et al. 2000). Is it that my arithmetic is faulty, or the cited neonatal prevalence of microcephaly exaggerated, or the condition little persists into childhood? Probable also is the transience of the facial dysmorphias; more of which is returned to below.

PKU, Ethnicity, etc.

Recent explorations of this hereditary disease have been at work along several tracks. Its underlying cause, as we saw, is complete or near complete deficiency of phenylalanine hydoxylase, but amazingly this gene locus has been discovered to house over 500 alleles (Scriver 2007), which must mean that virtually every individual with the disease has his or her own genotype if not phenotype. In Israel e.g. among a mere handful of patients, 43 different mutations were identified, often a unique mutation in a single family (Bercovich et al. 2008). Not clear about all this assortment is whether they all cause typical PKU.

Related to this allelic explosion may be the racial and ethnic diversity of PKU prevalence. An example of which was seen in peoples in southeast England in 1994–2004, with an estimate of 1.14/10000 live births in whites, 0.29 in Asians, and 0.11 in blacks (Hardelid et al. 2008). Such variability, not well explained, exists worldwide, ranging at one extreme from an estimate of about 1 per 70,000 in Japan (Aoki 2003), about what was found in US blacks – 1 per 50,000 (Hofman et al. 1991), to, at the high end, indeed the highest found, in Ireland and western Scotland, i.e. in those with Celtic ancestry, where it is 1 per 4500. Strangely about the same high rate occurs in Yemenite Jews (Avigad et al. 1990); in Ashkenazi Jews, being, on the contrary, extremely rare, as was noted years ago (Szeinberg et al. 1963) and confirmed later (Cohen et al. 1986).

All of which allelic and populational diversity means that much about the developmental ramifications of this disease is yet to be told.

Diabetes Mellitus

Diabetes is an old disease, going back at least to the Egyptian pharoahs. But only with the discovery of insulin early in the twentieth century (Banting and Best 1923) did it become possible for diabetic individuals to reach sexual maturity and advanced age. Also opened was the door to later harmful complications of the disease and unforeseen health problems (as was the case with some other newly overcome medical problems – is there always a price to pay?). Which for diabetic women included proneness to a number of detrimental outcomes associated with pregnancy – spontaneous abortion, perinatal mortality, large babies, and congenital malformation.

It must be explained that with respect to pregnancy, as it became definitively recognized only decades later, the disease has two forms, gestational diabetes, which occurs during pregnancy, and early onset diabetes, which usually begins during childhood or adolescence. The latter, pregestational diabetes mellitus, known as type 1 or insulin-dependent diabetes – which is the main concern here – is a major medical problem, especially because – in distinction e.g. to PKU – it is relatively common, with a prevalence of about 5/1000 population.

Pregestational Diabetes Mellitus

I begin here with the pregestational form of the disease, with its several associated consequences, one by one; and then interject a discussion of gestational diabetes, an entity explicitly recognized only years later.

Perinatal Mortality

Soon after establishment of the fertility-restoring benefit of insulin it became apparent that there was a price to be paid for this medical advance, when it was found that pregnancy in women with the early onset form of diabetes was associated with a high rate of offspring death. This was especially so during the earliest decades of the insulin era, when the rate of perinatal mortality, i.e. stillbirth plus neonatal death, was appallingly high, reaching 30–40%. In time, with improvement in maternal care and overall public health advances, it gradually abated, until at present the death rate of infants of diabetic women is approaching general population levels (Kalter 2000, 2011, Wylie et al. 2002, Penney et al. 2003, Evers et al. 2004, Pearson et al. 2007).

Spontaneous Abortion

The frequency of spontaneous abortion (SAB) also sometimes seemed to be increased. The more pressing problems presented by diabetic pregnancy in the earlier years overshadowed this question, but even the little attention it received was confused and contradictory. Nor is this to be wondered at since previous medical generations were often but poorly aware of how common SAB usually is. It later became well established that it varies from about 12% to over 30%, depending on the means of its discovery (Warburton and Fraser 1964, Leridon 1976, Wilcox et al. 1988).

The earliest studies of diabetic pregnancy noted relatively low SAB frequencies; not surprisingly since at that time women were often first seen by physicians at later stages of pregnancy. This was taken into account in a survey of articles that mentioned SAB in diabetic pregnancy, published over a wide span, by considering pregnancies that were seen from a relatively early stage separately from the others (Kalter 1987). In the former the mean SAB frequency was 12.7%, about the level usually found in studies of general populations; which led to the conclusion that diabetes is not associated with an excess rate of SAB in recognized pregnancy.

While this may seem to conflict with the high perinatal mortality rate it does not follow that they should be related, the one being a phenomenon of the earliest weeks of pregnancy and the other of the latest weeks; a distinction which became clear as perinatal mortality in diabetic pregnancy decreased over the years while SAB remained at the background level (Dorman et al. 1999, Penney et al. 2003, Evers et al. 2004, Temple et al. 2006).

Congenital Macrosomia

Early and even later in the postinsulin decades diabetic women often had big babies, usually weighing 4 kg or more. The great increase in weight, as well as in fetal length, was the result not of prolonged gestation, but of fetal overgrowth, especially in the last trimester, producing as the modern parlance puts it 'babies large for gestational age.'

The occurrence of such babies declined over time, but very slowly. In the 1930s–1940s they were nine or more times as frequent in diabetic as in nondiabetic pregnancies, in 1950–1960 down significantly, to 25–40%, but still a great problem; and even more recently big babies were occasionally still being born, even to metabolically well controlled diabetic women. But the story of big babies had further ramifications, told below.

Gestational Diabetes

The narrative is here temporarily interrupted. Let's first define gestational diabetes, as a noninsulin dependent state consisting of mild temporarily impaired glucose tolerance occurring during late pregnancy. Thus pregnancy may occur in a woman who is already diabetic and diabetes may occur in a woman who is already pregnant. The latter, called gestational diabetes since O'Sullivan (1961) coined the term, in time was accepted as a "part of normal pregnancy" (Freinkel et al. 1985). These authors were latecomers, however, since students of the subject were aware in the 1940s and 1950s that carbohydrate metabolism was frequently mildly disturbed during pregnancy, a situation called the diabetogenic effect of that state, but they were unsure whether or not it was merely a physiological by-product of pregnancy itself.

By chance the entity had been discovered even earlier, through a series of interlocked revelations. It began with the frequent birth of big babies to diabetic women; and then to finding that women gave birth to very large babies [first trumpeted years ago with a recording of the very first big baby known to be born to a diabetic mother, Bennewitz 1824] – not only after but also in the years before developing the disease, the so-called prediabetic period; a finding enabled by increasing numbers of diabetic women living to mature years. But the story did not end there.

The big baby phenomenon, upon being studied intensely, was seen to occur at a constant level throughout this precursory period, suggesting that big babies were a constitutional feature. Which then drew attention to other factors in their occurrence, a noticeable one being maternal obesity (Pomeranze et al. 1959).

Doubts mounting about the validity of these findings – all having been ascertained retrospectively – investigators turned to the prospective approach; which led to the discovery that many mothers of large babies were themselves obese; and in the upshot that obese women often responded abnormally to a glucose tolerance test. Although these interrelations were difficult to tease apart, there emerged from this complex the unexpected revelation that a considerable proportion, 1–5%, of *all* pregnant women have aberrant responses to the glucose tolerance test, i.e. have what might be interpreted as 'gestational diabetes.'

Gestational Diabetic Outcome

It must be underscored that no harmful pregnancy outcome is attached to gestational diabetes, aside from a predisposition to fetal macrosomia and neonatal complications (Persson and Hanson 1998, Johns et al. 2006). It has never been seriously claimed that gestational diabetes is associated with increased SAB, an improbability because this form of diabetes appears predominantly in the later months of pregnancy beyond the time it could be responsible for embryonic loss (Kalter 1987); nor was it ever held responsible for an increased occurrence of any other untoward outcome.

As for the apparent increase of perinatal mortality, reported in the years after the discovery of this form of the disease, it was probably due to selection bias (Kalter 2000). And last, publications with analyzable data regarding major congenital malformations in gestational diabetes revealed that it is not teratogenic (Kalter 1998), which again is not surprising, since malformations originate in early pregnancy, before appearance of this malady. (Convenience forces me to document these assertions solely with my overviews and analyses of the relevant publications;

interested readers can decide the question for themselves by looking into the copious references given in the several publications cited.)

The final word about this condition. Though impartially entitled "gestational diabetes: a non-entity?" this essay decisively demolished it, leaving no doubt as to the author's very justified skepticism of its existence (Jarrett 1993).

This decisive pronouncement came – perhaps as sign of frustration – near the end of a period during which the term gestational diabetes had been subjected to meaningless elaborations (see Kalter 2000, p. 56 et seq); a process that has continued into the present century, with no advance in understanding of any aspect of the matter.

Pregestational Diabetes Mellitus Resumed

Now to return to the main subject. The possibility that various harmful outcomes of insulin-dependent diabetic pregnancy were related to elevated level of maternal blood glucose made ongoing assessment of the latter necessary, even from before conception; such surveillance made by measuring long-lived blood components, the glycosylated hemoglobins, especially HbA_{1c} (for a history of these components in the diagnosis and management of diabetes see Kalter 2000, p. 11 et seq). Their significance for congenital malformations will be recounted below.

As for SAB, by the beginning of the present century many summaries of the subject had been made with no consistent finding of an increased frequency in diabetic pregnancy (Kalter 1987, 2000, Temple et al. 2006).

Major Malformations in Diabetic Pregnancy

Assessing the teratogenic potential of insulin-dependent diabetes is another matter. In time, as many facets of detrimental outcome in diabetic pregnancy were vanquished a persistent one – congenital malformations – came increasingly to the fore. Though isolated instances of congenitally malformed children of diabetic women had been reported previously (Kalter 1993), it was not until the 1960s that coherent evidence was presented of an excess of congenital malformations in diabetic pregnancy (Mølsted-Pedersen et al. 1964); soon supported by an international survey (Kučera 1971). After which it became almost the universal belief that maternal insulin-dependent diabetes is teratogenic, i.e. is associated with an increased frequency of congenital malformations (e.g. Mills 1982, Greene 2001).

Glycosylated Hemoglobin

The newly discovered blood components, already mentioned, especially glycosylated hemoglobin HbA_1, provided a more valid guide to glycemic state than previous methods of its measurement; and led to a considerable number of studies of the relation of this component to congenital malformation in the last 20 years of the twentieth century. Initial findings suggested that excessive glycemic level was associated with maldevelopment, but those of later more definitive studies were non-confirmatory, with glycemic level in early pregnancy and malformation rate almost uniformly found to be unrelated. Although the analysis was often contaminated by the frequent inclusion of defects not major and the presence of diabetic women not pregestationally insulin dependent, there was little unambiguous evidence of an association of glycemic level and congenital malformation (see Kalter 2000, pp. 178–9).

Mortalities Versus Survivors

A further matter complicating malformation statistics was lumping perinatal mortalities and surviving infants, classes with very different malformation prevalence history. The illegitimacy of admixing these subgroups is illustrated by general population data, in which over time the level of congenital malformations in perinatal mortalities grew as other causes of mortality were overcome (Kalter 1990), such that by the 1980s major malformations were present in over 30% of offspring dying perinatally, while in survivors it was 3%, as it has been historically (Kalter 1991).

The same is true of diabetic pregnancies. Let me give the reader an example of the situation in diabetic pregnancy, by describing the findings of a study made in northwest England in 1990–4, in which, most uncharacteristically, detailed information regarding malformations was imparted (Casson et al. 1997). Of the 462 recorded diabetic conceptions 5.2% were therapeutically aborted (37.5% of which were malformed), 16.9% were spontaneously aborted (1.3% malformed), of the term offspring 15.7% were stillborn or died neonatally (50% malformed), and the 9.4% of the remainder, the live births, were malformed.

Thus the total malformation frequency in the non-livebirths was 15.2% and in the survivors 9.4%. But the vast majority, 82%, of the defects in the survivors were of a trivial nature, while in the others the frequency was 1.7%, mostly cardiovascular as you will see.

Individually the malformations were skimpily designated. None in the abortuses was named. Cardiovascular malformations (type unspecified) apparently predominated; over half of them occurred in the stillbirths and neonatal deaths, apparently accounting for their total malformation rate; the others were seen in the survivors, in whom the overall frequency, as noted above, was 1.7%. The only other defects named were renal and skeletal abnormalities and hypospadias, not otherwise described and not all major defects, and of a prevalence not greater than in the general population – a total said to be 10 times that of the general population of England and Wales, a gross exaggeration.

If we contrast the Casson report, so replete with detailed findings as to allow full quantitative analysis, with another (Temple et al. 2006), we find that, as was usually the practice, in this report the defects in mortalities and survivors were not considered separately, making analysis impossible.

The detailed information supplied in the Casson report enabled it to be revealed that the frequency of malformations in offspring that perished during pregnancy was far larger than in survivors, the two in essence being separate populations; which makes it clear that a valid analysis requires that mortalities and survivors be looked at individually.

This was not a new revelation. It was seen in the outcomes of many earlier reports of insulin-dependent diabetic pregnancy: first that the malformation frequency in mortalities of diabetic pregnancy was not significantly different from that in mortalities of nondiabetic pregnancy; and second that the malformation frequency in surviving offspring of diabetic women was about the same as that in the general population (Kalter 2000, pp. 102–9, 231–4). See below for specific defects. [A summary and full critique of this topic is found in Kalter 2007, pp. 175–80]. With certain provisions, noted below, it may thus be concluded that type 1 diabetes does not appear to be teratogenic.

This conclusion is subject to recent studies of the question, an updating of which is here attempted. Since the turn of the century there has been a miniavalanche of reports of the outcome of type 1 diabetic pregnancy, a dozen and a half at least having been identified (Suhonen et al. 2000, Wylie et al. 2002, Boulot et al. 2003, Penney et al. 2003, Evers et al. 2004, Jensen et al. 2004, 2009, Roland et al. 2005, Sharpe et al. 2005, Galindo et al. 2006, Macintosh et al. 2006, Nielsen et al. 2006, Temple et al. 2006, Cundy et al. 2007, Guerin et al. 2007, Correa et al. 2008, Persson et al. 2009, Peticca et al. 2009).

Most of them it pains one to say are an unimaginative array of me-toos, with no redeeming quality. While dutifully reporting malformation number in toto, only a few thought to note the outcome in deaths and survivors separately; which are detailed here:

In a French multicentric study the frequency of major congenital malformations in perinatal mortalities was 42.1%, in liveborn offspring it was 4.7% (Boulot et al. 2003). In Scotland it was 22.2% in abortions and 3.4% in live births (Penney et al. 2003). In the Netherlands it was 33.3% in perinatal mortalities and 4.8% in survivors (Evers et al. 2004). In one Danish study it was 18% in perinatal deaths and 2.0% in survivors (Jensen et al. 2004), and in a later one, 16.1% in perinatal deaths and 2.0% in survivors (Jensen et al. 2009). And in England it was 33.8% in terminated pregnancies, stillbirths, and infant deaths and 6.6% in liveborn – said to be major as defined by Eurocat, but probably more inclusive than that (Bell et al. 2008). In every one of these instances, then, survivors were less often malformed than the nonsurvivors.

Despite this clear quantitative distinctiveness there may still be a doubt whether the array of malformations in "diabetic deaths" is different than in 'nondiabetic deaths.' Evidence regarding this matter, though not abundant, indicates that the congenital malformations in offspring that perish during pregnancy are similar in amount and type in diabetic as in nondiabetic pregnancies (MacHenry et al. 1979, Poland et al. 1981). With the era of such investigations giving every sign of largely passing, few more can be expected.

Thus, to reiterate, it is probable, being judicious, that type 1 diabetes is not teratogenic, since offspring of diabetic women that do not survive and those that do survive each resemble in degree and form of maldevelopment offspring of nondiabetic women.

Minor Abnormalities

With one possible exception. To make a clean sweep of it there is the question of minor defects. In and of themselves – though being relatively trivial physical divergences from the typical, and of little or no medical or cosmetic consequence, and thus having no intrinsic significance – they have nevertheless been thought to act as signals of major prenatal maldevelopment (Ekelund et al. 1970, Janz 1982, Pinsky 1985, Leppig et al. 1987, Koch et al. 1992).

If this be so, and – to play the devil's advocate – if maternal diabetes causes major malformations, minor defects, according to this argument, should be increased in diabetic pregnancy. Few reports of the outcomes of diabetic pregnancy listed major and minor aberrations separately and fully. And even fewer contrasted a nondiabetic group, important for examining the frequency of minor defects, because there is little agreement about what is to be considered a minor defect. A comparison of offspring of diabetic and nondiabetic pregnnacies in these handful of reports found almost complete accord that there was no difference between cases and controls in their frequency (Kalter 2000, p. 111). That should sweep this subject into the historical dustbin.

Specific Major Malformations

But we're not done yet, let's follow this trail to the end. Partly considered above were malformations in general. But what about particular ones? It has sometimes been held that, in addition to being increased overall in offspring of diabetic pregnancies, specific types of malformations were especially overrepresented, namely cardiovascular, neural tube, and certain sacral defects, with the last constituting a specific diabetic embryopathy. These claims were deeply scrutinized through a close reading of the world literature (Kalter 2000). The reader interested in the complete account is directed to this publication. The following summarizes the details.

Caudal Dysplasia

Many surveys of the malformations in offspring of diabetic women have found that they formed no particular pattern (Driscoll 1965, Neave 1984, Pedersen 1977, Holmes 1992). Opposed to this was the view that a particular syndrome of malformations is present.

I must note first that the subject began with an article, strangely enough, devoid of any relation to diabetes (Duhamel 1961). In it were described certain vertebral, urinary, genital, and lower limb defects whose frequent association led them to being considered a unit, named the syndrome of caudal regression.

The syndrome then, taken out of the original hands, became associated along the way with maternal diabetes. Thought at first to consist of specific defects of the upper leg (Lenz and Maier 1964, Lenz and Passarge 1965), and later also of sacral defects, led the name of the combined defects to be changed from "caudal regression" to "caudal dysplasia," and redefined as hypoplasia of the femur and absence of the sacrum (Passarge and Lenz 1966). A further refinement excluded femoral defects, and the entity made, finally, to consist of lower vertebral abnormalities alone (Lenz and Kučera 1967). Sacral abnormalities thus came to be considered a cardinal expression of diabetic teratogenicity (Kalter 1993), and accepted as a characteristic of the offspring of maternal diabetics (e.g. Benirschke 1987).

But this notion is troublesome. Sacral defects are rarely diagnosed in neonates, usually only being recognized months or years later, when the condition presents urological and other problem. Its relation to maternal diabetes is problematic therefore, especially since aside from a handful of instances no such spinal malformations were mentioned in records of several thousand offspring in series of diabetic women. In fact, data in the few publications dealing with population frequency of sacral absence in perinatal mortalities indicate the unlikelihood of this malformation being more common in the offspring of diabetic women than in newborns generally (Kalter 2000, Table 14.1, pp. 165–6).

Central Nervous System Malformations

Also said to occur in excess in offspring of diabetic women are anencephalus and spina bifida aperta. Because anencephalus is the most conspicuous and unmistakable of the NTD it has gotten the most attention, and extensive information regarding its worldwide distribution and prevalence is available. Since the condition is invariably lethal before or soon after birth a clear-cut record of its frequency at term or thereabouts is derived from perinatal mortalities.

Information was extracted from reports of series of diabetic pregnancies in which perinatal mortality was explicitly noted and in which all other malformations occurring were named as well. Not surprisingly there was an inverse relation between the frequency of anencephalus and that of perinatal mortality, with anencephalus increasing relatively as mortality decreased over time (Kalter 2000, p. 169).

Knowledge of the primary frequency of anencephalus, i.e. in embryos and early fetuses, comes predominantly from studies of the general population; which though sparse indicates that some large proportion of conceptuses with NTD, particularly of anencephalus, is spontaneously aborted (MacHenry et al. 1979, gave a figure of 41%), leaving relatively few to survive to birth. The same is undoubtedly true of diabetic pregnancy, with a greatly reduced frequency present at birth, as older and recent findings have shown (Kalter 2009, p. 180 et seq, Ray et al. 2004, Macintosh et al. 2006, Nielsen et al. 2006, Cundy et al. 2007, Guerin et al. 2007).

Cardiovascular Malformations

Congenital malformations of the heart and great vessels are the most common of all human malformations in the overall population, once having a frequency of

3–5/1000 live births diagnosed by 1 year of age (Hoffman 1990), and later about 8–10/1000 (Kalter 2007). The most common of these malformations is ventricular septal defect (VSD), which also increased from one quarter to one third of them all in earlier years to half or more later; the overall rise almost certainly due however to refined, more intense and prolonged diagnosis, enabling recognition of small septal defects not always evident or detectable at birth. In fact the later increase in the total level of cardiovascular malformations may have been due to the great increase in that of VSD alone. Most of these small defects (70%) however have no physiological significance and close spontaneously by 1 year of age.

Here the main concern are cardiovascular malformations in neonates, since it is at this age that children of diabetic women have mostly been examined for congenital defects. However difficulty of recognition in infancy makes perinatal mortalities the necessary source of information. Also the recently changed diagnostic methods require considering previous and later findings separately. It is encumbered by these qualifications that cardiovascular malformations in perinatal mortalities are to be assessed.

The information, based hence mostly on mortalities, combining much data, yielded frequencies in pre-1980 diabetic pregnancies of about 8%, about the same as in the 7% population estimate; while in later years it was over 14% in diabetic pregnancies, the increase mostly due to intensified diagnosis (Kalter 2000). It must be noted that these figures conceal the great difference between stillbirths and neonatal deaths, the frequency in the latter being manyfold greater than in the former (difficult in itself to explain), the gap closing over the years however.

Finally, although sparse, cardiovascular malformations in surviving neonates up to 1 year of age from diabetic pregnancies in the final quarter of the last century are of interest. Such diagnoses found a total frequency of perhaps 1.8%, which was about four times that in the overall population (Kalter 2000); about the same differential as later studies found (Loffredo et al. 2003, Wren et al. 1999, 2003).

Thus, with respect both to mortalities and survivors, the observed increased frequency of cardiovascular malformations in diabetic pregnancy must be tempered by the increase in these malformations in more recent years as perhaps the product of overdiagnosis.

Specificity of Teratological Response

The claim that maternal diabetes is associated with an increased frequency of particular congenital malformations was analyzed and found wanting, as noted above. What is left is the possibility of a general increase in malformation incidence without a preponderance of any one type of abnormality.

Does diabetes conform to teratological principles? Especially pertinent to the question are time and dose relations. By definition pregestational diabetes is present prior to and throughout pregnancy; and the question becomes whether such protractedness affects time- and dose-effect relations. As it was once expressed, the rationale for a nonspecific outcome is that "the diabetic state persists throughout pregnancy; thus one could reason that nonspecific anomalies of all organ systems would be expected, in contrast to the specific group of anomalies characteristic of drug- or virus-induced teratogenesis" (Simpson 1978).

That is not the way things work. Chronic or constitutional maternal states that are teratogenic offer examples. Fetuses may be infected with rubella throughout gestation, and yet the teratogenic effects will be related to particular stages of gestation, those at the time or times of maternal infection (Ueda et al. 1979). Another example is phenylketonuria, present throughout pregnancy, yet whose embryopathic effects show organ and time specificity (Koch et al. 1994).

Since diabetes was not associated with a specific constellation of congenital malformations, whose analysis could assign particular ones to different susceptible periods of early gestation, it departs in this respect from known human teratogens, and thus does not adhere to the specificity tenet required for honest to goodness teratogens.

Dose-Response Relation

Early in the study of the outcome of diabetic pregnancy attempts were made to relate various features considered to indicate severity of the diabetic state to the high perinatal mortality rates of the day, such as maternal age at onset of the disease, length of time the disease had endured, required insulin dosage, pathological complications, sex hormone imbalance, etc. None of them was clearly tied to infant death. Later, closely supervised maternal care and normal or nearly normal blood glucose levels were found to be related to improved survival.

The great decline in the death rate, accelerating in the 1950s, turned attention to the most prominent remaining lethal factor, malformations, and here also attempts were made to find maternal correlates that would give insight to prevention. Taking a hint from the beneficial effects of maternal glycemic control, studies were made of the possible association of glucose level in early pregnancy and malformation frequency.

If congenital malformations were caused by maternal diabetes, mediated by maternal carbohydrate imbalance, the degree of this imbalance, gauged by level of glycosylated hemoglobin, should have been correlated with the intensity of the effect, as appears to be the case for adult diabetic complications (Viberti 1995). This proposition was tested by determining the frequency of congenital malformations in diabetic women with different mean levels of this blood component early in pregnancy. But clear-cut, consistent findings supporting this contention eluded investigators (Kalter 2000, p. 180).

As noted, in diabetic pregnancy the perinatal mortality rate greatly decreased over time, but the malformation frequency in surviving offspring remained constant. Conforming to the first paradigm, ceteris paribus, the malformation rate should have increased, according to the second it should have decreased. It did neither. This is further indication that maternal diabetes is not teratogenic.

Conclusion

The belief that maternal diabetes is teratogenic has almost been axiomatic. But numerous matters deny it credence. The majority of ongoing studies of diabetic pregnancy outcome were conducted in diabetes clinics in large metropolitan hospitals, whose subjects were unrepresentative of the state of the disease in general. Evidence of which was the high glycemic levels found by them, far greater than in other sorts of studies of diabetic women (Hanson et al. 1990). Faulty conclusions have stemmed from admixing findings in surviving and nonsurviving offspring, as noted above – the record for congenital malformations in perinatal mortalities clearly finding, in each decade since early in the insulin era, that the frequency of congenital malformations in mortalities was not different from that in perinatal mortalities in the general population.

When years ago autopsies of perinatal deaths were pointed to as a means of exploring the effects of maternal diabetes on pregnancy outcome only three reports of such data were identified, with a mean malformation frequency in the 93 specimens of 22.6% (Rubin and Murphy 1958). Time has added many such studies, but has not changed the outcome. A summary gave the total number autopsied as 815, with a malformation frequency of 20.2%. In the four studies containing controls the total frequency of 17.4% further indicated that maternal diabetes does not lead to an increased level of malformations (Kalter 2000, p. 106). A recent prenatal detection study found NTD not significantly increased in diabetic pregnancy (Ray et al. 2004).

A full account of the few matters presented in abbreviation in this section, as well as information about many additional topics relating to the effect of diabetes on the outcome of pregnancy, is found in comprehensive works (Kalter 2000, 2011).

Fever

But Is It a Human Teratogen?

There is no doubt that raised maternal temperature causes congenital malformations in laboratory animals, as was demonstrated under intense experimental conditions (Warkany 1986, Graham et al. 1998). Whether fever, or more generally hyperthermia, in the states under which increased maternal temperature occurs in humans has caused or is capable of causing malformations is not yet settled.

This raises a ticklish question. This is because while many chemicals and other factors can cause malformations in mammals under experimental conditions, few of them are human teratogens (Schardein 2000, Shepard 2001). And this is so, not because humans might not react similarly if they were to experience the extreme conditions imposed on laboratory animals, but because humans rarely experience such drastic conditions. So it is necessary to be realistic, and understand that only when malformations result under conditions experienced by humans that the inducing agent can be accepted as a human reratogen. Is this true of hyperthermia?

It has been claimed that the temperature elevation in humans caused by febrile illness, sauna, and hot tub duplicates that achieved under experimental conditions. Teratogenic threshold levels of hyperthermia have been difficult to establish in animals. But it appears that temperatures must be raised 2°C or more above what is normal for the species, for some length of time, at periods of susceptibility, to cause malformations, while still higher ones are embryolethal (Graham et al. 1998). How do these compare with human temperature elevations said to be teratogenic?

Earlier studies had been contradictory (Kalter and Warkany 1983, Warkany 1986). A later one was instructive (Chambers et al. 1998). It concerned women calling a teratogen information service regarding a fever episode of different origins in a current pregnancy. About half reported temperatures of 38.9°C (102°F) or more for at least 24 h, the others less than 38.9°C or lasting less than 24 h. Offspring exposed to high fever had no higher malformation frequency than the others, but two of the 34 or 5.9% had anencephalus compared to none in the other groups. Which seemed to indicate that high fever lasting 24 h within the first 4 weeks postconception increased the risk for major malformations, especially anencephalus.

The report is laden with inconsistencies and doubtful aspects. Recall is all important because the trustworthiness of the only significant finding, the two occurrences of anencephalus in the high fever group, rests on the time the elevated fever occurred, i.e. whether it was during the critical embryonic period. This was not established (in fact the high fever group entered the study at a time beyond that of increased risk for anencephalus). Another possible factor, malformation recurrence, was suggested, but hardly substantiated, by the frequency of previous spontaneous abortions being significantly greater in the high fever than in the control women.

Sauna- and hot tub-induced hyperthermia was also accused of being teratogenic (described by Warkany 1986). This "hypothesis was met with considerable scepticism in Finland, where the sauna is an important constituent of everyday life, including during pregnancy...and where the incidence of anencephalus is among the lowest reported" (Saxén 1983). Another point. It must be remembered that rubella is also accompanied by fever, but as discussed above, neural tube defects were not among the varied malformations and defects associated with this infectious disease.

The question resurfaced as of late with a report of so-called characteristic multiple congenital abnormalities, including NTD and orofacial clefts, due to what was dubbed maternal high fever-related diseases in the second and later months of pregnancy (Czeizel et al. 2008b). The incriminated diseases included influenza, common cold with secondary complications, tonsillitis, herpes with high fever, etc.; the final judgment: 2.8% of cases were perhaps associated with high fever. Flogging an old horse.

A teratogen update on hyperthermia (Graham et al. 1998) began with a bang. It stated that "[H]yperthermia was the first teratogen in animals that was subsequently proven to be teratogenic in humans;" exhibiting how multiply wrong one simple sentence can be.

Environmental Hazards and Disasters

Discussed in this chapter are conditions that are pandemic, i.e. widespread and hence impact large numbers of people at more or less the same time; as distinct from those that people are exposed to one at a time. Most such potential dangers are controllable, but one, organic mercury, presents a risk that is not wholly avoidable. Only one is a 'natural' happening, as distinct from the others, which are the products of human action.

Iodine Deficiency

When I was a graduate student at McGill University in the early 1950s, for my master's degree project, under the direction of F. Clarke Fraser, I was given the task of exploring the connection between phenylthiocarbamide (PTC), also known as phenylthiourea, and goiter. Sensitivity to the bitter taste of PTC had been discovered in the 1930s to be a recessively inherited human trait (one of the first discovered); and insensitivity to the chemical was later found possibly to be related to the goitrogenic properties of brassicae, in which it was a natural ingredient. [How I went about the task is a long story. See my master's thesis – Kalter 1951.] But in essence it was my initiation into the world of scientific investigation, and a salutary lesson since I learned a lot about approaching people in clinical and exploratory settings, about statistical and sampling procedures, about searching the literature, and also to be a realist since my efforts turned out to have negative results.

My next exposure to goiter, in a sense, came with my association with another great figure in teratology, Josef Warkany. It is to his fascination with this disease, in fact, that experimental mammalian teratology owes much of its beginnings (Warkany 1988a). As a youth in Austria during the first world war, venturing into remote Austrian mountain valleys he saw a curious sight, people with huge goiters and their mentally retarded children, the outcome of lack of a vital nutritional ingredient, iodine.

H. Kalter, *Teratology in the Twentieth Century Plus Ten*,
DOI 10.1007/978-90-481-8820-8_12, © Springer Science+Business Media B.V. 2010

The Story of Iodine

Iodine deficiency has long been known to be associated with endemic goiter. Its story is fascinating. Over long geological eras iodine was leached out of soil and removed to the oceans, where it is now largely found. Its paucity in soil leads to its dearth in plants, particularly in areas away from seacoasts and at higher altitudes; and it is this lack of iodine, required for the synthesis of thyroxine, that is the major reason for the occurrence of endemic goiter in many parts of the world.

When pregnant women are deficient in thyroxine and thus become hypothyroid their fetuses suffer since they derive thyroid hormones in part from mom. Human fetuses begin synthesizing thyroid hormones at 10–12 weeks of gestation, and fetuses of other species at developmentally corresponding times. Failure of the fetal thyroid gland to produce adequate amounts of hormone leads to fetal hypothyroidism, which if not identified at birth and soon treated results in permanent mental retardation. But if in addition the mother is hypothyroid the problem is compounded and the infant will also suffer various neurological deficits (Utiger 1999).

Cretinism

Iodine deficiency is not an ordinary teratogen, since its only morphological consequence is widely separated cranial sutures and large fontanelles; but it does lead to pre- and postnatal growth deficiency and neuropsychological maldevelopment. Severe maternal iodine deficiency and consequent hypothyroidism combined with fetal hypothyroxinemia, especially in early second trimester, thus lead to physical and mental retardation, which continually worsen, with dwarfism of a type called cretinism ensuing.

While not every goitrous woman has children who are cretins, every cretin has a goitrous mother. The majority of cases of endemic goiter, in the past and still today, in many areas is the outcome of iodine deficiency. For example, in central Africa and remote parts of China, even today endemic goiter is common and is often accompanied by endemic cretinism and mental retardation.

Only at the end of the nineteenth century was iodine discovered to be an important constituent of the thyroid gland, and not till the first years of the twentieth century was its deficiency concluded to be the main cause of goiter and iodine administration found to reduce the size of the hyperplastic thyroid.

Endemic Goiter

Iodized salt was introduced into wide use in the 1920s; but in the many parts of the world where iodine supplementation has not yet been adopted or its consumption is insufficient it has been estimated that billions of individuals may still be subject to its consequences (Lewis 2009, Zimmermann 2009). The World Health Organization estimated in 1990 that 20 million people had some degree of brain damage due to iodine deficiency experienced in fetal life; its deficiency therefore still the most common preventable cause of mental retardation (Hollowell and Hannon 1997).

There is a crucial difference between the neuropsychological effects of iodine deficiency and the behavioral and cognitive effects said to be associated with certain other in utero exposures, e.g. to alcohol and antiepileptics. And that is, that the postnatal consequences of the former have been clearly identified as stemming from molecular abnormalities in the brain (Oppenheimer and Schwartz 1997), while a neurological basis for the putative effects of the others has not been described.

A recent summary and review of numerous facets of the question of iodine deficiency brought the subject up to date, but without really adding anything new (Lee et al. 2009).

Organic Mercury

Organic mercury is the only environmental factor for which there is clear evidence of being responsible for prenatal damage in humans. Events in several parts of the world, especially Japan and Iraq, proved that such damage can be caused by maternal consumption during pregnancy of foods contaminated with various forms of organic mercury, especially methylmercury (Nelson et al. 1971, Myers and Davidson 1998).

The Epidemic in Minamata

This danger was first revealed by the occurrence of several abnormal children about 8 years after the end of the Second World War in Minamata City, a small seaside community on the southwestern coast of the Japanese island of Kyushu. In hindsight, as it was later realized, the earliest sign that something odd was afoot was the strange behavior of the area's cats – violent convulsions and excessive salivation, and similar symptoms in other cats not long after being brought to Minamata.

Soon thereafter came an instance of the condition in a human, with the birth in 1953 of an infant with nonspecific neurological symptoms; the meaning of which became clear when similar signs showed up in other children (Harada 1968, Harada 1978, Tsubaki and Irukayama 1977).

It was soon deduced that the condition, given the name Minamata disease, was due to maternal consumption of fish and shellfish contaminated with methylmercury, stemming from the effluent of an industrial plant located along the river flowing into Minamata Bay. A second Japanese outbreak, of lesser extent, occurred in 1965 in Niigata, a coastal town in a northwest region, and its cause also identified as consumption of seafood contaminated with methylmercury.

Congenital Minamata Disease

Forty years or so later, by 1995, the number of cases of organic mercury poisoning had risen to over twenty-two hundred (Harada 1995); though the number of congenital occurrences, i.e. of strictly prenatal origin, was unclear (Igata 1993). The few facts suggested that these were infrequent (5.3%), though greater in Minamata than elsewhere. The youngest patients were 6–12 months old when clear signs of the condition, diffuse cerebral palsylike symptoms, first appeared. But it was unclear in particular instances whether exposure was in utero or later, from contaminated breast milk.

Numerous brain pathologies were found, including general atrophy, hypoplasia, and abnormal cytoarchitecture of the cerebral cortex, and various changes in the cerebellum. Most prenatal damage was apparently due to disturbances in later months of gestation, since the frequency of major malformations was not greater than usual. This is uncertain however because of inadequacies of neonatal diagnosis (Harada 1995).

Source of the Methylmercury

The basis of the problem was seafood, the end of the chain containing the accumulated toxin; and the nutritional practice of the population of the region, an important proportion of whose diet, especially of fishermen and their families, consisted of fish and shellfish. But looking back, the pollution of the area waters, fished by the seaside inhabitants, originated with the effluent from a factory operated by the Nippon Chisso Company, a leader in Japan's chemical industry, using massive amounts of mercury as the catalyst in the production of acetaldehyde.

The region had benefited economically from the factory, but reaction set in as the illness was suspected of being related to its presence. Questioned, the company and members of the local Kunamoto University, who were in possession of the facts, were at first uncooperative. Only after governmental intervention did the company admit its negligence in withholding critical information and in continuing to dump waste and provided compensation for the damage.

It was estimated that 70–150 tons of mercury had been discharged into Minamata Bay, removal of which by dredging, starting in 1977, took 13 years to accomplish (www.enu.edu/minamata). The announcement, as noted in the New York Times on July 30, 1997, by the governor of the Kunamoto Prefecture, declaring "the fish in the bay to be safe to eat," it would seem put finis to this extraordinary episode in human toxicology.

The Epidemic in Iraq

A similar epidemic took place in Iraq, but though much briefer than in Japan was more devastating, causing microcephaly, seizures, mental retardation, and cerebral palsy, as well as less severe neurological outcomes (Bakir et al. 1973, Greenwood

1985). It followed consumption of home-made bread using imported grain that had been treated with a fungicide containing methylmercury. The grain arrived in September and October 1971 and was distributed until January of the following year, when the authorities issued warnings about it. By March 1972 over 6000 cases were reported, but the frequency of prenatally damaged infants appeared to be low (Amin-Zaki et al. 1979).

Matters of Environment and Dose

Inorganic mercury is ubiquitous in air, soil, and water. In lakes and oceans it is converted to methylmercury by common bacteria and moves up the food chain into edible seafood species. Ordinarily its concentration in marine life is low, but in areas of man-made pollution, as in Minamata Bay, it can reach 40 parts per million (ppm), and in hair of patients up to 700 ppm (Harada 1995).

Study of the fetal dose-effect relation was made difficult in Japan by the initial extremely high prolonged level of exposure, and the relatively low one for a long time afterward; while in Iraq and elsewhere the high level and limited duration of exposure facilitated such analysis. In a small sample of mother-infant pairs, the methylmercury content of maternal head hair was weakly related to certain neurodevelopmental features in children (Marsh et al. 1987). But a dose-response analysis suggested that as little as 10 ppm of mercury could be harmful (Cox et al. 1989), compared to an average of 1 ppm or less in the US population (Khera 1979).

Studies Elsewhere

Studies in fish-eating populations in several parts of the world have been negative or inconclusive. A series of reports concluded that a diet high in ocean fish posed no threat to developmental outcomes of young children in the Republic of Seychelles (Davidson et al. 1998), "current evidence does not support the hypothesis that consumption of even large amounts of fish during pregnancy places the fetus at neurodevelopmental risk from MeHg exposure" (Myers and Davidson 1998), "data do not support the hypothesis that there is a neurodeveopmental risk from prenatal MeHg exposure..." (Myers et al. 2003), and "there is no convincing evidence for an association between prenatal exposure and child development..." (Davidson et al. 2006). Such pronouncements have not quelled anxiety.

In the Faroe Islands, whose population is said to be highly dependent on seafood, including pilot whale, a small sample of infants were tested at about 2 weeks of age and found to have deficits in certain functional abilities in association with a 10-fold increase in cord-blood mercury concentration (Steuerwald et al. 2000). When tested at 7 years of age, children exposed prenatally to mercury showed negligible IQ deficits (Axelrad et al. 2007).

In young Inuit children in Greenland, another population exposed to methylmercury through fish eating, there were no exposure-associated neurobehavioral deficits, while neuropsychological ones were of doubtful statistical significance; results that were in accord with other evidence of prenatal or early postnatal exposures to methylmercury causing trivial effects hardly of personal or medical importance (Weihe et al. 2002).

An extraordinary instance is to be noted, for what it is worth. A young woman, following an attempted suicide by intravenous injection of liquid mercury, became pregnant, and despite high levels of mercury in various tissues had a child that was without "immediate adverse health effects" (Pugach and Clarkson 2009).

What should the public be told? I can do no better than cite the conclusion of an article looking deeply into the problem (Oken and Bellinger 2008). "Recent publications . . . evaluating the overall effects of maternal prenatal fish consumption suggest that there is on balance no harm, and perhaps some benefit, from greater maternal fish consumption."

Agent Orange

Agent Orange, a herbicide used by the US military during the war in Vietnam in the 1960s, was claimed to have many harmful effects, among them congenital malformations in children of military veterans.

Agent Orange – the name came from the orange stripes on the 55-gallon drums it was stored in – was a mixture of 2,4-D (2,4 dichlorophenoxyacetic acid) and 2,4,5-T (trichlorophenoxyacetic acid), the latter contaminated with an impurity, TCDD or dioxin (2,3,7,8-tetrachlorodibenzo-*p*-dioxin). For defoliant purposes it was dispersed mostly from aircraft and exposed to it were an unknown but probably relatively limited number of US, Australian, and New Zealand military personnel and a larger but also unknown number of Vietnamese soldiers and civilians.

Male Exposure Toxicity

Studies in several animal species of the developmental and reproductive effects of the chemicals in Agent Orange found them not to be teratogenic (Institute of Medicine 1994). This is irrelevant however since there is no parallel between exposure of pregnant female animals in experimental studies and exposure of men prior to fathering children. Nevertheless, as was pointed out, offspring sired after paternal exposure may have been harmed through chromosomal aberrations or gene mutations induced in spermatogonial elements, or by transfer of chemicals or metabolites to seminal fluid or spermatogonia (Brown 1985, Trasler and Doerksen 1999).

These possibilities were invalidated by various facts: the chemicals are quickly metabolized and removed from tissues; male rats administered dioxin did not sire abnormal offspring; previous exposure of men to toxic substances was not associated with increased frequency of malformation, stillbirth, or spontaneous abortion (Pearn 1983, Wolfe et al. 1995, Schnorr et al. 2001, Lawson et al. 2004). All of which led an Institute of Medicine document to state that there was "inadequate or insufficient evidence of an association" (Institute of Medicine 2006). Nevertheless allegations based on meta-analysis held that certain malformations occurred in excess, although no concrete support for the contention was presented (Ngo et al. 2006).

A footnote: apparently only one study has ever been made of female military personnel who served as nurses and non-nurses in Vietnam (Kang et al. 2000). It was learned by telephone interview, long after the period of service, that among the very limited number of the children of women who responded that no unusual reproductive consequence occurred (spontaneous abortion, stillbirth, etc.), except for a slightly elevated rate of "moderate-to-severe" birth defects in livebirths appearing up to 2 years of age; by which was meant chromosomal abnormality, malignancy, genetic disease, malformation, etc. The only malformation named was spina bifida, which occurred in 15.7% of children of those who served in Vietnam and 14.3% in non-Vietnam veterans, extraordinarily high frequencies difficult to interpret and impossible to accept.

Exposure of Vietnamese Nationals

Anecdotal reports in the early 1970s of unfavorable pregnancy outcomes in Vietnamese women carried little weight. Broad scale studies were taken more seriously, but comparing pregnancy outcomes in 22 Vietnamese hospitals in 1960–5 and 1966–9, years of light and heavy spraying respectively, revealed no difference in malformation and stillbirth frequencies (Cutting et al. 1970). Ascertainment difficulties, record unreliability, imprecise exposure information, etc. however left the matter undecided.

A report of an increased number of children with cleft lip in a Saigon hospital (Herbicide Assessment Commission 1970) was deceptive, since it was based on surgical cases and not malformation occurrences (Fraser 1990, personal communication 2002). Reports of certain types of malformations associated with presumed exposure of Vietnamese men (Constable and Hatch 1985) were contradicted by most findings in US and other studies, noted below.

Exposure of US Military

Although the overall frequency of congenital malformations was not increased in the children of US Vietnam veterans, two defect types, spina bifida and cleft lip with or without cleft palate, were increased in those whose fathers had possibly been more heavily exposed (Erickson et al. 1984). These positive findings, out of the over 90 anomaly entities examined, may well be explained by data dredging: "multiple comparisons are *by chance alone* [sic] likely occasionally to yield a positive finding" (Hatch and Stein 1986). Regarding exposure data, noted by the Centers for Disease

Control, serum dioxin levels in Vietnam veterans who served in a heavily sprayed area were virtually the same as in non-Vietnam veterans (CDC 1987).

Hospital records of major malformation and stillbirth frequencies in pregnancies in 1977–80 of wives of Vietnam veterans failed to reveal an association with military service; nor was there a clear difference in such outcomes in pregnancies of wives of men with military service in Vietnam and those whose husbands were non-Vietnam veterans (CDC 1989, Aschengrau and Monson 1990).

Ranch Handlers and Reproduction

US Air Force personnel directly involved in the spraying missions in Vietenam were perhaps the most exposed of any US troops to Agent Orange (Lathrop et al. 1984). Their children did not have a greater frequency of serious congenital abnormalities than those of personnel not exposed to the agent (Hall and MacPhee 1985, Hatch and Stein 1986, Sever et al. 1997).

Paternal serum dioxin level, measured years after the time of exposure, was associated with over a dozen abnormalities, but formed no consistent malformation pattern; many of which, e.g. hypospadias and Down syndrome, were not major malformations or were etiologically irrelevent (Wolfe et al. 1995). These weak data were nevertheless mistakenly accepted as possible evidence of a relation of exposure to an increased risk of some defects (Institute of Medicine 1998).

Studies specifically directed to the question of spontaneous abortion stated explicitly that the risk for this outcome was not increased (Stellman et al. 1988, Aschengrau and Monson 1989, Schnorr et al. 2001).

Exposure of Australian Military

There was no evidence that children of Australian army Vietnam veterans had an increased frequency of congenital malformations overall or of any of about 100 specified ones (Donovan et al. 1984). The study was based on all the soldiers and not the exposed ones only, which it was thought might have obscured possible positive findings (Reznik et al. 1984). In addition, epidemiological study provided evidence of the teratological ineffectiveness of male exposure to herbicides (see argument and references in Hall and MacPhee 1985). For comments on these appraisals, which called them into question (Steele et al. 1990), see below.

Telephone interview of Tasmanian veterans of the Vietnam war noted more children with minor and major abnormalities than in families whose fathers were nonveterans, but the evidence was confusing and inconclusive (Field and Kerr 1988). A close look at it found that this "flawed and deceptive study" contained many misleading statements (Coombs 1988, p. 315).

Nonmilitary Herbicide Exposure

Children of professional 2,4,5-T sprayers in New Zealand had no increased congenital defects and miscarriages (Smith et al. 1982); as was also true of children of men exposed industrially to chlorinated dioxins in Michigan (Townsend et al. 1982).

Children of men in British Columbia sawmills exposed to chlorphenates, dioxin-contaminated wood preservatives, had increased frequencies of a disparate set of defects, cataracts, genital anomalies, and NTD, but not stillbirths and neonatal deaths (Dimich-Ward et al. 1996).

Occupational exposure to pesticides and various effluvia was sometimes associated with spontaneous abortion and various congenital malformations in offspring; and sometimes not (see Olshan and Schnitzer 1994, Kristensen et al. 1997, Blatter et al. 1997 for many relevant references).

Herbicide Exposure and Sex Ratio

Among the numerous reproductive characteristics studied as possibly affected by herbicide exposure – sperm count, time to conception, etc. – almost incidentally, was that of sex ratio. This is a field of contemplation that has long fascinated many students of human biology and often entrapped many a thinker. Some investigators of reproductive consequences of Vietnam service, not able to resist the lure, however found no sex ratio disturbance, i.e. only the usual slight male excess (Stellman et al. 1988, Michalek et al. 1998). But that doesn't end the story.

An explosion in a trichlorphenol plant near Seveso in 1976 in northern Italy released large quantities of TCDD or dioxin. Its aftermath will be gone into below. Considered here is the deficiency of males in live births in the years soon after the explosion, ascribed to exposure of fathers, but not mothers (Mocarelli et al. 1996, 2000). The allegation was weakened however by the sex ratio discrepancy already existing several years before the explosion – another instance of an attractive story demolished by a nasty fact.

Sex ratio was not disturbed after exposure to dioxinlike chemicals in Taiwan (Rogan et al. 1999); sex ratio variation was not associated with exposure to extremely high amounts of TCDD in US factories producing Agent Orange (Schnorr et al. 2001).

Other findings were less clearcut. A declining sex ratio in the last several decades of the twentieth century in some industrialized countries was taken as alerting to health hazards, especially from dioxins (Davis et al. 1998). Numerous writers, taking issue with this interpretation, noted that small fluctuations in the sex ratio, such as these had occurred before sometimes long before the age of industrialization in several regions of the world (see citations in Kalter 2003).

Permit me a small interruption to consider this matter per se. Almost without exception slightly more boys are born alive than girls. Cycles have been discerned in which the number of male births fluctuated a bit (James 1998), but generally the ratio at birth has remained close to 1:1.

Fisher (1958), explaining the close equality, said that since the total reproductive value of males in a population is exactly equal to that of the females, it follows that the "sex ratio will so adjust itself, under the influence of Natural Selection, that the total parental expenditure incurred in respect of children of each sex, shall be equal." In other words, when there are fewer members of one sex, the other sex has a higher fitness for producing individuals of the next generation, and the system stabilizes at a 1:1 ratio.

This proposition seems to be substantiated by the increase often noted in the sex ratio during and immediately following wars (MacMahon and Pugh 1954, James 1981); and the negative correlation between sex ratio at birth and that in adults of reproductive age (Lummaa et al. 1998, James 1995, 2000).

What has been considered above is the secondary sex ratio, i.e. the ratio in live births; which is usually about 1.05, or 51.2% males. This is probably not identical to the primary sex ratio, that in conceptions, which is conjectured to be about 120, but may be much greater. For example, in abortuses induced for medical or social reasons, and therefore expected to be representative of conceptions generally, it was 164 (Kellokumpu-Lehtinen and Pelliniemi 1984), and in anatomically normal previable spontaneously aborted fetuses 151 (Honoré 1988).

The initial male predominance – for which there is no credible explanation (an unlikely one, related to sex hormone concentration, was proposed by James 2002) – persists throughout pregnancy; even though it diminishes as more males than females die at all stages, the greatest during organogenesis (e.g. Byrne and Warburton 1987). Thus males outnumber females throughout, even in perinatal deaths (e.g. McKeown and Lowe 1951, Naeye et al. 1971, Feitosa and Kreiger 1992, Møller 1996, Sakamoto et al. 2001).

To return to the question at hand. The examination mentioned above of a possible sex ratio discrepancy in offspring of Vietnam military personnel exposed to dioxin, not being considered definitive, alternative explanations were sought. A most likely one, differential death of males in utero, failed the test since spontaneous abortion was not increased in pregnancies of wives of Vietnam personnel (e.g. Aschengrau and Monson 1989). A number of others were proposed, which will not be described since they are hardly worth considering (those interested can find the full account in the prior edition of this work: Kalter 2000). Refusing to let it lie, the flogged horse is still being subjected to attempted revival (James 2008).

The Political Dimension

With Agent Orange the science of teratology found itself at the center of a sociopolitical storm. Soon after the war Vietnam veterans charged that exposure to herbicides had caused an increased rate of a wide variety of health disorders including malformations (Coombs 1988). The controversial negative findings of an investigation of one such claim (Donovan et al. 1984, Hall and MacPhee 1985) was in turn challenged and then defended (Jacobs and McNamara 1996, Coombs 1987, Steele et al. 1990). For the full sorry tale see Kalter (2003, p. 205).

As was bitterly commented in the London Sunday Times of August 29, 1999, "public opinion is replacing scientific studies" in product safety assessment.

The Aftermath

The matter of Agent Orange and its health consequences has not yet come to an end. It was highly emotional anecdotal accounts of abnormal children of Vietnam veterans continuing to be published (Li and Johansson 2001) that prompted a meeting in March 2002 in which a joint US and Vietnam Agent Orange research program on the impacts of the pesticides on human health was announced (Young 2002). Agreeing to do the research is the easy part, a participant said, "the more difficult task will be to develop research studies that are definitive and address the underlying causes of disease in Vietnam." Key areas for research announced include spontaneous abortion, miscarriage, congenital malformation, neurological disorder, and cancer.

As of 2009 research was confined to exploring the "lingering social, political and environmental effects of the defoliant's use during the Vietnam War..." (University of California, Riverside Newsroom, 4/15/2009). But outcries still fill Google sites, aided and abetted by the US Department of Veterans Affairs (2009).

Seveso

The twentieth century saw many industrial, environmental, and ecological calamities, not only Minamata and Agent Orange, but also Chernobyl, Three Mile Island, Love Canal, Bhobal, Sellafield, Seveso, and others. Seveso presented some especially worrisome elements.

On July 10, 1976 there was an explosion in a factory producing trichlorophenol near the town of Seveso, 20 km north of Milan, Italy. Vapors containing various chemical products, primarily sodium trichloropenate and a toxic by-product, TCDD, were discharged into the atmosphere and dispersed by wind over an area of over about 1800 densely populated hectares southeast of the plant (Hay 1977). To examine the possible harmful effects of the TCDD, soil samples were measured and three zones of contamination demarcated, the most polluted, with 773 inhabitants, one less polluted, with about 4700 inhabitants, and a surrounding one, slightly and unevenly polluted (Reggiani 1980). The frequency of major malformations was not increased, even in the area of heaviest pollution (Bisanti et al. 1980, Mastroiacovo et al. 1988), but the number exposed was thought too small to be definitive.

Because of fears of malformations a number of public and private services were alerted and urged to report all such occurrences beginning January 1977 when women 3 months pregnant or less at the time the accident happened would be delivering. Such an alert was issued because malformations were known to have been greatly underreported previously (Abate et al. 1980).

Underreporting continued in 1977, when in all the cities of the entire region the malformation frequency was 1.4%, about half that usually expected, and continued in 1978 when it was 1.6%. Still these frequencies were larger than those previously reported, but only because of the even poorer record prior to 1976. But in neither year was it greater in the more polluted than in the less polluted area.

The low frequency it was reasoned may have resulted from increased spontaneous abortions or, because fearing a bad outcome, some women had their pregnancies interrupted. The former was dismissed because no such trend was detected through 1980 (Fara and Del Corno 1985), but the latter was supported by the birth rate decrease in the area in the 6 months of 1977 (Pocchiari et al. 1979); a decrease however that was a continuation of one begun a year or more earlier (Bisanti et al. 1980).

Examination of abortions found no clear evidence of an increased malformation frequency (Rehder et al. 1978); and defects in abnormal births were unhelpful, a hodgepodge without a semblance of a pattern (Bisanti et al. 1980. Reggiani 1988).

The final word was the finding of no untoward frequency of malformations in all deliveries in 1977–82 to women who were residents of the area in July 1976 (Mastroiacovo et al. 1988). All of which clearly indicated that exposure to the levels of TCDD experienced was not teratogenic.

Queries made in the following 20 years into the relation of other birth outcomes and TCDD serum levels shortly after the explosion in women living in the most contaminated areas found nothing more than trivial decreased birthweight and gestational age (Eskanazi et al. 2003). More extended observations noted increased mortality, from a variety of unrelated diseases, certain neoplasms, diabetes among females, etc. (Consonni et al. 2008).

Sellafield

The Sellafield matter may be briefly summarized. It began with an allegation in 1983 in a Yorkshire television program that a cluster of cases of childhood cancer was associated with a nuclear reprocessing plant near Sellafield in Cumbria, a county in northwest England. The matter was reignited some years later when the then editor of Nature commented that "[f]or the past decade, there has been persistent anxiety about the reasons why most of the conspicuous clusters of early-onset leukaemia and lymphoma in Britain should be near the nuclear plants at Sellafield and Dounreay..." (Maddox 1990).

While the question of carcinogenesis remained unresolved other concerns arose, namely whether babies born in 1975–85 in a town near the Sellafield plant had deleterious outcomes due to proximity to the plant, including stillbirth and congenital malformation; but with no evidence of such occurrences the question expired of itself. But was later again inflamed, when it was charged that congenital malformations were increased around a Cardiff plant that makes radioactive materials, but again with no clear evidence of a link with pollution from the plant. The full excruciating details of these events are found in the previous edition of this work Kalter (2003).

Chernobyl

In April 1986, almost 10 years after Seveso, what has been called the world's worst nuclear power accident happened at Chernobyl in the then USSR, now the Ukraine, linked to failure to carry out numerous required safety procedures. The result in time was the radiocontamination of water, air, and soil in regions of the former Soviet Union and western Europe. One clear association – besides the widespread psychological distress in workers cleaning up the debris and in others in and around the contaminated areas – to be taken seriously, was the approximately 30-fold increase in the incidence of thyroid cancer in children exposed to the fallout, especially in those younger than 1 year of age at the time of the explosion (Wiersinga 2001).

The fear aroused in pregnant women in exposed regions of the Soviet Union and elsewhere, it turned out, was unwarranted, since the radiation from the fallout was estimated to have been in the range of the natural background for most European countries and never reached teratogenic levels (Castronovo 1999, Hoffmann 2001).

In fact neither were unfavorable pregnancy outcomes increased, despite scattered reports of certain congenital malformations in children conceived immediately after the fallout (Haeusler et al. 1992). Nonetheless emotions carried the day and thousands of elective abortions were performed and the birth rate declined in several countries, even in areas remote from the accident.

A well reasoned explanation of the overestimation of the initial medical consequences of the disaster, including limited means of thyroid tumor diagnosis, was given by Pastel (2002) and Jarvin (2009).

Polychlorinated Biphenyls

The harm environmental contamination by industrial products may cause in humans was early drawn attention to by situations involving polychlorinated biphenyls (PCB). PCB are mixtures of synthetic organic hydrocarbons with properties that favored their use in hundreds of industrial and commercial applications. By the early 1970s toxic effects of PCB were appearing in agricultural animals and marine life; but it was their discovery in foods and human tissues and especially their persistence in the environment that – upon public outcry of course – prompted the US

Congress in 1976 to enact the Toxic Substances Control Act, which all but prohibited the manufacture, processing, and distribution in commerce of PCB. But though banned since then in most western countries the dangers they pose persist because of their presence in the environment.

Cola-Colored Babies

An early indication of human PCB toxicity came from Japan and Taiwan in the summer of 1968 in the form of an epidemic of severe acne, chloracne, in children. The disease was soon traced to cooking oil contaminated with PCB, to which children had been exposed prenatally through its consumption by mothers. Other signs included intrauterine growth retardation and irregularities of the skull. The telltale sign, present in most of the infants, was a deep brown pigmentation of skin and nails, due to increased melanin, giving them the name cola-colored babies.

The pigmentation largely cleared in some months and the impaired growth diminished after a few years; but neurological examination of older children gave an impression of developmental delay in some of them (Miller 1971b, 1985, Rogan 1982, Rogan et al. 1988).

PCB-Exposed American Children

Decades after production of PCB ceased their long half-life insured continuance in surroundings, and presence in human milk and tissues of residents of industrialized countries, concern abut which led to studies of prenatally exposed children in various US regions (Jacobson and Jacobson 1997). For never explained reasons children in US southeastern areas were often found to have been exposed to higher PCB levels than elsewhere, but without associated effects on birth weight and head circumference (Rogan et al. 1986).

Other types of findings, behavioral and neurological deficits in infancy and early childhood in some US areas, were difficult to understand, since maternal PCB body content in those locales was similar to general US population levels; raising the question whether epidemiological studies are able to detect subtle associations between exposure to low levels of environmental contaminants and their possible effects. Hutchings (1993) called attention to the problem that arises "from the difficulty of according biological significance ... to the term [behavioral teratogen] when it is used in such a way as to give equal weight to central nervous system teratogens and to agents that produce only subtle or ambiguous changes in behavior (e.g. altered activity) and are linked only by inference to damage in the central nervous system."

A writer not long ago was tempted to aver that "neuropsychological effects are among the most common human birth defects," and that "the organochlorines as a class are recognized as bona fide human teratogens" (Dietrich 1999). It is, as O. Wilde recognized, a human failing to be able to resist anything but temptation.

Notwithstanding, though having decreased in the last 30 years, PCB population background levels remain elevated, which it is felt may pose long term health effects despite difficulty of interpreting what these may be (Walkowiak et al. 2001, Hopf et al. 2009).

Love Canal: A Study in Political Teratology

The following chronology it is hoped will help grasp the events in what came to be the Love Canal tragedy.

Love Canal was a multifold disaster, technological, political, legal, sociological, psychological, toxicological, not least teratological. Named after William T. Love, an investor, the canal was dug in 1893 between the upper and lower Niagara River, bypassing Niagara Falls, to supply cheap hydroelectric power, but was abandoned when technological advances made it redundant. Filled with rainwater it became a giant swimming hole in the summer and an ice skating rink in the winter.

In 1942 it came into the possession of the Hooker Chemical and Plastic Corp. which drained it and dumped thousands of tons of its manufacturing chemical by-products sealed in metal drums into it. The dump was then filled in and the land sold for $1 to the city of Niagara Falls, which built housing and an elementary school on the site.

In 1976 a Niagara Falls newspaper reported that materials, later learned to have leaked out of the drums, had been seeping into basements of homes in a part of the area, and that illness and injuries to human, animal, and plant life were rumored. The paper then reported that chemical analysis of residues near the Love Canal dumpsite indicated the presence of organic chemicals, including chlorinated hydrocarbons, which were being carried through city storm sewers into the Niagara River.

Toxin presence in soil was confirmed, the contamination was called a serious health threat, and the school at the site was ordered closed and pregnant women and children under aged 2 evacuated.

In 1979 Beverly Paigen, a cancer researcher at Roswell Park Memorial Institute, Buffalo, reported a high rate of birth defects and miscarriages among Love Canal families.

State investigators found traces of highly toxic chemicals around the school and ordered more tests to determine the extent of contamination.

The EPA approved $4 million for remedial work at the Love Canal. The City received $1 million in Federal Disaster Assistance Administration funds to help pay debts incurred at the Love Canal. Later in the year the New York State Legislature extended property tax exemptions to another 300 families in area of Love Canal.

In August the City of Niagara Falls Board of Education closed the school pending the outcome of further studies of chemical contaminants, and students were transferred to various schools throughout the city.

In September the State Supreme Court ordered extension of state-funded relocation of residents to allow residents time to submit statements signed by physicians that they are unable to live in their homes due to remedial work.

The first Love Canal lawsuits naming Hooker and three public agencies were initiated, and the State Supreme Court rejected a $2.5 billion lawsuit filed on behalf of 900 Love Canal residents.

In October Jane Fonda and husband Tom Hayden visited the Love Canal.

A US House subcommittee recommended relocation of another 140 families after reviewing research by Dr. Paigen that said chemicals from the Love Canal were migrating.

Over 800 lawsuits had been filed naming Hooker, the city, county, and Board of Education totaling $11 billion.

A federal report indicated the odds of Love Canal residents contracting cancer was as high as 1 in 10.

The Federal Justice Department initiated a $124 million lawsuit against Hooker in connection with chemicals buried at four sites in the city.

The next February the EPA announced it found chemicals suspected of causing cancer in air samplings at Love Canal.

In April the State filed a $635 million lawsuit against Occidental Petroleum and two of its subsidiaries, Hooker Chemical Corp. and Hooker Chemicals and Plastics Corp., claiming they were responsible for Love Canal disaster.

The EPA announced chromosome damage was found in 11 of 36 residents tested in the Love Canal region.

President Carter declared Love Canal a national emergency paving way for relocation of another 710 families.

This account reflects the mounting hysteria in the locality, because of the pronouncements that cancer was being caused and that babies were being harmed. What were the adverse effects on prenatal growth and development alleged to be due to the chemical pollution of Love Canal? A statement labeled "Health Hazards at Love Canal," in which these allegations were detailed was presented by Paigen to the US House of Representatives Subcommitte on Oversight and Investigations in March 1979. In it she said that her survey had shown an increased frequency of spontaneous abortion, birth defects, and low birth weight in pregnancies close to the areas in question.

The survey found that 12% of children born in "wet areas" and 5% in "dry homes" (i.e. areas that were more or less contaminated) were malformed – the quality of the judgments indicated by the assertions that "some . . . were minor . . . such as webbed toes, an extra toe, or extra or unusually spaced teeth. Others . . . included a deaf child, 5 children with mental retardation, 6 with kidney abnormalities, and 3 with heart defects."

This information had apparently been obtained through anecdotes related by the area's residents. No other listing of the defects, to this writer's knowledge, has been published, not even in a paper stating that the prevalence of malformations was higher in exposed than controls for all years 1966–76 (Goldman et al. 1985). Paigen expanded her efforts by adding to the argument the ethical dimensions of scientific conflict (Paigen 1982).

Meanwhile information from the state health department indicated an excess of spontaneous abortion in the "historical water area," 12% vs. 8.2%; and an excess of congenital defects in one area adjacent to the Canal, but not another (Vianna 1980).

All of these assertions were disputed by epidemiological consideration that judged that there was no clear increased risk for these pregnancy outcomes in women living next to the canal (Heath 1983).

Love Canal and Chromosomes

That did not close the story. A different chapter began with a 1980 EPA report of possible damaged chromosomes in Love Canal residents, who thus were at increased risk of cancer and malformations. The report was soon discredited, but not before President Carter declared a state of emergency and ordered relocation of residents. The political/scientific events leading up to and following the EPA report, and the psychological harm the residents suffered, were recounted in detail (Holden 1980, Kolata 1980b).

These dangers were repudiated by an analysis that revealed no increase of cancer in the Love Canal area, and no different levels of cytogenetic aberrations than in controls (Janerich et al. 1981, Anon 1983a).

The Last Word

The final word appeared in a review of a book about the Love Canal inquiry: "The most upsetting aspect of this saga, and its aftermath, is that there was never any creditable science done. No scientifically believable study, contemporaneously or subsequently, can be pointed to with any confidence, and not having learned anything of scientific substance or future value about this situation is a real tragedy" (Frank 1998).

The present author can add little more. A tempest in a teacup, engendered by parental fears abetted by naive scientists, simpleminded journalists, and overzealous governmental paternalism. He can do no better than note a Wikipedia item from 1998, which quoted a Dr. Elizabeth Whelan as saying that "people were not falling ill due to exposure to chemical waste, but from stress caused by the media."

Disease Medication and Teratogenesis

Modern medical advance and pharmaceutical innovation have wrought much to be thankful for. The difference between the barely efficacious means my father had to alleviate his severe asthma and those at my disposal that deal so effectively with the condition he passed on to me and I to my children is but one small example of what we today have to be thankful for in this regard. But not in every particular has progress been benign. There are some with drawbacks.

Retinoids

One thing led to another. Studies in animals of vitamin A deprivation and excess brought to light the importance of this nutrient for normal prenatal development (Hale 1933, Warkany and Schraffenberger 1944b, Cohlan 1954). Motivated by this discovery teratolgic experiments were made (Kalter and Warkany 1959, Geelen 1979), which led to studies of analogues of the vitamin (Soprano and Soprano 1995), and from them to recognizing the basic role of the metabolite retinoic acid in embryogenesis (Hofmann and Eichele 1994, Mendelsohn et al. 1994, Lohnes et al. 1995, Maden 2000).

A Human Teratogen

It had been suggested earlier that large amounts of vitamin A might be a human teratogen, suggestions supported by retrospective studies, flawed however, of the vitamin A content of serum of mothers of infants with NTD (Gal et al. 1968, 1972). Then came reports of various malformations in children of women taking huge doses of vitamin A during pregnancy (Rosa et al. 1986, 1994), followed by misguided alarms and warnings about permissible levels of vitamin A consumption (Pauling 1986, Shepard et al. 1986).

Several studies looked into these matters. Retrospective studies of the maternal consumption of multivitamins (see chapter on folic acid and malformations for details) found equivocal or negative associations with NTD frequency

H. Kalter, *Teratology in the Twentieth Century Plus Ten*,
DOI 10.1007/978-90-481-8820-8_13, © Springer Science+Business Media B.V. 2010

(Mulinare et al. 1988, Milunsky et al. 1989). Another examined vitamin A intake by mothers of infants with malformations of cranial neural crest origin, because such defects had been found in infants prenatally exposed to synthetic retinoids, as discussed below (Werler et al. 1990).

A prospective study of maternal consumption of multivitamin supplements and vitamin A-containing foods concluded that more than 10,000 international units (IU) of vitamin A daily was a potential teratogenic risk (Rothman et al. 1995). However, the overall 1.5% malformation frequency found was lower than the usual background level, and even in the highest retinol intake group it was only the usual 3%.

The allegation nevertheless led to a reexamination of previously collected population-based data. In one instance the rate of cranial neural crest-derived and other congenital malformations was not increased in infants of mothers who took fairly large doses of vitamin A and multivitamins (Khoury et al. 1996). And in another orofacial clefts and conotruncal heart defects – parts derived from cranial neural crest cells – were not associated with maternal use of vitamin A levels probably exceeding 10,000 IU daily (Shaw et al. 1996).

In a variation, vitamin A consumption, as supplements or in fortified cereals, by mothers of malformed and nonmalformed infants was not different from each other – incidentally, it was surprising that so few women consumed supplements at all (Mills et al. 1997).

The offspring of a selected group of women, with various disease conditions, consuming large doses of vitamin A, 10,000–300,000 IU daily for at least a week during the first 9 weeks of pregnancy, had no increased rate of spontaneous abortions and only a low frequency of major congenital malformations, of kinds not associated with retinoids (Mastroiacovo et al. 1999).

Studies with cynomolgus monkeys (*Macaca fascicularis*), to the extent we can rely for such comparisons on our primate cousins, offered assurance of the teratologic safety of the usual doses of vitamin A taken daily during human pregnancy (5000 IU in most commercial vitamin preparations). Based on the similarities in the teratogenic susceptibility of humans and monkeys to the synthetic retinoid 13-*cis*-retinoic acid, preliminary estimates of no-effect levels of vitamin A in human pregnancy were calculated to be much greater than this, about 25,000–37,000 IU/day (Hendrickx et al. 1980, 2000).

Synthetic Retinoids

Vitamin A, an ingredient essential for normal growth, vision, and reproduction, was first described in 1909 and given its present name in 1929; its chemical structure determined in 1931, and its pure form synthesized in 1947. Rosa (1993) noted that as long ago as the sixteenth century explorers knew of the harmfulness of eating polar bear liver (containing tremendous amounts of vitamin A, 20,000 IU/g), a fact the native peoples of the north, avoiding it, must have long been aware of.

Deficiency of the vitamin was discovered to cause xerophthalmia in rats in 1922 and cutaneous abnormalities in humans in 1931. Which led in the 1940s to treating severe acne and other skin disorders with very large doses of it, 100,000–300,000 IU daily for months. In the 1960s the vitamin A derivative all-*trans*-retinoic acid was synthesized and became available for oral and topical use for these disorders. But its use was limited by its side effects.

Isotretinoin

An isomer largely free of side effects, 13-*cis*-retinoic acid, synthesized in 1969, was soon used to treat skin disorders (Hartmann and Bollag 1993). Called isotretinoin, with the commercial name Accutane, it was registered in the US in 1982 for oral treatment of severe recalcitrant nodular acne, a common ailment of adolescence and older ages. Because retinoids were known animal teratogens women were urged not to use Accutane except when taking a contraceptive before beginning treatment.

Neverthless, within a few months its inadvertent use led to congenital malformations (Rosa 1983, Lammer et al. 1985, 1988). In pregnancies retrospectively ascertained the frequency was 84% and prospectively 18%. The latter, confirmed by the 20–30% figure in additional pregnancies, proved to be the more realistic evaluation. But even the lesser ones may still have been overestimates, resulting from the high rate of preemptive elective abortion – in none of which malformations was detected however (Lammer et al. 1985, Dai et al. 1992).

As of mid-1989 more than 400 reports of isotretinoin use during pregnancy had been made (Dai et al. 1992), which more than doubled by the end of the century (Atanackovic and Koren 1999). At which time it was estimated to be the most widely used teratogenic drug in the US, with about 2.5 per 1000 women of ages 14–45 taking it – but with few further occurrences of deformity identified, the result of modified use of the drug, etc. (Honein et al. 2001a).

The Population at Risk

As a leader writer expressed it "...any girl who is old enough to have acne is old enough to be fertile" (Anon 1985). By 5 years after the drug was introduced it was estimated that more than 300,000 women had used isotretinoin (Stern 1989). A US market survey found that about 540,000 girls and women aged 12–44 had done so in 1982–8 (Dai et al. 1992). Because it was also prescribed for mild cases of the disorder it was estimated that its use was 15–20 times greater than the number of women with severe acne in the US, and 6–8 times greater in Sweden and the UK (Rosa 1993). It was estimated that in 1999 alone 810,000 women of reproductive age received prescriptions for isotretinoin in the US (Honein et al. 2001a).

The number of exposed pregnancies is uncertain and conflicting. By 1988 433 were reported to the drug's manufacturer (Dai et al. 1992); an estimated 1200 pregnant women using the drug during the first trimester in that year (Faich and Rosa

1989). By the end of 1990 the number of infants in the US known to have the characteristic isotretinoin embryopathy had risen to 92, but elsewhere only five had been identified (Rosa 1993).

These numbers are probably a minimum, since precautions to avoid exposure may not have been sufficiently adhered to, despite the advertised dangers of the drug (Autret et al. 1997, and others). A Boston survey e.g. found that there were about 900 pregnancies in 454,273 women treated with isotretinoin in 1989–99 – a number estimated to be 38–40% of women taking the drug (Anon 2000b). An even later updating of isotretinoin use, in 1998–2006 in selected countries, still found malformations in some small number of term babies (Garcia-Bournissen et al. 2008). A comprehensive and ongoing examination of these questions however it seems has not been well thought through.

Retinoic Acid Embryopathy

The congenital malformations caused by isotretinoin constitute a characteristic syndrome, given the name retinoic acid embryopathy syndrome (Lammer et al. 1985, Rosa 1993). The structures most often affected are the cranium and face, heart, thymus, and hindbrain, and the single most common abnormality microtia or anotia, often accompanied by atresia or stenosis of the external auditory canal. The hindbrain malformations consist of cystic dilatation of the roof of the fourth ventricle and partial or complete absence of the cerebellar vermis, simulating the pattern in the sporadically occurring Dandy-Walker malformation.

It is a curiosity that cerebellar malformations greatly similar to these human instances have been seen in laboratory animals with diverse etiology: in a hereditary hydrocephalus in mice and experimentally caused in mice by the riboflavin antagonist riboflavin and in pigtail monkeys by all-*trans*-retinoic acid (Benda 1954, Bierwolf 1956, Kalter 1959a, Fantel et al. 1977).

A side by side comparison revealed the similarity in the pattern of malformations in animals and humans, perhaps unique to the retinoids (Hendrickx et al. 1983, Rosa et al. 1986). For example, those caused in mice by large doses of vitamin A consisted of an array of defects, including ear and mouth abnormalities (Giroud and Martinet 1956b, Kalter 1960, Kalter and Warkany 1961) and others that were analogous to a human malformation picture, the third to fourth pharyngeal pouch complex or DiGeorge syndrome, a sporadic condition of heterogeneous etiology (DiGeorge 1968, Warkany 1971, p. 740, Lammer and Opitz 1986). What links many of the retinoid-induced malformations, in experimental animals and humans, as noted above, is their apparent stemming from disturbances of cranial neural crest cell activity in early embryogenesis (Brockman and Kirby 1984).

In contrast with these similarities are the great differences in the dose of isotretinoin required to cause teratogenesis, far smaller in humans than in other animals: 0.5–1.5 mg/kg vs. 2.5–5.0 mg/kg in cynomolgus monkeys (Hummler et al. 1990), and 10–100 mg/kg in various small animal species (Kochhar and Christian

1997); specificities perhaps explained in part by metabolic and transport differences (Collins and Mao 1999, Tzimas and Nau 2001).

Further Threat: Etretinate

Etretinate, an analogue of retinoic acid, was first marketed in Europe in 1973, for treatment of severe skin disease, several years before its release in the US, and all reports of its teratogenicity up to the present writing are of non-US origin (Happle et al. 1984, Orfanos et al. 1987, Hopf and Mathias 1988, Lambert et al. 1988, Rosa 1993). However, almost none of the malformations resembled any of the cardinal features of the retinoic acid embryopathy (Rosa 1993, Geiger et al. 1994, Briggs et al. 1998).

Also reported were malformations in children conceived some time after women ceased taking the drug, which were attributed to its prolonged presence in maternal tissue (see review in Collins and Mao 1999). But none of the known cases had malformations typical of the retinoic acid syndrome. For example, the detailed description of one case belied the allegation that the pattern was "identical to that of the retinoic acid (isotretinoin) embryopathy," since the external auditory canals were enlarged, not stenotic or atretic, a characteristic defect (Lammer 1988). The same teratogenic nonspecificity was true of experimental studies of etretinate: various malformations were induced by it or its metabolite etretin, but apparently not of the ear or thymus (Hummler et al. 1981, Kamm 1982, Kistler and Hummler 1985, Agnish et al. 1990), perhaps except for a relatively limited study with confused results (Turton et al. 1992).

The question arises of course whether instances of malformations after etretinate cessation were in fact due to the drug or were sporadic occurrences the drug was conveniently blamed for. Isotretinoin, in contrast, with its relatively brief half life in humans, was never thought to present a long-term threat; which a prospective study clearly established (Dai et al. 1989).

Acitretin

Since acitretin, another synthetic retinoid congener, has a relatively short half life it promised a decreased teratogenic potential. This was deceptive however, since it was unexpectedly found to be transformed to etretinate (Collins and Mao 1999). Thus it is uncertain whether the malformations in a pregnancy in the Netherlands, with those of the ear being the only sign of the retinoic acid embryopathy, were due to the administered drug or to its transformation product, or indeed not associated with the treatment at all (de Die-Smulders et al. 1995). The same uncertainty pertains to a recent case from Argentina (Barbero et al. 2004).

World-wide pregnancy data reported from the time acitretin became available in 1983 brought the record up to date (Maradit and Geiger 1999). There were 123 pregnancies, the majority within 2 years of therapy discontinuation, 83 of which were reported prospectively. Apprehension led to many abortions (undescribed), but none of the 47 neonates had a major congenital malformation.

It is clear thus that etretinate and acitretin differ from other retinoids by not producing the full-blown retinoic acid embryopathy; and perhaps by not being teratogenic at all.

Topical Tretinoin Use

Reports of malformed infants born to women treated topically with tretinoin were contradictory. Two infants had unilateral ear and external auditory meatal defects, a sign of the retinoic acid embryopathy (Camera and Pregliasco 1992, Selcen et al. 2000), whereas others had various malformations not of the typical variety (De Wals et al. 1991, Lipson et al. 1993, Rosa et al. 1994). Additional studies, retrospective and prospective, also cleared the drug of teratogenic risk (Jick et al. 1993, Shapiro et al. 1997, Loureiro et al. 2005). Such negative findings are in agreement with findings of plasma tretinoin concentrations after topical treatment which indicated a low teratogenic risk (Nau 1993, Chen et al. 1997).

Like thalidomide, with its new therapeutic applications, the beneficial effects of retinoids guarantee its continued and wide use. Only stringent adherence to stipulations that have been formulated to prevent or greatly limit the possibility of their inadvertent use during pregnancy will see the disappearance of the grave fetal damage they can impose.

In sum, again like thalidomide, we see here the hopeful disappearance of an external source of embryonic maldevelopment.

But concern still continued. An account from Berlin of pregnancies exposed to isotretinoin in 1993–2008 noted that three-quarters of them were aborted, anxiety the cause, without confirmation of fetal damage, and none of the live births had retinoid embryopathy symptoms (Schaefer et al. 2009). In a three-nation survey of pregnant women exposed to isotretinoin in 1998–2006 a low malformation rate indicated no cause for alarm (Garcia-Bournissen et al. 2008). And similarly, a population-based evaluation in 1984–2002 of tretinoin use in pregnancy in Canada was again not supportive of teratogenicity (Bérard et al. 2007). Single case reports complete the roster (Giannoulis et al. 2005). The danger it therefore seems has been averted.

Antiepileptic Drugs

Epilepsy, the falling sickness, is as old as mankind. A Babylonian tablet in the British Museum, recording the oldest detailed account of the disease, is dated from at least 2000 BCE. As is true of so many frightening and inexplicable phenomena faced by human beings in the past epileptic seizures were long ascribed to demonic forces and dealt with by exorcism – as described e.g. in the New Testament (Luke 9:39). Another 2000 years of superstition had to elapse for Hippocrates' view of epilepsy as a disorder of the brain to be regained in recent times.

Opposing the impression of epilepsy as impairing mental function the condition has afflicted some of the most notable individuals in every age and in every walk of life, from Alexander the Great to Agatha Christie.

Epilepsy is not only one of the oldest but is also one of the commonest chronic diseases of humankind. The CDC reported that there were 1.4 million persons with epilepsy in the US in 1996; and that it afflicts 5.2 per 1000 women under 45 years of age, a number that hasn't changed in recent years (Niswander and Gordon 1972). Thus, based on the current nearly 4 million annual US births over 20,000 American women with epilepsy have babies each year.

Therapy of Epilepsy: Fetal Consequences

It was not until the 1850s that medieval methods of potions and sorcery gave way to a rational means of controlling epileptic seizures, when a chance observation led to the use of potassium bromide for this purpose (Bodendorfer 1978). This medicament became widely used in the US and Europe during the second half of the nineteenth century; and opened the door to other drugs for the treatment of epilepsy, mainly phenobarbital introduced in 1912 and hydantoins in 1938. Many newer ones were discovered subsequently; but all, old and new, brought problems in their wake (Meadow 1991).

Over 50 years elapsed between the introduction of the first of these drugs and 30 after the second before suspicion arose of their teratogenicity, despite incidental observations of congenital defects in children of epileptic persons along the way (Müller-Küppers 1963). This delay is not hard to understand, since none of the defects had the severity or singularity that e.g. allowed the more rapid recognition of thalidomide's prenatal damage.

Thus while the frequency of congenital malformations in children of women attending an epilepsy outpatient clinic was within the expected range, it seems to have been overlooked that one of the malformations, orofacial clefts, had a frequency rather greater than the control level (Janz and Fuchs 1964). The suspicion of its association with the disease grew with the specific finding of cleft lip and palate in children of epileptic mothers, mostly receiving phenylbarbitone (Meadow 1968, 1970).

Fetal Hydantoin Syndrome

The association of antiepileptics and malformations was strengthened in time, the particular defects and combinations sometimes varying according to which of the many antiepileptic medications had been taken (Kelly 1984b, Källén et al. 1989,

Friis 1990, Yerby et al. 1992). For example, one combination, cardiac malformations and cleft palate, was associated with trimethadione (German et al. 1970), a type of medication long since replaced by newer antiepileptic ones.

The first of many case-control studies reported 4.4% congenital malformations, which though moderate was considered to be a greater frequency than expected and to support the suspicion of the teratogenicity of such drugs (Speidel and Meadow 1972). Affected children, exposed mostly to phenobarbitone, had a miscellany of unspecified abnormalities, comprising a specific though variable pattern. Consisting especially of minor craniofacial and peripheral skeletal anomalies, they were classified as minor, since their impact on viability was minimal and they were frequently transitory.

Soon as further described the anomalies included distal digital hypoplasias, which combined with other features were judged to represent a complex of defects, termed the fetal hydantoin syndrome (Barr et al. 1973, Loughnan et al. 1973, Hill et al. 1974). The complex, becoming more sharply defined in time, was depicted as consisting of prenatal-onset growth deficiency, craniofacial abnormalities, nail and digital hypoplasia, and mental retardation (Hanson and Smith 1975, Hanson 1986).

Let me review and critique the items said to compose the complex. First, reduced birthweight and head circumference, seen in one study (Hvas et al. 2000), was not present or hardly present at birth in others (Yerby et al. 1992, Holmes et al. 2001); and in which at 1 year of age exposed and control children did not differ significantly (Leavitt et al. 1992). A long-term Swedish study of nearly 1000 children concluded that "the influence on fetal growth of treatment during pregnancy seems to be of minor clinical importance" (Wide et al. 2000).

Next, minor facial defects: were not significantly more frequent in prenatally exposed children when examined at 5–6 years of age than in controls (Gaily et al. 1988). Also since such features were common in the children of unmedicated epileptic mothers, and with epileptic mothers themselves also possessing them, it was suggested that many of them are inherited. Because of this the syndrome as enunciated by Hanson and Smith (1975) was questioned; and the opinion expressed that the fetal effects of antiepileptic exposure consisted only of hypertelorism and digital anomalies (Gaily et al. 1988).

The anomalous craniofacial features were also transitory, many (upper lip and nose length, inner canthal distance), while present in neonates, upon postnatal examination were not different in case and control subjects (Leavitt et al. 1992). Severe instances, following massive doses of phenobarbitone, were exceptions in persisting for several years (Seip 1976).

Diagnosis of distal digital hypoplasia depended on the quality of the evaluation, and thus was not an objective sign of exposure. One study found a larger frequency in case infants than in controls (Hill et al. 1974); others noted that it varied in severity (Andermann et al. 1982). But overall the frequency was not large; publications summarized by Hill et al. (1974) reported hypoplastic digits and nails in 0.8% of a large number of infants exposed to antiepileptic medication; but controls were lacking.

The frequency of hypoplastic digits appeared to decrease with age (Kelly et al. 1984, Koch et al. 1992, Yerby 1992), even by 3 months (D'Souza et al. 1990).

Radiographs of 5 year old children revealed distal phalangeal hypoplasia – "irrelevant for the well-being of the child" – only in those exposed to the largest doses of phenytoin, in this case 9% of the sample (Gaily 1990); and in prenatally exposed children and adults only various "subtle" (meaning no doubt trivial) digital changes were present: stiffness, increased nail area, increased dermatographic arch patterns, smaller distal phalanx of digit four (Lu et al. 2000).

Thus while antiepileptic exposure may be suggested by nail hypoplasia at birth and older ages, and perhaps only or especially when of more severe degree resulting from polytherapy (discussed below), its fading in time revealed it to be a form of localized growth retardation, often reversed by catch-up growth. Distal digital hypoplasia, also seen after exposure to other antiepileptics, e.g. valproate (Gaily and Granström 1992. Yerby et al. 1992, Moore et al. 2000), was an unreliable indicator of exposure. In one sizable prospective study it was not seen at all (Canger et al. 1999).

A family history of malformations may have predisposed to the production of severe malformations by certain antiepileptics, a relation rarely noted in studies of exogenous teratogens. In fact, susceptibility to the teratologic effect seemed to have a genetic component, as indicated by family studies in which treated epileptic mothers who already had an affected child were more likely to have another than were those whose first-born child was unaffected (Van Dyke et al. 1988).

Another sign of retarded development, undescended testes, was found in 15% of newborn boys exposed to phenobarbital, but in only 2.8% exposed to phenytoin, the same as in controls (Dessens et al. 2001). The gestational age of the boys at birth was 30–43 weeks, a relevant fact since undescended testes is a normal feature before 37 weeks of gestation. But no infant with the condition was small for gestational age and all were born at term. So prematurity may not have been responsible; except that the condition was absent at follow-up. It is strange however that it was increased after exposure to phenobarbital but not phenytoin, which is contrary to antiepileptics of whatever kind usually causing the same array of minor defects.

Finally, the nonspecificity of the minor defects was emphasized by the same constellation being seen after phenytoin, carbamazepine, and phenobarbital (Yerby et al. 1992, Holmes et al. 2001); and also, as was mentioned by Seip (1976), by their "remarkable similarities to the fetal alcohol syndrome," as others did also (e.g. Hill et al. 1974, Robert and Källén 1994, Moore et al. 2000).

Recognizing Minor Defects

It has been maintained that, although these minor midface and distal digital defects individually can present diagnostic difficulty, and in fact may not be pathognomonic, when they occur together they form a distinctive pattern and a readily recognized picture. Since major malformations occurred relatively infrequently in such cases the diagnosis of the fetal hydantoin syndrome depended primarily on the far commoner minor defects. For example, in one study the minor defects were eight times more frequent than major ones (Hanson et al. 1976).

Few objective analyses of the recognition of the minor symptoms of antiepileptic exposure have been made. One that did make such an assessment had some pertinent findings (Carlin et al. 2000). The facial and hand/feet features of three groups of infants – one exposed to antiepileptics, another whose mothers had a history of seizures, and the third a population-based control – was evaluated by two sets of examiners, specialist physicians and assistants trained in classifying antiepileptic indicators, and their assessments compared.

The frequency of the anomalies was increased in all three groups of children, no doubt due to close scrutiny; and although the assessment of the two sets of examiners seemed to be in good agreement it was considered only fair to moderate. The overall conclusion was that "it is difficult to assess subtle physical anomalies in newborns in a reliable and reproducible fashion," even when made by individuals trained specifically to recognize conditions supposed to be associated with a particular prenatal exposure.

What is to be made of this provocative evaluation? Additional studies of such outcomes are always desirable; but meanwhile when judging the reality of the subtle effects of prenatal exposure to these and other medications discussed throughout this work, it cannot be neglected.

Is Epilepsy Teratogenic?

It is not clear sometimes whether abnormalities in children exposed prenatally to a drug are due to the drug or to the disease, wholly or in part, for which the drug is prescribed. Thus in the matter of epilepsy it was of interest to inquire whether it was the treatment itself that had teratogenic potential, perhaps connected with maternal seizures during pregnancy, or genetic factors associated with tendency to the disease.

This question was raised by extensive American and Finnish data suggesting that fetal damage previously attributed to antiepileptic drugs was due to the disease itself (Shapiro et al. 1976). A Mayo Clinic study seemed to agree, despite conflicting evidence (Annegers et al. 1978). Also in agreement was a study that concluded that "major malformations were linked to epilepsy itself, i.e. its genetic background," though numbers were inadequate for statistical certainty (Koch et al. 1992).

Many studies have explored this subject, but consistency has eluded them. Increased major malformations were found in children of drug-treated epileptic women, but not children of untreated epileptics (Nakane 1979, Dansky et al. 1982); in children of both treated and untreated epileptic mothers, and epileptic fathers (Majewski et al. 1981, Friis and Hauge 1985); in children of epileptic mothers but not fathers (Dansky et al. 1982); and in children of epileptic fathers (Koch et al. 1982); and the same mixed findings were true of minor defects (Beck-Mannagetta et al. 1982, Majewski et al. 1981, Rating et al. 1982).

Parental epilepsy may have been an etiological element in some studies of oral clefts. Cleft lip with or without palate, but not isolated cleft palate, was associated with epilepsy in population studies of mothers of children with the defect (Kelly et al. 1984, Kelly 1984a). A later analysis was contradictory: the frequency of both cleft lip with or without palate and isolated cleft palate was increased in children of epileptic mothers; but genetic factors seemed of little importance because the frequency did not exceed expectation in the children of epileptic fathers and siblings of epileptic patients (Friis 1989).

Strong evidence of the innocence of epilepsy was supplied by multistudy comparisons of malformation frequencies in children of treated and untreated epileptic mothers, which found that the level in the former almost always exceeded that in the latter (Hanson and Buehler 1982, Finnell et al. 1995).

Recent analyses, which perhaps settled the matter, found that physical features, including facial characters, and IQs of children of epileptic mothers who had not taken antiepileptic drugs during pregnancy were not different from those of children of nonepileptic mothers (Holmes et al. 2000, 2001, Holmes 2002). All in all, despite the confusion and inconsistency, the preponderant evidence indicates that epilepsy per se is not teratogenic.

To the extent that animal studies can answer the question, a study of reproduction in mice possessing the neurological mutant gene quaking can be mentioned (Finnell and Chernoff 1982). Quaking is an autosomal recessive gene, homozygotes for which have spontaneous tonic-clonic seizures. Phenytoin administered to pregnant mice of this genotype greatly reduced the frequency of the seizures but did not prevent malformations, indicating that the drug and not the genetic condition was responsible for fetal maldevelpment.

Epilepsy and Spontaneous Abortion

No single environmental agent or medical circumstance is conclusively known to increase the frequency of spontaneous abortion (Rushton 1985); and this is true as well of epilepsy and its pharmaceutical therapies (Nelson and Ellenberg 1982, Annegers et al. 1988, Battino et al. 1992, Oguni et al. 1992, Martin and Millac 1993, Schupf and Ottman 1997, Canger et al. 1999). The subject of the etiology of spontaneous abortion is discussed in various sections of this work.

Major Congenital Malformations

The overall frequency of congenital malformations in offspring prenatally exposed to antiepileptics has varied greatly (see tables e.g. in Hanson and Buehler 1982, Finnell et al. 1995), though the mean overall rate, 6.0%, was considered to be about twice that in the general population (Friis 1990).

The variation undoutedly stemmed partly from differences in the abnormalities included in the count, which though not usually made explicit consisted of both major and minor defects as well as of those that could not be attributed to the antiepileptic drugs at all (e.g. Fairgrieve et al. 2000). In the few cases where abnormalities were listed it was possible to verify this supposition of heterogeneity (e.g. Canger et al. 1999). In another study at most only one-third of the overall frequency of 14% were major malformations, the remainder minor defects (Kaneko et al. 1988).

Differences between studies in the amount and composition of polytherapy may also have been responsible for the variation. Polytherapy, the use of more than one antiepileptic per patient per pregnancy, sometimes resulted in higher frequencies of maldevelopment than monotherapeutic regimes; e.g. 6.5% after single drug treatment and 14% after multiple treatment (Kaneko et al. 1988); or particular combinations of drugs may have been more teratogenic than others (Samrén et al. 1999). In contradiction was a multihospital study of single or multiple drug therapy with malformation frequencies not statistically significantly different from each other (Holmes et al. 2001); but in opposition later was a UK register study finding an increased risk when the polytherapy included valproate (Morrow et al. 2006).

Most of the increase in major malformations was accounted for by a small number of abnormality types (e.g. Bjerkedal 1982). The dimension of this fact was especially shown by a French study in which the vast majority of the unequivocally major malformations were cardiovascular and neural tube defects, the last mainly after valproate exposure; of which more below (Dravet et al. 1992).

Early studies had noted an increased occurrence of certain malformations, especially cleft lip, as well as cardiovascular malformations (Fedrick 1973, Annegers et al. 1974). An Oxford linkage study, of mostly phenytoin-treated women, noted a diverse assortment of defects, most of which however could hardly be considered major malformations (Fedrick 1973). In a Mayo Clinic study, although the overall level of malformations was not increased, there was a significant excess of facial clefts and cardiovascular malformations, that of the latter perhaps owing to postnatal follow-up (Annegers et al. 1974). In this as in other retrospective studies cleft lip with or without cleft palate appeared to be the most common of the major malformations, with a frequency about nine times that expected.

The relatively minor role of antiepileptics in the overall malformation picture must be stressed. A calculation found them responsible for but 3.3% of nonsyndromic oral clefts in a population (Abrishamchian et al. 1994); i.e. of the order of recurrence expected in affected families (Curtis et al. 1961). A broader analysis, putting the teratogenicity of antiepileptic drugs into perspective, found that children with malformations due to such drugs accounted for 1.3% of all malformed children born in the US in 1981 (Kalter and Warkany 1983).

Carbamazepine

Studies of particular antiepileptic drugs must be noted. The teratogenic potential of carbamazepine has been in doubt. It was first approved for use as an antiepileptic in the US in 1974, and later, considered safe, became the drug of choice in various areas (Morrison and Rieder 1993, Wide et al. 2000). The first suspicion otherwise arose with anecdotal findings of a pattern of minor craniofacial and digital defects in children of postneonatal ages (Jones et al. 1989). Similar studies found more minor anomalies but no increased major malformations (Ornoy and Cohen 1996, Nulman et al. 1997a).

The possibility of more serious consequences was uncovered by a survey finding a risk of about 1% spina bifida (Rosa 1991), supported, but far from proven, by studies mostly with a limited number of subjects (Källén et al. 1989, Gladstone et al. 1992, Källén 1994, Samrén et al. 1997, 1999). A sizable prospective study noted an increase in a diverse assortment of malformations (not all major), but with no apparent defect specificity, and an overall frequency not much larger than the background one; and, at variance with some but not all studies, a rate no different with polytherapy than monotherapy (Samrén et al. 1999, Canger et al. 1999, Holmes et al. 2001).

The findings, mainly of minor anomalies and spina bifida, emerging from studies with numerous limitations, still await, as was commented years ago, definitive proof of the drug's culpability – from "large, prospective, multicentre. . .investigations" (Anon 1991). The teratogenic potential of this drug thus seems at present not to be conclusively established. It is also pertinent to note the low teratogenic potential of carbamazepine found in various experimental studies (Robert and Källén 1994).

Valproic Acid

Valproic acid (VPA) caused a teratogenic picture distinct from that of other antiepileptics. The first warning of its possible harmfulness came from experimentally based calculations that indicated it to be a more potent teratogen than phenytoin (Brown et al. 1980). Several individual instances of defective children prenataly exposed to VPA (Lammer et al. 1987) were hardly proof of the drug's danger; but a French monitoring system supported the animal prediction by finding a strong association with spina bifida (Robert and Guibaud 1982, Robert et al. 1988), almost all located in the lumbosacral region (Lindhout et al. 1992b).

Further evidence came from several European areas, yielding risk figures of 1–3% after exposure in early pregnancy (Bjerkedal et al. 1982, Lindhout and Schmidt 1986, Samrén et al. 1999). Larger frequencies were discovered by prospective studies (Omtzigt et al. 1992, Lindhout 1994, Canger et al. 1999), which also revealed a clear dose-response relation, with risks beginning at 600–1000 mg/day (Koren et al. 2006). (An incidental note: a dose-response relation was found for various malformations in a sizable antiepileptic Israeli study, but with no instance of NTD; an absence perhaps connected with the infrequency of these defects in some Jewish populations – Kalter 2009, p. 161.)

Remarkable in the light of the close epidemiological connection between spina bifida and anencephalus is the rarity of the latter malformation after VPA, for which there is no good explanation. As for other major malformations, there is no clear evidence of such an association with VPA (Källén et al. 1989, Rodríguez-Pinilla et al. 2000). [Experimentally, the predominant malformation induced by VPA is exencephaly (the prenatal forerunner of anencephalus), though careful timing of drug administration to pregnant mice also induced spina bifida (Nau et al. 1991, Ehlers et al. 1992).]

VPA monotherapy did not reduce birth weight or head circumference (Jäger-Roman et al. 1986, Ardinger et al. 1988, Martínez-Frías 1990); but caused facial dysmorphias (DiLiberti et al. 1984, Winter et al. 1987, Ardinger et al. 1988) as well as other minor malformations when used alone or combined with other antiepileptics (Jäger-Roman et al. 1986, Martínez-Frías 1990). The pattern of aberrant facial features, almost all retrospectively discovered in cases of older children, was similar in many respects to that due to hydantoins, and like the latter seemed to lack specificity.

The validity of the minor defects, as of the similar ones associated with carbamazepine, will remain unclear until they are prospectively identified in neonates and followed up to older ages. Two siblings exposed to VPA montherapy helped fill this requirement; both were born with digital contractures, which persisted in one of them to a later age; facial aberrancies however were noticeable only at older ages (Kozma 2001).

VPA is very different chemically from other antiepileptics, being a fatty acid, 2-propylpentanoic acid, with a relatively short half life in humans. But the most extraordinary fact about it is that it is the only known discrete environmental agent that causes a neural tube defect in human beings, spina bifida aperta, one of the commonest of major congenital malformations, and that the effect appears to be restricted almost entirely to the posterior neural cord.

Finally, an indication of a hereditary susceptibility should be mentioned. Despite the usual low proportion of VPA-affected children, multiple-sib occurrence was noted in several families (Malm et al. 2002).

Altered Antiepileptic Use

The pattern of antiepileptic prescription changed over time. The great variety of drugs and their use in combination was reduced, while monotherapy, of valproate and carbamazepine in particular, increased (Lindhout et al. 1992a, Oguni et al. 1992, Wide et al. 2000). This change may not have led to a significant reduction in the overall rate of malformations, but may possibly have changed the array of the defects.

Meanwhile numerous newer antiepileptic drugs have been developed in the hope of discovering ones with no or reduced fetotoxic potential, among them felbamate, gabapentin, lamotrigine, topiramate, and tiagabine (Morrell 1998). For few, however, has sufficient information as yet accumulated to be able to comment with certainty on their fetal safety (Nulman et al. 1999, Reiff-Eldridge et al. 2000, Dean et al. 2002). An inkling of information about lamotrigine had emerged in the new millenium, one international registry finding monotherapy nonteratogenic (Cunnington and Tennis 2005), as did a study in Croatia (Miskov et al. 2009) but others an increased risk of fetal maldevelopment (Brodie 2006, Holmes et al. 2009). Stay tuned.

Later Writings

Recent years have seen a barrage of commentaries on the potentialities of antiepileptics for causing prenatal maldevelopment. But equally remarkable was it that few presented new facts or ideas, whether regarding the effects of monotherapy versus polytherapy, etc. (e.g. Juárez-Olguín et al. 2008); the conclusion being as yet, and again, that the only such medications of the many available today with proven teratogenic risk are carbamazepine and especially valproic acid, as was already known years ago (Samrén et al. 1997, Nguyen et al. 2009, Ornoy 2009, Velby et al. 2009).

Psychological Effects

To the time of this writing there have been at least 20 studies of behavioral and cognitive outcomes in children of various postnatal ages associated with in utero exposure to antiepileptic drugs. See Christianson et al. 1994 for an early report. I will spare the reader the tedium of a detailed summary of them, and try to extract the sense of their findings. To wit, various neurodevelopmental deficits were associated with exposure in some cases but not others. Positive findings were more often found by prospective than by retrospective studies; but most were usually of little practical consequence for life's struggles. An exception may be a recent finding of significantly dose related reduction of IQ in 3-year-old children exposed prenatally to valproate, but not to other such drugs (Meador et al. 2009). Two authors did the writer the service of reviewing the literature and finding that indeed retrospective views tend to exaggerate in this respect (Moore et al. 2000, Kozma 2001).

Contradictions abounded. In one study exposure to carbamazepine and phenobarbital was associated with poor performance, but not exposure to phenytoin; while in another phenytoin elicited an association but not carbamazepine; and in a study of various drugs the only significant finding followed exposure to carbamazepine. One study found no difference in IQ between children exposed to monotherapy and polytherapy, while the latter often led to more severe effects of other sorts. And in many other cases no cognitive, behavioral, or other psychological effect was found at all.

Postnatal environmental factors that may confound associations have largely been neglected, as were the possible effects of maternal IQ and socioeconomic variables. Also poorly answered was whether medication – prenatal or postnatal – or maternal illness was primarily if not entirely responsible for the nonphysical abnormalities reported. Cognitive dysfunction was associated with maternal seizures during pregnancy as well as with exposure to antiepileptic drugs. Antiepileptics necessary for maternal health and well being do not stop being taken at birth, and can be excreted into breast milk, extending the exposure into infancy. In the face of these uncertainties and inconsistencies it must be concluded that there is as yet no clear evidence that trivial and inconsequential psychological effects are associated with prenatal exposure to antiepileptic drugs alone, especially since all were no doubt tainted by confounding variables.

Summary

Almost four decades of study of the teratogenic risks run by the offspring of treated and untreated epileptic women yielded great understanding of the question. But along the way much misapprehension and premature pronouncements have had to be corrected. An earlier view stated that there was no "clearcut agreement that any one of the four major drugs used for the treatment of seizure disorders (phenytoin, phenobarbital, valproic acid, and carbamazepine) is more teratogenic than others" (Malone and D'Alton 1997). This assertion is no longer true, nor was it entirely true then.

At present only VPA appears to fit this description, in its propensity to cause the serious malformation spina bifida. The usual earlier estimate of the risk of this condition was approximately 1% in offspring coming to term, higher than occurs in the overall population; but at present it is believed to be of the order of 3% or even greater. But this needs clarification, especially when it is recalled that NTD prevalence per se is affected by many variables, which have as yet been little looked into in studies of antiepileptic teratogenesis; and that risk figure approximates the rate of recurrence of the condition in previously affected families in the overall population, an observation which demands that family history be an integral component of such studies. And as well, as has been commented, there may be "a pharmacogenetic component to sodium valproate's teratogenic and neurodevelopmental effects. Future research may enable us to identify those women whose offspring may be affected" (Duncan 2007). This writer, further, has cogently pointed out the dilemma posed by valproic acid, that "it is the most effective drug in the treatment of idiopathic generalized epilepsy. . ."

Animal Studies

A last word must be given to experimental studies with these drugs. A vast review of the experimental literature dealt exhaustively with this topic (Finnell and Dansky 1991). Numerous pages of tables listing species used, drugs, dosages, human dosage equivalents, significant findings, etc., led to the summarization that "the experimental literature has been extremely beneficial in validating the teratogenicity of selected antiepileptic drugs. . ." The mountain brought forth a mouse.

It might be asked, what is the purpose of experimental teratology studies. Fraser (1959) once said they don't have much relevance "to the bulk of human malformations..." employing as they do degrees of severity hardly ever to be experienced by people, but that they are "useful tools in the analysis of abnormal development." Nau (1987) said "It is very unlikely that an animal species will ever be found [because of metabolic considerations] suitable for laboratory experimentation," an opinion perhaps modified by later considerations (Nau 2001). But the uses of animal studies go far beyond the mere purpose of weeding out possibly toxic pharmaceuticals during the premarketing process, and many examples of the "lessons about human dysmorphogenesis to be learned from experimental teratology," were described by Fraser (1977) in his contribution to the Handbook of Teratology.

Discussion of what this means and its importance are discussed elsewhere in this work, in certain examples under the headings genetics and individual responses, multifactorial/threshold concept, and a new concept of embryotoxicity. Major insights that derived from experimental studies, with direct relevance to human studies, depicted in his usual masterful way by Wilson (1973, 1977), are also outlined in pages above.

Lithium

Its Discovery

Lithium presents a cautionary teratological lesson. This element was discovered in 1817, not too long ago, by the Swedish chemist Johann-August Arfvedson. Despite its name, which means stone, from the Greek lithios, proposed by the chemist Berzelius, it is the lightest of all the solid elements at atomic number 3 and weight 6.94. Soon after its discovery it was put to medical use in the treatment of diseases such as gout; and in the latter part of the nineteenth century Sir Alfred Garrod, of inborn errors of metabolism fame, detailed its therapeutic merits in treating mood disorders, which he believed were caused by "gout retroceding to the head." (This information, as so much in these days, was gleaned from the Encyclopedia Britannica and PubMed, combining old and new sources.)

It is fascinating that after the drug (it turned from an element to a drug with its medicinal applications) lost its many claimed healing powers around the onset of the twentieth century, these should have been rediscovered and lithium found by an Australian psychiatrist (Cade 1949) to be of value in treating acute mania, in a return to Garrod's prescription for mood disorders. It continues to be prescribed for the manic stage of bipolar disorder and also as an antidepressant. Reports of its toxicity when used as a salt substitute for patients on low sodium diets prevented the drug from receiving FDA approval until 1970.

Lithium came into early clinical use in Europe; but cautiously since it had been found to be teratogenic in experimental mammalian studies, large doses causing various malformations in some mice and rat stocks, but not in other lines of rodents or rabbits, pigs, and monkeys (Smithberg and Dixit 1982, Moore 1995).

Lithium is frequently prescribed for maintenance therapy of recurrent manic-depressive disorder, an illness affecting 1–2% of the population, women more often than men. Women of reproductive age are thus among those prescribed the drug, presenting a vast teratologic threat (Dinan 2002).

Its Teratogenicity

With the potential teratogenicity of lithium in mind a registry was initiated in 1968, to which the outcomes of many if not most pregnancies of women receiving the drug it was hoped would be reported as a means of early detection of malformations. A few instances of women taking the drug throughout pregnancy and having normal babies (Fries 1970, Silverman et al. 1971) gave hope that humans, unlike sensitive animals, would not be vulnerable. A preliminary record of 18 women given lithium over the entire course of pregnancy with no recorded malformations supported this likelihood (Fries 1970).

But later the same year the registry received information about two children with malformations (Schou and Amdisen 1970). Recognizing the bias adhering to the method used to gather these data, the investigators were not entirely convinced and urged that registries be informed of the birth of normal as well as malformed children. Nevertheless they implored women of reproductive age to take contraceptive measures while being treated. On the whole optimism continued as the collection grew to 60 children with no greater malformation frequency than expected (Schou and Amdisen 1971).

With the first full report of the registry the optimism evaporated (Schou et al. 1973). The information, gathered from physicians and other sources in various countries, concerned 118 children of women given lithium in the 1st trimester or throughout pregnancy. Nine of the children had structural abnormalities, including in six of 12 neonatal deaths. Most of the malformations were serious, six of the nine cardiovascular malformations, two the "Ebstein type." The authors, as though still reluctant to admit the drug's harmfulness, again noted that the way the information had been collected may have given a false picture of the outcome of lithium-treated pregnancies and implicity urged caution against its overinterpretation.

Ebstein's Anomaly

The tendency in lithium exposed pregnancies to Ebstein's anomaly and perhaps other cardiovascular malformations appeared to be substantiated by their occurrence in several children (Nora et al. 1974). Ebstein's anomaly is a severe malformation in which the tricuspid valve, the valve between the right atrium and the right ventricle, is defective and displaced into the right ventricle. It is also rare, only 300 cases having been noted in the literature (Warkany 1988b) in the more than 100 years since its description by Wilhelm Ebstein in 1866. Nora et al. (1974) gave its incidence as about 1 in 20,000 total births, and thus the two occurrences they reported were unlikely to have happened by chance. A later epidemiological study of cardiovascular malformations in the Baltimore area corroborated this figure, finding an occurrence of 2.5 per 100,000 livebirths (Correa-Villasenor et al. 1994). A review of the condition noted it accounted for about 1% of all heart defects (Attenhofer et al. 2005).

Updates of the lithium registry augmented the number of cardiovascular defects, noting a total of 10 out of 13 malformed babies in 143 medicated pregnancies (Weinstein and Goldfield 1975). Further updates solidified the impression that, while biased recognition might exaggerate its frequency, with regard to specific malformations, where there was smoke there must be fire (Schou 1976, Weinstein 1979). What seems to have been the last report from the registry recorded major malformations in 21 of 212 infants, including six with Ebstein's anomaly and 11 with various other cardiovascular malformations (Weinstein 1979).

A different approach to the question was taken in a Swedish study (Källén and Tandberg 1983). By linking the records of birth and gynecologic registries 97 women with manic-depressive disease were identified who had received lithium and/or other psychotropic drug treatment during pregnancy. Six of their children had congenital malformations including four with serious heart defects, though none was an Ebstein's anomaly, while two of the children of the 190 control women had heart defects (one a Down child, a condition in which heart defects are frequent). This was a statistically significant difference and therefore added to the evidence that lithium medication during pregnancy increased the risk of heart disease (Källén et al. 2006).

A detailed review of the lithium studies prompted Warkany (1988b) to state that the increased risk of Ebstein's anomaly posed by such treatment contraindicated the use of the drug in manic-depressive women of reproductive age unless pregnancy was prevented. But, intimating doubt, he also cautioned that questions remained that could only be resolved by prospective studies.

The Retraction

A brief report foreshadowed the start of a countertrend (Källén 1988). It noted that case-control data collected through the International Clearinghouse for Birth Defects Monitoring Systems had found that drugs were not major contributors to tricuspid malformations and that the association between lithium and Ebstein's anomaly was weak.

The retreat gained momentum with several case-control studies. One conducted by the CDC Birth Defects Monitoring System, a nationwide program monitoring about 30% of the newborns in the US, identified 76 infants with possible Ebstein's anomaly in 1981–4 (Edmonds and Oakley 1990). For 34 of the cases sufficient information was provided by physicians and hospital staffs to meet malformation definition criteria, and for none of them did the mother have a history of manic depressive disease or of lithium use during pregnancy.

A Canadian case-control study found 59 instances of Ebstein's anomaly diagnosed in 1971–88 with complete medical and drug exposure history (Zalzstein et al. 1990). None of the mothers of the children had manic-depressive disease or been treated with lithium before or during pregnancy; nor was the rate of lithium exposure during pregnancy in the case mothers different than in the controls. It was pointed out however that even these impressive findings could not rule out a random association, since at a lithium intake of 1 per 1000 pregnancies and an anomaly frequency of 1 in 20,000 the likelihood of the anomaly occurring spontaneously in a pregnant woman taking lithium is 1 in 20,000,000, but can do so with a great deal of confidence (Zalzstein et al. 1990).

A Hungarian case-control study of over 10,000 patients, in which the number of malformation instances was very small, found no association between lithium and malformations (Czeizel 1993a). In an epidemiological study in which parents of cases with Ebstein's anomaly were interviewed correlations were found with several environmental factors, but not with lithium (Correa-Villasensor et al. 1994).

The Finale

More than 20 years after the first suspicion arose of the prenatal harmfulness of lithium, after it had been expressed again and again that registry-collected cases were unreliable indicators of teratogenicity, what appears to have been the first prospective study of this difficult problem, for a while all but drove the nail into the coffin and absolved lithium for all practical purposes of being a teratogen (Jacobson et al. 1992).

It took a large multicenter effort to gather 148 lithium-using women – whose size is told by the lengthy list of coauthors – and to find that the cases and matched controls had close to the usual frequency of congenital defects, 2.8% and 2.4% of livebirths respectively, and one each a cardiovascular defect, an Ebstein's anomaly in the exposed group, aborted at 16 weeks of gestation, and a ventricular septal defect in the controls.

Some commentators thought it improbable and others "noteworthy" that so rare a malformation can have occurred by chance in so small a number of offspring (Ferner and Smith 1992, Cohen et al. 1994). According to a set of evaluators, because of concerns over differences between study centers in the enrollment of the subjects, etc., the pregnancy outcome data were "very difficult to interpret" (Moore 1995). A recent review of relevant cases, evaluating prospective versus retrospective studies, also expressed such caution, holding that "lithium does not seem to be a significant teratogen" (Yacobi and Ornoy 2008).

It seems that the statement made years ago (Kalter and Warkany 1983) that lithium was among the drugs no longer considered teratogenic in customary use – while it may have been premature at the time – seems to have been well vindicated.

However, at the same time that a much relied-upon medical resource assessed the teratogenic risk of lithium as "small" it continued to advise prenatal diagnosis in lithium-exposed pregnancies (Friedman and Polifka 2000, p. 393). Such advice, as well as that of avoiding pregnancy, if stringently followed, would preclude the teratologic innocence of lithium ever being established with certainty.

An Addendum

Bringing the record up to date, as far as such matters can ever be resolved, exposure in early pregnancy to a number of antidepressants, either tricyclic antidepressants or selective serotonin reuptake inhibitors (SSRI), was found to be free of teratogenic risk (Chambers et al. 1995, Nulman et al. 1997b, Alwan et al. 2007, Davis et al. 2007, Louik et al. 2007, Einarson et al. 2009, Wichman et al. 2009), or unclear because of confounding factors (Wogelius et al. 2006, Bar-Oz et al. 2007). In contradiction were possible associations of SSRIs with low frequencies of septal defects and other undescribed cardiovascular malformations (Diav-Citrin et al. 2008, Pedersen et al. 2009).

It may be of interest to interject that 60 nonmalformed children prenatally exposed to lithium, when examined at 5 years of age or greater, and judged against siblings not prenatally exposed, had no increased frequency of physical or mental impairment (Schou 1976).

Antihypertensives

Hypertension occurs commonly during pregnancy, and its treatment is problematic. Antihypertensive medication in the later months of pregnancy is known to be associated with fetal deleterious effects (Quan 2006); but its effects in earlier pregnancy are controversial.

A report of such a consequence has been questioned. An increased frequency of major congenital malformations was reported in offspring of nondiabetic women exposed to angiotensin converting enzyme inhibitors in the first trimester (Cooper et al. 2006). The risk of defects was about doubled and consisted mainly of cardiovascular malformations. These findings were questioned as perhaps affected by ascertainment difficulties, and as of now the matter in unresolved (Ray et al. 2007).

Folic Acid and Congenital Malformation

All the known or conjectured environmental causes of congenital malformations in human beings consist of positive factors, chemical, infectious, etc. (Kalter and Warkany 1983) – with two exceptions. The only two instances of malformations being attributed to negative conditions are deficiency of iodine, described earlier in this work, and deficiency of the vitamin folic acid. The latter, a major subject, is told here.

The effects of folic acid deficiency on human pregnancy outcome have long been debated. Megaloblastic anemia, one such sign, was once relatively common in pregnancy. In addition to this incontestible sign various complications of pregnancy have been claimed as associated with folic acid deficiency, including placental abruption, spontaneous abortion, perinatal mortality, and congenital malformations (Pritchard et al. 1970), of which the last has received much attention.

Maternal Folic Acid Status and NTD

Of the innumerable discrete environmental factors that have been alleged to cause the dysraphic congenital malformations of the central nervous system known as neural tube defects (NTD) only valproic acid and deficiency of folic acid have been considered credible. But the advent of these two beliefs was very different from each other. The connection with valproic acid was fortuitous whereas the connection with folic acid evolved historically.

Over the years the relation of folic acid to fetal maldevelopment has been investigated in two ways; an older one examined its association with maternal folic acid status, and a newer one, the association with maternal vitamin usage.

The earlier focus consisted of examining the pregnancy outcome of women probably folic acid deficient, or the folic acid status of women with malformed children, i.e. prospectively or retrospectively, as epidemiologic parlance describes it. A study of the former sort from Canada seemed to find a positive connection, i.e. malformations in some children of women diagnosed perinatally with hematologically diagnosed megaloblastic anemia (Fraser and Watt 1964). The authors were cautious in imputing the defects to folic acid deficiency, however, feeling as is so often

H. Kalter, *Teratology in the Twentieth Century Plus Ten*,

DOI 10.1007/978-90-481-8820-8_14,

the case that it "should be investigated further." Especially, they might have added, because the anemia usually developed late in pregnancy, beyond the embryonic sensitive phase.

In a similar study from England women with chemically diagnosed folic acid deficiency had children with a greater frequency of major congenital malformations than those of women with normal folic acid levels (Hibbard et al. 1965). The validity of such comparisons, one need not be reminded, depends on the circumstances of the ascertainment of the subjects and the diagnosis of the malformations, in controls no less than cases.

A retrospective study found that a larger proportion of mothers of malformed infants had reduced folic acid levels than did control mothers; true of children with malformations generally as well as of those with central nervous system defects (Hibbard et al. 1965, Hibbard and Smithells 1965). Several caveats were voiced or were implicit. Matters of ascertainment and confounding – e.g. alcohol and antidepressant use – may have intruded; the folic acid status was discovered perinatally, and may not have been indicative of the condition earlier; the assay test used was not always accurate (Kitay 1969). Smithells (1994) later said of these preliminary communications that they "sank without so much as a ripple."

As often happens these positive indications were followed by negative ones. None of the newborn infants of women with severe folic acid deficiency had serious malformations (Pritchard et al. 1970); no folate-deficient pregnant women had a malformed infant (Daniel et al. 1971); mothers of severely malformed infants had no lower serum folate levels than mothers with uncomplicated pregnancies (Emery et al. 1969, Scott et al. 1970). These studies were all retrospective, but neither did a prospective study have positive findings (Hall 1972, 1977).

Later Concentration Studies

Studies of folate concentration in pregnancies made in more recent years were no less contradictory. In a prospective study from Ireland, a high NTD risk region, serum folate concentration in early pregnancy was not significantly different in women delivering infants with NTD, almost all of whom were not folate supplemented, than in control women (Molloy et al. 1985). A contrary finding came from Scotland, another high risk region, where women with two or more previous offspring with such defects, refraining from dietary folate ingestion in the three immediate preconception months, had red cell folate concentrations significantly lower than controls (Yates et al. 1987).

In Finland, a lower risk area (Saxén 1983), a prospective study saw no difference in serum folate concentration between women bearing NTD offspring and controls (Mills et al. 1992). This was true as well in a British high risk area, with blood folate concentration and folate dietary intake no lower in women with a previous NTD pregnancy than in controls (Wild et al. 1993); contradicting incidentally an earlier study (Schorah et al. 1983).

The latest and probably the last study of this type was a case-control study of 15-week pregnant women in three Dublin hospitals in 1986–9, which found red blood cell folate level and NTD risk inversely related, such that those with the lowest level had a malformation rate of 6.6 per 1000 and the highest 0.8 per 1000 (Daly et al. 1995); from which findings calculations were made of the vitamin supplementation strategy needed to prevent NTD to one extent or another.

These findings and calculations were suspect, however, because the overall NTD birth prevalence in the sample, 1.9 per 1000, was far lower than that found by the European Registration of Congenital Anomalies, which reported prevalences in Dublin in 1980–6 and 1987–92 of 3.45 and 2.72 per 1000 respectively (EUROCAT Working Group 1991, Anon 1995), as well as a continuously falling NTD prevalence during those years.

The underestimate perhaps resulted in part from the exclusion of gestations of less than 23 weeks (Kirke et al. 1993); but mostly it would seem, judging from the discrepancy between the mean annual pregnancy numbers in the EUROCAT reports and that in the cited papers, it was due to underrepresentation.

Such inconsistencies were not clarified by estimation of midpregnancy folate concentration, since maternal levels in pregnancies with and without NTD were not different (Holzgreve et al. 1991). But the authors were cautious in interpretating the results, merely stating that the results cast doubt on an easy explanation for the positive findings of folic acid prevention of NTD occurrence.

Using another variation – amniotic fluid vitamin B_{12} content – folate, etc. were measured at midpregnancy in women delivering offspring with NTD (Gardiki-Kouidou and Seller 1988, Dawson et al. 1999). Both studies found a lower vitamin B_{12} content, but not of folate, raising among other things the question of the interrelations of these two vitamins, a matter of some complexity.

The Connection Develops

Of the environmental aspects possibly implicated in the connection between folic acid deficiency and NTD the one that loomed largest was nutrition, because of the close relation that had been found years before between anencephalus and social class. As early as 1939 a clear connection had been drawn between social class and stillbirth and infant mortality, and the basis of this relation, at least in part, imputed to be quality of diet (Baird 1947).

In Great Britain, according to the classification of the British Registrar General, individuals were divided into five classes, I to V, based on the occupation of the head of the household, class I professional, II mangerial, III skilled, IV partially skilled, and V unskilled. Applying these divisions Edwards (1958) found a marked correlation between class and anencephalus and spina bifida, for the former rising in a straight line from 0.9 per 1000 total births in class I to 3.6 in class V. Similar trends were later found in several populations in different parts of the world (Golding 1982). (But not everywhere, Vrijheid et al. (2000) finding only a slight gradient, if any, of risk for NTD with socioeconomic deprivation in UK in 1986–93; perhaps due to its prevalence declining and the instituting of selective abortion.)

It seemed to make sense, then, that the connection between social class and NTD, with its suspected nutritional basis, should be followed up by a focus on various nutrients, including folic acid. A study begun in 1969 in Leeds, England strove to do this, and the early findings were promising (Smithells et al. 1976). First-trimester levels of several vitamins – folate, vitamin A, vitamin C, and riboflavin – were all lower in women of classes III and above than in the others.

Question arose however. The central nervous system malformations were associated only with decreased levels of vitamin C and red blood cell folate but not serum folate, and there was a social gradient for red blood cell but not for serum folate, although both were significantly correlated (Wild et al. 1993). But especially problematic, the women were volunteers and perhaps unrepresentative in numerous ways of the Leeds population.

Diagnosis also presented a difficulty. All central nervous system defects were classified as NTD, but only four so qualified – three anencephaluses and a myelomeningocele, the others being meningocele and microcephaly. Furthermore none of the four infants with qualifying defects was borne by women of the lowest social classes. That such erroneous classification continued was indicated by the later inclusion of cranial meningocele, meningocele, and iniencephaly among the NTD (Smithells et al. 1981, Seller and Nevin 1984). [Iniencephaly has sometimes been classed as a NTD and sometimes not; only in some small but unclear proportion does this abnormality seem to involve a defect of neural tube closure (Howkins and Lawrie 1939, Warkany 1971, pp. 234–6, Lemire et al. 1972).]

Maternal Vitamin Usage and NTD

Most studies of the vitamin usage of women bearing offspring with NTD were passive, i.e. comprised interviews of women to discover the patterns of consumption of vitamins in the several months preceding and following conception. An early retrospective study of this sort found that a high proportion of mothers of infants with congenital abnormalities took vitamins including folic acid during the first months of pregnancy (Nelson and Forfar 1971). Other studies, characterized as active, asked women to take vitamins of one sort or another from before planned conception through the embryonic period of vulnerability to NTD. The first such studies were of NTD recurrence, i.e. of women who had previously borne NTD offspring.

Folic Acid and NTD Recurrence

Earlier studies of this nature had results that were sufficiently suggestive for vitamin deficiency to be proposed as a factor in the genesis of central nervous system defects, and for the hypothesis to be tested through their recurrence, i.e. a study of women who already had children with central nervous system defects.

A multicenter intervention project of this sort was conducted in which women planning a further pregnancy but not yet pregnant were given a commercial multivitamin tablet (each including 0.36 mg folic acid) or an iron preparation daily for a month or more before conception through a time following neural tube closure. Controls consisted of women not vitamin supplemented who also previously had children with central nervous system defects, but who were already pregnant or declined to take part in the study (Smithells et al. 1980). The results were highly significant: only 1 of the 178 examined offspring of the supplemented women had a NTD while 13 of the 260 examined children of the control women had NTD (0.6% vs. 5.0%).

Various matters must be considered. A considerable fraction of the recognized abortions in both groups were not examined, 45% and 34%, respectively, leaving open the possibility of overlooked occurrences of NTD. In addition the spontaneous abortion rate in both groups of women was low, just under 10% (even though the unsupplemented ones were already pregnant at the study's onset), meaning that some further spontaneous abortions may have been missed.

Next, not all of the supplemented women were screened prenatally and the number of the unsupplemented women that were so screened was not noted. Which may be relevant since the sole NTD offspring of the supplemented women was discovered in this way, while 11 of the 13 NTD offspring of the unsupplemented group were so detected and aborted.

Perhaps therefore only NTD prevalence at birth should have been the basis of the comparison: namely none in 182 supplemented pregnancies and 1 in 265 unsupplemented pregnancies. But even the former instance is suspect, consisting as it was of a "skin-covered lesion," and hence not a true NTD. A discussion of NTD birth prevalence versus so-called total prevalence, i.e. in neonates plus elective interruptions, will be found below.

Other questions were also raised. Pointed out as the most troublesome matter was that the study was not randomized or double blind. Also troubling was the wide geographical distribution of the supplemented and control mothers, which included high and low risk areas, possibly prejudicing the recurrence risk (Stone 1980). The question turned on the relation of occurrence and recurrence risk. This was answered by a survey of different geographical areas that found them to be positively correlated (James 1981). Even more pertinent, this was also true of different social classes within a given population (Nevin et al. 1981). It is thus possible that the findings of Smithells et al. (1980) were distorted by including a larger number of control women than supplemented women from high risk areas.

Findings of new studies, replying to these criticisms, confirmed that the vitamin supplmentation was associated with a greatly reduced rate of NTD recurrence (Smithells et al. 1981). Combining results from the earlier and a later cohort further strengthened the impression that vitamins had a preventive effect, with the recurrence rate in supplemented and unsupplemented women after one and more than one previous NTD child being 0.5% and 4.2%, and 2.3% and 9.6% respectively (Smithells et al. 1983a). But if recognized spontaneous and therapeutic abortions were omitted the outcome was not impressive.

Skeptical reactions were again forthcoming. There was concern about possible self-selection of low-risk women in the supplemented groups, especially in the second cohort (Emanuel 1983). The rejoinder to this was that with few exceptions the case and control women were referred by medical personnel (Smithells et al. 1983b); which in itself however would not seem to guarantee a balanced ratio of case and control women.

There was also suspicion of the reliability of the referral diagnosis of the women as having previously had a child with a NTD, as indicated by the fact that two of seven previously affected children, as detailed in one publication, did not have a NTD (Seller and Nevin 1984).

A reiterated criticism was that the study was not randomized and controlled (Oakley et al. 1983, Wald and Polani 1984), to which the testy reply was given that the authors were aware of this and it need not be "laboured further." It was true nevertheless, and prevented unqualified acceptance of the findings. It was also pointed out (Oakley et al. 1983) that the recurrence rate found in the supplemented and control women combined (2.9%) was just about what was being found in the period studied in genetics counseling centers in the UK, a period of reduced occurrence and consequently of reduced recurrence, which suggested that supplementation had been of little or no protection.

A final report from the Leeds group (Smithells et al. 1989) summarized its 1976–87 findings: 1 NTD recurrence in 187 fully and partly supplemented women, and 18 such recurrences in 320 unsupplemented women (0.5% vs. 5.6%), a statistically significant difference. The individual results for the three cohorts (1977–80, 1981–4, 1984–7) told a different story. First, the overall recurrence rate decreased, from 5.4% in the first interval, to 2.3% in the second, to 1.7% in the third. The three intervals therefore were not homogeneous and should not have been examined for risk differential as a unit. Most troubling however, individual statistical tests showed that in none of them was there a significant difference between the supplemented and unsupplemented groups.

Randomized Trials

A study of the sort that was recommended to avoid various inadequacies had in fact already been conducted (Laurence et al. 1981). Women in two districts of south Wales who had previously given birth to children with NTD and were planning to have further children took part in a double-blind randomized control trial, one group taking 2 mg folic acid twice a day and the other a placebo. The outcome was positive, the first group having no NTD child in 44 pregnancies, and the second four in 51 pregnancies. But because an appreciable number of the folate group did not comply with the protocol and the overall number of women was relatively small the results were inconclusive. The outcome was also weakened perhaps by inclusion of non-neural tube central nervous system malformations.

It is unclear whether the same weakness was also true of a study conducted in Havana, one that had the advantage however of avoiding the imponderabilities of multivitamin ingestion, by patients taking only 5 mg folic acid daily (Vergel et al. 1990). Details of patient ascertainment, treatment compliance, etc. were not described; only that women registered in the Provincial Genetic Department as previously having had a pregnancy complicated by a NTD were included in the study. Again the unsupplemented women had a higher NTD recurrence rate than the supplemented ones, but the limited number prevented the difference from being significant.

MRC Trial

It was believed that the various doubts entailed in other studies would be removed by well designed trials, which would anticipate future criticism by being large as well as randomized and double blind. A study of this sort, conducted under the auspices of the British Medical Research Council, insured the first by involving 33 centers (17 in the UK and 16 in other countries – Hungary, Israel, Australia, Canada, the then USSR, and France). It began in 1983 and continued until 1991, when, after sequential analysis, the findings were considered definitive enough to justify bringing the study to an end (MRC 1991).

Women who had previously given birth to children with a NTD were randomly assigned to groups taking daily 4 mg folic acid, other vitamins, both, or neither from before conception to the 12th week of pregnancy. The results, vindicating expectations, indicated that folic acid, but not the other vitamins, protected against NTD, which recurred in 1.0% (6/593) of those taking folic acid, and 3.5% (21/602) in the others, thus affording about a 72% recurrence protection.

Several comments are to be made about the study. How women came to participate was not stated; ostensibly they were "recruited," but whether and how much self-selection was involved was not noted. To counter past criticism the UK women, who composed 44% of the overall number of subjects, were randomly allocated to treatment group according to social class. Whether this was true of women from other areas is unclear. The largest segment of them, 41%, were from Hungary, and even in a supposedly classless society, as it was then, it cannot be taken for granted that socioeconomic inequalities did not exist. This possibility, to my knowledge, has not been addressed (Czeizel and Fritz 1989, Czeizel and Dudás 1992, Czeizel 1993b, Czeizel et al. 1996).

Regardless, two circumstances – the no doubt diverse socioeconomic status of subjects, and the temporal decline in the frequency of NTD, which began even before the practice of selective abortion was instituted or folic acid therapy came onto the scene – cannot but have confused the outcome.

Next, it was posited that most of the reduced NTD recurrence happened in high risk populations (Smithells et al. 1991), and by extension folic acid supplementation may be less effective or even ineffective in low risk ones. Hungary in the 1980s was a relatively low-risk area, the total prevalence of NTD, i.e. in all births and prenatally diagnosed and terminated fetuses, being about 1 per 1000 (Czeizel 1993b). Thus the findings of the MRC trial may have been prejudiced by intermixing subjects of various risk status.

Last, although the recurring NTD were true defects of neural tube closure – anencephalus, spina bifida aperta [all except encephalocele about which there is a question – see comment below] – as often before, no assurance was given that the same was always true of the defects that led to the women being included in the study. As in the study cited above other central nervous system defects may also have inadvertently occasionally done this duty. Thus there remain a number of unresolved uncertainties about the MRC study.

[The classification of encephalocele, and the related meningocele, is confused. These localized, later arising, i.e. postclosure, defects, relatively less frequently occurring, consist of external saccular protrusions of meningeal tissue through skull or vertebral interruptions. A generic term for these protrusion defects is spina bifida cystica, which have been erroneously grouped with spina bifida aperta; but in essence are not NTD.]

Irish Double Blind Trial

A randomized double blind multicenter trial was also carried out in Ireland in 1981–90 (Kirke et al. 1992). Women who previously had a NTD baby were randomly allocated to receive periconceptionally only folic acid (0.36 mg/day), multivitamins plus folic acid, or multivitamins only. The results failed to uphold the hypothesis that folic acid is protective since the NTD recurrence frequencies in the test and control groups were not significantly different (0/89 vs. 1/172). Socioeconomic variables played no part in this since the groups had approximately equal proportions of women of classes I and II. The earlier intervention study (Smithells et al. 1980) using this folic acid dose therefore was not supported.

It was almost inevitable that this trial would be uninformative. Declines of almost 65% in the NTD birth prevalence occurred during its course (strangely, this was said to have been unanticipated), and almost none of the appreciable number of spontaneous fetal deaths was examined pathologically.

Texas Trial

We move west now to the story in Texas; which began with a brief note (Pemberton 1992). It reported that several cases – what has come to be termed a cluster – of anencephalus, 30 in this instance, occurred in a localized region, a city and the surrounding area along the Texas-Mexico border, in a relatively short period of time. For want of a better explanation the ‘epidemic’ was blamed on toxic waste from sources across the border in Mexico. At the same time, rather incongruously, it was acknowledged that the frequency of this malformation is higher in Hispanic people, which it was said may explain why Texas has a NTD frequency twice the national average; a disparity also true of other areas along the border (Canfield et al. 1996a). This seems to be where the episode came to a rest for some time. (It is apropos here to mention that aggregation – temporal and geographic clustering – of medical phenomena, malformations, etc. – is common enough for a service to have been set up for the purpose of reporting and investigating such occurrences – Eurocat Cluster Advisory Service 1993).

The story continued, in a slightly different form, some years later when a survey of a small number of high risk Hispanic women in an area along the US-Mexico border found that folic acid supplementation was associated with reduced NTD recurrence in term births (Anon 2000a); but as the accompanying editorial stated the limited nature of the study made the results suggestive at best.

These results were not confirmed by a multisource case-control study of this population mostly during 1995–8 in 14 Texas counties along the Mexican border (Suarez et al. 2000). From interviews of Mexican-American women with and without NTD children (identified at birth or prenatally in abortions) it was learned that approximately the same proportion of cases and controls (2.0% and 2.5%) had consumed vitamin supplements containing folic acid daily from 3 months before to 3 months after conception. Supplementation therefore had not reduced the frequency of the defects in this population.

The authors seeking an explanation for a finding so out of line with many others noted first that it may have been because of the small number of women taking the vitamin supplement. Or, they speculated, it may be that Mexican Americans, because of their higher NTD prevalences, require greater folate intake for prevention. The mandated inclusion of folic acid in grains, discussed below, may in time have had an impact on NTD prevalence in this population. This likelihood is considered below; as are other studies with Hispanic populations in Mexico and California. A full description and discussion of the tale as it unfolded in Hispanics in Texas appears in Kalter (2009).

It may also be mentioned that a similar failure to find a significant NTD reduction after mandatory folic acid food fortification in a population in northeastern Brazil was also reluctantly not accepted, for a number of reasons, insufficient follow-up period, insufficient period of consumption of fortified products (Pacheco et al. 2009).

Folic Acid and NTD Occurrence

Folic acid supplementation having been shown to the satisfaction of some to reduce the NTD recurrence rate, or because of the realization of its drawbacks, several attempts were made to learn whether the same could be achieved for first time occurrence of these abnormalities. Studies of the latter sort in fact would have wider and more general significance, and also be greatly aided by the majority, perhaps as much as 95%, of these conditions being first occurrences, thus affording far larger numbers of subjects for study than recurrence studies.

Atlanta Study

The first study of this type, conducted as part of the Metropolitan Atlanta Birth Defects Program, compared the periconceptional use of multivitamins by women who from 1968 through 1980 had newborn infants with anencephalus or spina bifida with such use by mothers of babies without malformations or with serious malformations other than NTD (Mulinare et al. 1988).

Such usage was reported by about 14% of women overall, but was twice as great in case as in control women. Supporting previous findings of such protection perhaps being more effective in high than low risk groups (and noted elsewhere in this work), the study found NTD frequency reduced in whites but not in other races, remembering that the NTD frequency in blacks, the predominant nonwhite population of Atlanta, is perhaps one-tenth that of whites (Mellin 1963, Altemus and Ferguson 1965, Erickson 1976).

The question was asked, in commenting on the results of the Atlanta study, whether NTD would disappear were all women to take vitamin preparations at the time of conception. The answer given was "probably not," but the reason for it was only briefly considered (Holmes 1989). The question will be discussed further below.

The Atlanta study had several weaknesses, especially that it relied on parental recall of long-past events; a potentially troublesome matter that has frequently bedeviled epidemiological studies (see citations below). Details of its design and other features were described by Erickson et al. (1984). Information regarding many aspects of maternal health before and during the index pregnancy was obtained by telephone interview some years after the births. Not all parents were located or cooperated, and for only about 70% of the eligible individuals in each group were interviews completed. Participation was lower for nonwhite than for white mothers, leaving a gap since the former contributed a sizable proportion of all births. Difficulties may have been introduced by using birth certificates of children "without defects" to select one group of control mothers (the accuracy of the recording of malformations on such documents has frequently been found wanting); and by case infants including stillbirths as well as live births, the controls only live births.

NIH Study

A study organized by the National Institutes of Health hoped to avoid the imponderables encountered in probing maternal recall after extended periods of time (e.g. see Gittelsohn and Milham 1965, Mackeprang et al. 1972, Hexter and Harris 1991), by interviewing women by telephone regarding multivitamin or folate use within 5 months of delivery (Mills et al. 1989). The women were recruited in 1985–7 at two centers, in California and Illinois, through various sources. The periconceptional use of the vitamins by women with an infant or fetus with a NTD diagnosed prenatally or postnatally was compared with that of control women with infants without major congenital malformations or with various pregnancy complications, including non-NTD abnormalities (the latter intended to provide a group with as good recall as women with abnormal offspring are presumed to have). The upshot: there was no significant difference between case and control mothers in the rate of use of multivitamin or folate-containing supplements.

Several explanations were given for the discrepancy between this finding and that of the Atlanta study, namely secular reduction as well as geographic variation in NTD frequency, genetic differences between groups, recall bias. It may be noted that some significant fraction of the case offspring had central nervous system malformations that were not abnormalities of neural tube closure.

Boston Study

In a prospective study women in the care of over 100 participating obstetricians mostly in New England were asked at midpregnancy, at amniocentesis or serum alphafetoprotein screening, whether in the 3 months preceding pregnancy they took a folate-containing vitamin preparation (Milunsky et al. 1989). The NTD frequency in those replying negatively was 3 per 1000 and positively 0.9 per 1000, a reduction of 70%, the pregnancy outcome learned of from the delivering physicians or the mothers. The sample of women was treated as a unit, with no consideration given to ethnicity, race, and socioeconomic status – factors of relevance to NTD occurrence – especially of those from the Boston area, who composed one-third of the group.

The Debate

A heated exchange followed upon publication of the discrepant conclusions reached in the last three studies described. One group (Mills et al. 1990) defended itself against the charge by another (Milunsky et al. 1989) of recall bias or other miscalculation; and countercharged that their accusers misclassified women as to vitamin usage, as well as overlooking subject motivation. The latter replied (Milunsky et al. 1990) that the risk variations were of no importance. Another critic (Shapiro 1990) weighed in to the effect that a relatively large proportion of women in the Boston study were aware of their pregnancy outcome, which may have prejudiced recall. A partly reconciling solution was offered (Anon 1990, Seller and Nevin 1990), which suggested that perhaps the contradictory findings were due to the studies having been done in areas with different NTD prevalence.

At this point, aside from difficulties of ascertainment and maternal memory bias, it seemed possible that the main reason for the discrepant outcomes were differences in NTD prevalence among populations, such that folate-containing vitamins prevented NTD only or especially in high risk groups. This consideration is not without contradictions of its own, as will be seen below.

Two Widely Separated Studies

Further confusing the record were two studies also with divergent findings, one from Australia the other from Hungary. The former was based on questioning mothers some years after they had children with NTD regarding food and vitamin consumption before and during pregnancy. Earlier analysis of the data indicated that dietary intake of folate in early pregnancy protected against the occurrence of NTD (Bower and Stanley 1989); but detailed restudy of the vitamin data concluded on the contrary that there was no definitive evidence of an association between periconceptional vitamin supplementation and such defects (Bower and Stanley 1992)– perhaps exhibiting the complexities of uncovering the details of dietary intake.

The Hungarian effort, described in a "final" report, was an intervention study. Volunteers were randomly assigned to groups taking a daily multivitamin tablet, including 0.8 mg folic acid, or a trace element mixture, from some time before planned conception to the 12th week of pregnancy (Czeizel and Dudás 1992, Czeizel et al. 1994). In the final outcome there were no NTD in 2471 births in the multivitamin supplemented group and six in 2391 trace element births, indicating that the vitamin product afforded protection. (As discussed above, there is doubt about considering the spinal bifida cystica in the control group a true NTD.)

These reports were unclear on several points. Not all the so-called 'evaluated' pregnancy outcomes were included in the NTD analysis. This omission was not explained, except for the ambiguous statement that the analyzed material consisted of women with 'confirmed' pregnancies. The distinction between evaluated and confirmed was unexplained. The multivitamin supplemented group included, also without explanation, a substantial proportion that received no supplement.

Also troublesome was the fact that after receiving their final supply of tablets, at the 12th week of pregnancy, the subjects were referred to prenatal care clinics, which reported pregnancy outcome to the study directors. It may thus be that untoward pregnancy outcome during the first trimester was not monitored, and spontaneous abortion not taken into consideration (Czeizel et al. 1994). The conclusion of the study that "this primary preventive method can reduce the occurrence" of NTD therefore is thrown into doubt. [And incidentally a doubt also cast upon was the outcome of the MRC trial described above, since over 40% of its subjects were Hungarian.]

Other Recent Studies

Boston Case-Control Study

Women with and without NTD offspring were interviewed within 6 months of delivery or therapeutic abortion (the proportion of the latter was unmentioned) in three metropolitan area hospitals in northeastern US and Canada in 1988–91, regarding dietary practice and frequency of vitamin supplement use, with or without folic acid, in the month preceding conception and throughout pregnancy (Werler et al. 1993).

Only in the 7.8% of the case women who used folic acid-containing vitamin supplements daily was there an apparent NTD protection, a relative risk of 0.5. In women able to estimate daily folic acid intake – those presumably most certain to have taken folate-containing supplements – intake varied from less than 0.4 mg to over 1 mg; but no clear dose-response relation could be discerned. Inaccurate recall may have been a factor since in many instances subjects were asked about events occurring 20 months earlier.

Dietary folate intake was estimated for the case women, over half of them, who took no supplements at all, with even less confidence than for the vitamin supplement data. The confounding effects of demographic and health behavior factors on the relation of vitamin supplementation and NTD were considered to be of negligible importance (Werler and Mitchell 1992).

A final matter relates to the unusual ratio found of anencephalus to spina bifida, 1:3.7, very different from the approximately 1.5:2 often found in the northeast US (e.g. MacMahon and Yen 1971); which raises the possibility that some spontaneous abortuses with anencephalus were undetected.

California Study

A similar case-control multicenter study (Shaw et al. 1995) was made at about the same time in California, an area of lower overall NTD risk than other US regions (Strassburg et al. 1983). Women were interviewed 4–5 months after the approximate time of term delivery regarding dietary practice and multivitamin usage in the months before and after conception. From this information folate intake was estimated, and vitamin usage found to protect against the occurrence of NTD.

Usage in early pregnancy was unprotective. But vitamin usage not begun till the 2nd trimester, beyond the time of neural tube vulnerability, seemed to be protective; a finding that either reflected poor maternal power of recall, or made the protective value of vitamins doubtful. Unfortunately the manner of presenting the data prevented independent calculation of the degree of protection and judgment of the findings.

The study also found that NTD prevalence was far greater in Hispanics than in other ethnic groups (as also found by several others, e.g. Strassburg et al. 1983, Canfield et al. 1996b, Feuchtbaum et al. 1999). Yet, contrary to some evidence that folic acid affords greater protection to high- than low-risk populations, Hispanics (mostly Mexicans, who formed almost half the subjects) were far less protected by vitamins than were others (Shaw et al. 1997, Hendricks et al. 1999, Kirby et al. 2000).

Interpreting the findings was also made problematic by the prevalence of the malformations in Mexicans varying according to migration and residence status (Harris and Shaw 1995). In Mexico the NTD rate was 3.3 per 1000 live births (close to the highest ever recorded), while in those born in Mexico residing in California it was 1.6 per 1000 and in California-born Mexicans 0.7 per 1000, the last close to that of US-born Caucasians in California. What accounts for the decreased risk in expatriate births, whether partly ongoing secular decline, etc., remains in question.

With respect to the extremely high NTD rate in Mexicans born and bearing children in Mexico expectation was frustrated by the dietary intake of folate in the periconceptional period being adequate, i.e. by absence of a relation to folic acid (see references in Harris and Shaw 1995). Similarly disappointing was the finding in Mexican-Americans discussed above (Suarez et al. 2000).

A later report from Mexico, on the contrary, noted a far lower NTD frequency, in hospital births in Nuevo Léon, in northeast Mexico, a NTD frequency in 1995–9 of 1.47 per 1000 births, similar to the level in acculturated Mexicans in the US (Hernández-Herrera et al. 2008). A number of possible reasons for which come to mind.

Emigration and Acculturation

We digress now for a moment. The gradient in the Mexican NTD rate once again raises an old question, with a novel wrinkle: whether, at a given moment in time, heredity or environment is predominant in the etiology of malformations. Does the situation in Mexicans and similar instances in other groups help answer the question – assuming that the answer is one or the other? On the one hand, the many instances of migrant groups retaining, at times for generations, the risk present in their ancestral communities argue for heredity. On the other hand, examples in which emigrants come to resemble the groups they live among, although perhaps not immediately, depending perhaps on adoption of new cultural patterns, seem to argue for the environment (Hobbs 1969, Leck 1969). In part the latter pattern seems to be true of the Mexican groups discussed here.

Other groups display other patterns. A pattern of persistence is well exemplified by the NTD prevalence in American blacks continuing after many generations to be significantly lower than in whites (Milic 1969, Kurtzke et al. 1973, Myrianthopoulos and Melnick 1987), which is true of blacks elsewhere as well (see below); while the opposite is supported by the NTD prevalence in Pakistani and Indian migrants to Birmingham (Leck 1969) and the Irish in Boston (Naggan and MacMahon 1967) coming to resemble that of the indigenous population.

Matings between populations with distinctly different NTD prevalences can sometimes refine or modify these arguments. For example in Cape Town, South Africa (where race was once officially documented) the NTD rate in whites and blacks was 2.5 and 0.95 per 1000 births respectively, while in those of mixed ancestry it was 1.05, and remained relatively constant during the 20 years of a study period (Buccimazza et al. 1994). Thus the mixed-ancestry rate, though intermediate, was close to that of the black group. But was this so regardless of maternal race? (Incidentally, here as has often been the case, whites were members of significantly more affluent classes than blacks, thus possible effects of nutrition cannot be discounted as interacting with those of genetic factors.)

An important piece of information, namely whether the malformation rates in children of mixed parentage differed according to the sex of the parent, may have provided an additional etiological clue. As it did in crosses between Europeans and Caribbeans, where though the rate in children of mixed ancestry was intermediate to those of the parental groups, it was significantly greater when the mothers were European than when they were Caribbean (Leck and Lancashire 1995). Thus in each instance the outcome was different from but close to that of the maternal group.

Leck and Lancashire (1995) from such revelations concluded that the difference in risk of neural tube maldevelopment between Europeans and Caribbeans (predominantly blacks) "seems more likely to have arisen from variations in the genotype of the conceptus than from environmental differences." This is not true of Caucasian groups with different NTD rates, where the rate in migrants eventually has become more like that in the native group, as with the Irish in Boston. In this instance it seems clear that environment is the dominant force.

But how do the Mexicans discussed above fit in here? Is the intermediate rate in Mexican-born California residents due to some degree of ethnic mixing or is it more probably a transitional state, a way-station on the road to complete similarity to the non-Hispanics, the result of as-yet partial acculturation? If the latter, as seems likely, length of residence should be in play; and should be investigated. Recent studies have confirmed these differences between residency groups, but invoking the nebulous concept of acculturation has not aided in their explanation (Ramadhani et al. 2009, Canfield et al. 2009).

In summary, *interracial* NTD risk is entirely or predominantly genetically determined; while *interethnic* risk has a large, but not necessarily wholly, environmental component (whatever the risk factor or factors may be). At the present these generalizations are obviously based on limited knowledge and need to be tested: by study of other migrant situations and of intermatings of other ethnic and racial pairs, black-Asian, etc.

Now, back to the matter at hand.

China Studies

The populations of some areas of China form a remarkable contrast with each other in respect of NTD frequency, certain northern provinces having high frequencies, some among them the highest ever recorded, and certain southern ones low frequencies.

The earliest known NTD records in Chinese groups, of populations of south China origin, noted frequencies at the relatively low end, of 0.7–1.3 (here and below, per 1000 births) (Wei and Chen 1965, Stevenson et al. 1966, Emanuel et al. 1972, Ghosh et al. 1981, Lau and Fung 1984). In contrast in northern regions the overall NTD frequency was 4.4, ranging from 3 in city areas to 10 in rural ones (Lian et al. 1987), differences sometimes considered to be related to nutritional availabilities. (Findings regarding urban-rural variations elsewhere however usually noted that urban areas had the higher NTD frequencies – Turnbull et al. 1977; thus inferences regarding nutrition must be approached cautiously.)

The NTD classification was somewhat confused, making analysis difficult; puzzling also was the absence of mention of iniencephaly, in the light of findings in northern regions of China of a larger presence of craniorachishisis and iniencephaly than seen elsewhere (Moore et al. 1997). Iniencephaly, however, as noted earlier, may seldom involve an open neural tube.

A China-wide survey of births in 1986–7, extending these observations, found a spread of NTD frequencies from 0.7 to 10.5, with a gradient that tended to decline from north to south and from east to west, with the Yangtze River the dividing line between north and south (Xiao et al. 1990). These findings confirmed those of various authors, writing in Chinese, cited by Lian et al. (1987).

No clear reason for this regional difference has been discovered, not climate, diet, ethnicity (Berry et al. 1999), except the possible greater availability of various folate-supplying foodstuffs in the south than the north (Melnick and Marazita 1998), contradicted by other findings, as noted elsewhere.

This great north-south difference led to a public health campaign in 1993–5 of periconceptional folic acid supplementation to reduce the level of these malformations (Berry et al. 1999). In a nonrandomized trial in two areas, the high risk northern province of Hebei and two low risk southern provinces of Zhejiang and Jiangsu, women were instructed to take a tablet, which they were to purchase, of 0.4 mg folic acid daily, from some time before conception through the first trimester of pregnancy, or some part of the latter, while other women were not to take such a tablet.

The outcomes were as follows. In the northern province the NTD frequency in offspring of 20 or more weeks of gestation of women who took the tablet at any time was 1.3, while in those not taking it the frequency was 6.5; and in the southern provinces the comparable figures were 0.7 and 0.8, i.e. a 79% and 16% reduction respectively.

The fabric of the study must be remarked upon. Infants with NTD were identified through a surveillance system set up shortly before the onset of the trial, the accuracy and completeness of which therefore had not been tested. Many possible uncertainties regard the consumption of the folic acid: women "were asked to purchase pills," which may have entailed financial and logistical difficulties; questions concern the diligence of the health workers whose task was to validate their ingestion. Diagnosis of abnormalities was made through photographs sent to a central location, but accessibility of photographic equipment seems not to have been verified.

Other matters of possible importance were not mentioned, the sex ratios of the affected infants and the ratio of the types of their malformations. These would have enabled certain comparisons in the case and control groups in the Berry et al. report with data found in a southern region (Lian et al. 1987). In the latter the overall ratio of anencephalus to spina bifida was 1.2:1; and the sex ratios for anencephalus and spina bifida were 0.36 and 0.65 respectively. This information might have revealed, e.g. whether these malformations were equally often prevented or not.

A nongenetic explanation of the regional difference was postulated, most Chinese being ethnically uniform. Although the dietary regimens in the two regions were not examined it was assumed that climatic and other advantages of the southern area, such as greater access to folate-containing foodstuffs, were responsible for the baseline differences between the regions. The far larger proportion of women who were farmers in the northern than the southern area was not considered in this hypothesis.

Changes in China aside from interventions must also be noted. NTD frequency decreased in Hebei in the north and Jiangsu and Zhejiang in the south between 1986–7 and 1993–4 (Xiao et al. 1990, Berry et al. 1999), being 7.7 earlier and 5.5 later in the north; and 1.3 and 1.0 in the south. Thus time itself, as in other parts of the world (e.g. Yen et al. 1992), and in various life situations, had healing qualities; which may partly explain the findings made by Berry et al. (1999).

All these reports from China dealt with offspring that survived to late pregnancy, information about induced abortion being omitted. This matter, however, so important in any consideration of NTD epidemiology, cannot be ignored, especially because, as it has been reported – see below – abortion induction has been in common use in China, for the purpose of sex selection.

As for the other side of the coin, spontaneous abortion, it was said to comprise only 2% of all registered pregnancies in several regions, an unrealistically small number, meaning without doubt that many abortuses, with and without NTD, eluded notice. How this may have affected birth prevalence can only be conjectured. One study found the plasma concentration of homocysteine and folate not significantly different in women who aborted spontaneously than in controls (Ronnenberg et al. 2002).

Other and Later Reports

A malformation survey, conducted in 1992–4 in concert with a Sino-American cooperative project, found major differences in NTD frequency in unnamed north and south regions, 4.56 and 0.71 respectively (Pei et al. 2003). [This article is in Chinese, as are others described below, and only bare details were to be gleaned from the summary in English.]

In 1996–2000 a birth defect monitoring network identified an overall NTD frequency of 1.13, almost twice as great in the north as the south, during which time the frequencies fell significantly (Dai et al. 2002). In four selected counties of Shanxi Province, near Hebei, located in the upper northeast, the NTD frequency in 2003 was an unbelievable 12.4 per 1000 still- and livebirths, with 50.0% of anencephaluses and 41.8% of spina bifidas terminated following prenatal diagnosis (Li et al. 2006). Only about 4% of women took folic acid during periconception.

A study of congenital malformations generally, in certain communities in Gansu province, in the northwest, found a NTD frequency in 2001 of 6.7 (Cheng et al. 2003). The high levels in the north thus continued. A counter to this high tendency was reported from Guizhou Province in the southeast, with a mean frequency in 1996–2004 of 1.22; the low level considered to be due to folic acid usage, but with no details (Liu et al. 2007).

A study of a relatively small number of women was made in 2000–2 in parts of four provinces, three in low prevalence east and southeast central areas and the fourth in the far northeast (Chen et al. 2008). In the southern areas the control frequency was 1.82, and in takers of multivitamin tablets 0.45; and in the northern 1.73 and 0.0, respectively. However, the NTD frequency was also reduced in those not taking the tablets, but eating large amounts of vegetables and fruits. Nothing conclusive came out of this study.

As noted, in all-China studies the background frequency of NTD declined steadily, from 2.36 to 1.10 in the brief span of 1986–2002. Regrettably, there was but sparse documentation of the separate declines in the north and south. Only in two later reports was this information forthcoming, namely that in 1992–4 it was 4.83 in the north and 0.71 in the south (Pei et al. 2003), and in 1996–2000 1.99 and 0.58 respectively (Dai et al. 2002). Thus in this 10-year interval, of 1992–2002, there was a 59% decrease in the north and 18% in the south. This conforms to what we've seen elsewhere: that the temporal decline is directly proportional to the starting level.

Thus, a temporal decline was occurring independent of multivitamin intervention, and what was uncovered is another example of what was seen elsewehere, namely, that folates seemed to prevent NTD only or especially in high risk groups; and thus that in the north, where this was so, what had occurred was simply a continuance of the temporal trend outlined above for China.

Several still later reports of NTD trends in folic acid supplemented pregnancies in different parts of China have not clarified the picture. In four southern provinces the NTD frequency in births in 2000–2 of women receiving perinatal multivitamin supplementation including folic acid was found to be decreased compared to the control (Chen et al. 2008). The provinces were widely separate geographically, with greatly varying NTD background frequencies, and the mean decreased frequency thus was little meaningful. Suspicion of the data given was aroused by one such background frequency, 2.45, in Henan Province, one of the four areas with a greater frequency, being much larger than it was said to be in a contemporary report, in which it was noted to be 1.47 in 1996–2005 (Zhou et al. 2008).

In sum, studies up to the end of the twentieth and early in the twenty-first century presented little incontrovertible, consistent, and persuasive evidence to support the hypothesis that folic acid is unambiguously associated with a reduced frequency of NTD.

Has Folic Acid Prevented NTD?

Folic Acid Food Fortification

The hypothesis that maternal deficiency of folic acid predisposes to NTD has been accorded much credence. So much was this so that in 1992 the Centers for Disease Control (CDC 1992) issued the recommendation that "all women of childbearing age in the United States who are capable of becoming pregnant should consume 0.4 mg of folic acid per day for the purpose of reducing their risk of having a pregnancy affected with spina bifida or other NTDs," and that women who have already had a NTD pregnancy should take 10 times this amount, 4 mg folic acid per day, starting when they planned to become pregnant (CDC 1991).

This was apparently considered an insufficient precaution, and it was further recommended that cereal grain products be enriched with folic acid to a level of 140 mg/100 g of product (FDA 1993). Other governments soon concurred (Wald and Bower 1995, Cornel and Erickson 1997), and elsewhere experts went even further, recommending not less than 0.8 mg/day (Van Allen et al. 1993).

The challenge was to determine how much folate would be needed to prevent NTD (Rush and Rosenberg 1992). Since in no study had folic acid been able to prevent NTD entirely reasoning had it that only some fraction was preventable by this means (Oakley 1993). By circular rationale the latter was estimated to be about half of all NTD in the US, and preventable by 0.4 mg/day (Oakley et al. 1994). Arcane discussions followed of the levels of folate needed to achieve particular amounts of NTD reduction (Daly et al. 1997, Wald et al. 1998). Calculations in New Zealand e.g. showed "emphatically" that 400 mg of folic acid prevented up to 70% of neural tube defects (Borman 2001), and analyses of the relation of folic acid intake and serum folate concentration led to the deduction that taking a 5 mg tablet daily would reduce the risk by about 85% (Wald et al. 2001).

The subject of "folic acid-preventable" NTD is still with us; but it seems still has along way to go, since it was estimated near the end of the first decade of the twenty-first century that a mere 9% of preventable ones are being prevented by current fortification programs, which to be accelerated requires more countries to adapt such programs (Bell and Oakley 2008).

It was not until 1996 that fortification of cereals and grains with 140 mg folic acid per 100 g of grain was authorized by the FDA (1996), and not until January 1, 1998 that this went into effect; an action that was credited with the increased serum folate levels in subjects in New England, California, and elsewhere in the US (Jacques et al. 1999, Lawrence et al. 1999, Anon 2000c). Parenthetically, European governments, especially in the UK, were castigated for not mandating universal fortification of flour with folic acid; accused of "committing public health malpractice" and allowing "a continuing epidemic of preventable human illness" (Oakley 2002).

One wonders whether malfeasant officials of Her Majesty's Government were then held criminally liable when British children were born with NTD, even of those ostensibly not preventable by folic acid fortification, as French doctors can be sued who did not inform pregnant mothers they were carrying NTD fetuses?

Obviously, the ultimate question is whether supplementation and especially fortification have led to continually lowering the frequency of NTD. Several preliminary attempts to answer the question proved negative. An early international birth registry study from mid-1987 to mid-1996 found no "...evidence that...any change in time trend was attributable to the introduction of national folate supplementatrion policies" (Rosano et al. 1999). Analysis of later time trends in Britain (Abramsky et al. 1999, Kadir et al. 1999) and North Carolina (Meyer and Oakley 1999) revealed a decreased NTD frequency in the time preceding folic acid fortification, but no further decrease afterward; and several explanations were offered for this failure – insufficient fortification, poor compliance with preconceptional folate supplementation, low folate intake, etc.

Other studies focused more directly on the assumed connection between folic acid fortification and changed NTD prevalence. A reported decline in England and Wales from the 1970s to 1997 was concluded to have been due at least in part to the increase in dietary folate (Morris and Wald 1999). A Netherlands pediatric survey, comparing prevalence in periods before and after 1996, the year periconceptional use of folic acid increased in that country, found that though there had been a decrease it was not statistically significant (Bruin et al. 2000). NTD declines, mostly of spina bifida, in South Carolina from 1992 to 1998 were coincident with the increased folic acid use, but no causal connection was offered (Stevenson et al. 2000). Information on US birth certificates for two periods recorded declines in NTD prevalence that were attributed to the folic acid fortification during the more recent years, although "the long-term downward trend in anencephalus prevalence that preceded folic acid fortification makes it difficult to interpret the...decline following fortification" (Honein et al. 2001b).

The US National Center for Health Statistics (NCHS), using nationwide birth certificate data, recorded a steady rate for spina bifida in 1991–7 and a slight dip in 1998–2001, with that for anencephalus fairly stable since 1994. The trend for both defects was downward in the most recent years, which if continued would be significant (Anon 2002); but data discussed below (Williams et al. 2002) disputed this rosy outlook.

This and other discrepancies (e.g. Rosano et al. 1999, Stevenson et al. 2000) may indicate that the retrieval methods used by Honein et al. (2001b), as well as by the NCHS, missed a significant number of NTD occurrences, probably because birth certificates greatly underrecord malformations (Watkins et al. 1996).

Further difficulty revolves around calculations demonstrating the efficacy of fortification (Honein et al. 2001b). According to birth certificate entries some women "received third-trimester only or no prenatal care." From this it was deduced that NTD birth prevalence in such pregnancies was less likely to be affected by any "changes in patterns of vitamin supplement use." Nevertheless the NTD prevalence in these pregnancies also declined, by larger rates in fact than those in pregnancies of all women. Can it be that part of the differences in rate between the subsample and all women reflect elective termination of some affected fetuses in women cared for from early in pregnancy, not indicated on birth certificate sources, despite the authors' denial of this possibility? And in part as well, may the decline have been part of the historic downturn?

This report again raised the question of birth prevalence versus total prevalence, and the discrepancies and confusion swirling about it. Is it possible e.g. that the relatively low NTD prevalence found by Honein et al. (2001b) was partly due to it having been based on live births only, with stillbirths and aborted NTD-affected conceptuses not having entered into the count? This possibility was considered "implausible," but no evidence was offered for the opinion. Data presented by others (e.g. Forrester and Merz 2000) that contradicted this assertion will be noted below.

To compensate for the deficiencies of standard birth certificates in not including prenatally diagnosed and electively terminated NTD cases, a population-based procedure was designed that collected malformation data from multiple sources (Williams et al. 2002). The programs, conducted in about half the US states, differed considerably from each other in the intensity and diversity of the collection methods; nine included prenatally diagnosed NTD and approximately half of them fetal death or elective pregnancy termination.

Interest focused on comparing the prevalence of spina bifida and anencephalus during three periods, 1995–6, before fortification, 1997–8, during the transitory period, and October 1998–December 1999, following mandatory fortification. Data from all the participating programs showed a decrease in the prevalence at birth of spina bifida, even in the transitory period, i.e. prior to mandatory fortification; whereas that for anencephalus while decreasing from the first to the second period fell no further afterward. The bottoming out of the decline continued thereafter, with the frequency at birth not changing in 1998–2001 – as the authors commented, "the period of optimal and mandatory folic acid fortification (Mathews et al. 2002)." However, not especially emphasized, was the fact that in programs with prenatal ascertainment the prevalence of both malformations did not decline at all from 1995 to 1999. Thus it is questionable once more whether folic acid fortification had any beneficial effect.

A follow-up overview of NTD prevalence in almost half of all US surveillance districts from 1999 extending to 2004 found, strangely, contrary to those of the study recounted immediately above, that it was spina bifida that barely shifted, while anencephalus significantly declined, all NTD combined however barely differing over the period (Boulet et al. 2008). Extending the findings to 2005 changed nothing, especially considering that the report pertained to births not conceptions (Boulet et al. 2009).

A similar asymptotic trend was depicted graphically in a Greenwood Genetic Center publication, which showed only the slightest NTD frequency alteration from 1997–8 to 2005–6 in children of women consuming folic acid daily (Anon 2007). This study thus found no consistent preventive effect of folic acid fortification; the authors even conjectured that "supplement use did not account for the. . .NTD prevalence decline in the observation period."

Similar trends were seen in Hispanics, whites, and blacks, upon the background of the usual frequency differences among them; with no thought having been given to the change simply being a reflection of temporal shift in all three (Williams et al. 2005). Among white non-Hispanic mothers however no significant decrease in spina bifida was seen in the recent postfortification period 2001–5 (Anon 2009).

Several additional reports dealt with NTD occurrence, some associated with maternal folic acid use. Data from Ontario, though not easy to interpret, indicated an increase in "total" NTD frequency in 1986–95, probably it was thought due to increased prenatal detection, and a decrease in affected live- and stillbirths afterward associated with the rise in the rate of therapeutic abortions (Gucciardi et al. 2002).

NTD frequency in Nova Scotia was compared in the years before and after dietary addition of folic acid, 1991–7 and 1998–2000, respectively (Persad et al. 2002). In the earlier period the frequency was fairly constant, about 2.43 per 1000 terminated pregnancies and live- and stillbirths, and in the latter declined to about 1.00. A decline however was already in evidence before the fortification program had been completed, to 1.33 and then further to 0.83. All of which might be interpreted as a continuation of the decrease already in progress.

A decline in NTD was also noted 1993–2002 in seven of the 10 provinces of Canada, from 1.37 per 1000 terminations and live- and stillbirths, in the years before folic acid fortification was mandated, to 0.72 afterwards (De Wals et al. 2007). However, that this downward pattern was part of the historic trend already in progress was shown by the decline in anencephalus and spina bifida in live- and stillbirths from 1.16 in 1979–81 to 0.91 in 1991–1993 in years prior to the fortification in Canada (Johnson and Rouleau 1997).

A study in 1995–2004 in Sweden of the general malformation rate, including NTD, found no reduction in infants of women who used folic acid early in pregnancy compared to all other women giving birth at the time (Källén 2007).

A study made in selected California counties found that the frequency of NTD at birth had declined in 1989–96, but had not continued, in 1998–2003, after folic acid fortification was instituted (Chen et al. 2008). A comparison in Western Australia of the trend in 1980–2006 was hampered by folic acid supplementation and fortification not having been mandatory, and its putative effects being judged by estimates of voluntary consumption (Bower et al. 2009). The figure (the only way data were reported) indicated that a decline in the NTD frequency started in the mid 1990s, but whether it was associated with folic acid was not clear.

And finally, a multicenter US study in 1998–2003, "found insufficient evidence for an association between maternal folic acid supplement use or dietary folate intake and neural tube defect occurrence. Reported folic acid supplement use was similar among women who had neural tube defect-affected pregnancies and women who had pregnancies not affected by birth defects" (Mosley et al. 2009). This negative finding was bizarrely interpreted to mean that "fortified food is probably providing sufficient folic acid to prevent folate-related defects" (Mills and Carter 2009). A recent elaborate meta-analysis has not altered the conclusion that the decrease in the frequency of neural tube defects has been related little or not at all to folic acid supplementation (Blencowe et al. 2010).

Nor to be neglected is a study of pregnant women with epilepsy, in which there was no difference in NTD frequency in children of those folic acid supplemented preconceptionally and those not, the almost apologetic explanation for the failure to find a difference, that "extrapolation from studies carried out in the general population to groups of women with epilepsy may be questionable" (Morrow et al. 2009)!

In sum, no consistent evidence has as yet appeared to support the hypothesis that folic acid ingestion is associated with a reduced frequency of NTD. Perhaps more time must pass to allow judgment. But meanwhile other factors cannot be ignored.

Secular Decline

Other elements known to affect the birth prevalence of NTD must be considered. One is secular shift in the frequency of anencephalus and spina bifida; and another is prenatal diagnosis and elective abortion of embryos and fetuses with such abnormalities.

It is an indisputable fact that as of this writing the last six or more decades have seen a sustained decrease in the frequency of NTD in many parts of the world; for the most part of no apparent cause. The decline, which has often been rapid and consistent, began as early as during the late 1930s (Gittelsohn and Milham 1962, MacMahon and Yen 1971) – hard upon the heels of its ascent in the several preceding decades (MacMahon and Yen 1971, Elwood 1973) – and which with variations of one sort or another in the onset and pattern of its trend (e.g. Elwood 1973, Janerich 1973, Källén and Löfkvist 1984, Czeizel and Karig 1985) has been of widespread occurrence ever since (Stone 1987, Elwood and Little 1992, Yen et al. 1992, Rankin et al. 2000).

Several reasons for the trend in frequency of NTD have been conjectured. Any part alcohol consumption may have had was not clarified by the frequency declining after the repeal of the Eighteenth Amendment to the US constitution (MacMahon and Yen 1971). Another possibility, not yet disproven, based on the hypothesis that socioeconomic deprivation in childhood predisposes women to bearing children with NTD, is that improvement in such conditions led to a reduction in their frequency (Anderson et al. 1958, Sever and Emanuel 1981, Emanuel 1993).

The one widely accepted actor in the decline is periconceptional supplementation with folic acid, credited with eliminating the so-called preventable fraction, which, as discussed above, has been reckoned to be about half or so of all NTD. Before this thesis can be accepted other facts must be remembered: first, that the decline in the frequency of NTD began years before folic acid supplementation programs were instituted. Nevertheless the preventive action of the vitamin afterward may still have complemented the 'spontaneous' decline. This possibility can be examined by analyzing whether the rate of decline accelerated in later years, as supplementation and fortification took hold. Preliminary efforts so far have failed to prove this, but may still be a valid consideration.

Terathanasia

A decrease in the frequency of NTD at birth may be a consequence in part of spontaneous abortion of affected conceptuses. Analysis of this question is impeded by several facts: abortion is common, occurring at a minimum in 15–20% of recognized human pregnancies; the frequency of malformations is several times greater in abortuses than at birth; and many malformed conceptuses are spontaneously aborted (Nishimura et al. 1968, Boué et al. 1976, Rushton 1985).

The last, the natural screening process, was given the name terathanasia by Warkany (1978b), who cited a number of studies in which offspring with anomalies were subject to preferential death and prenatal elimination (e.g. Kalter 1978, 1987). He asked whether this phenomenon can be recognized and even facilitated in humans, as a mean of reducing the malformation load. Was folic acid an answer?

Some years ago the NTD frequency at birth and the spontaneous abortion rate were found to be inversely related in regions of Wales, and area differences in the latter offered as an explanation for geographical variability in the former (Roberts and Lloyd 1973, Fedrick and Adelstein 1976). This implied that the primary frequency of NTD, i.e. that present in early embryos, was similar in all regions but came to differ owing to differences in rate of spontaneous abortion.

This interesting idea, however, was not substantiated by studies of variations in other areas and of other factors – e.g. social class, season, and birth order – which revealed not an inverse but a positive relation between NTD prevalence at birth and spontaneous abortion rate (Byrne and Warburton 1986, Little and Elwood 1992).

More recently the topic of an inverse relation between rates of spontaneous abortion and malformation was again broached when it was suggested that folic acid may reduce the NTD frequency by causing the abortion of a larger proportion of affected embryos in supplemented women than in nonsupplemented women (Hook and Czeizel 1997), a human example of terathanasia. The increased spontaneous abortion rates found in the two studies most convincing of an effect of folic acid supplementation on NTD prevention (MRC 1991, Czeizel and Dudás 1992) were relatively small, but as Hook and Czeizel (1997) pointed out, they were sufficient to account for the malformation reductions.

This proposal, not surprisingly, caused an eruption of reactions, some more reasoned, others less so, while offering alternative explanations (Burn and Fisk 1997, Hall 1997, Schorah et al. 1997, Wald and Hackshaw 1997). After a period of quiescence, the topic reemerged, when in a preliminary communication the radical proposal received some support from a study in California, which found a larger spontaneous abortion rate in folic acid users than in nonusers, on the order of that found in above-noted randomized studies (Windham et al. 2000).

Another example of an apparently high rate of spontaneous abortion in women taking folic acid preconceptionally was believed to be an artifact, the result of intense monitoring of pregnancies (Stevenson et al. 2000). Such findings were interpreted favorably by one author (Hook 2001), but dismissed as chance occurrences by others (Wald and Hackshaw 1997). Further question revolved around the slim possibility that folic acid enables more pregnancies to survive, postponing fetal loss to detection at a later stage, producing more live births and fewer malformations (Hook 2001). A bit of negative evidence on this score comes retrospectively from China, where it was found that daily folic acid supplementation did not affect the risk of spontaneous abortion (Gindler et al. 2001). The rate, however, while not appreciably different in cases and controls, was rather low, about 9%, and may reflect inadequate recognition of the event. A study of abortion in folic acid supplemented epileptic women found a reduced rate (Pittschieler et al. 2008). This is where this particular question rests at the moment.

Prenatal Diagnosis

Detection of embryos and fetuses with NTD and their prenatal elimination obviously reduces the number of these abnormalities seen at term. Clearly without knowledge of the contribution of this selective action to the decline in recent years in the prevalence at birth of NTD judgment of the part played in this outcome by folic acid programs will be unsound.

It was discovered in the early 1970s that elevated levels of alphafetoprotein in amniotic fluid, which occurred through its leakage from the open neural lesions, pointed to the presence of embryos and fetuses with anencephalus and spina bifida aperta (Brock and Sutcliffe 1972). Later, more convenient and accurate methods of detection came from using maternal serum alphafetoprotein, ultrasound, and other techniques (Wald et al. 1974, Campbell and Pierce 1983).

These procedures were soon put to widespread use to identify and terminate affected pregnancies (Leck 1983, Watson et al. 1983), with the inevitable consequence of reducing the prevalence at birth of these malformations. Numerous estimates were made in the following years of the amount of this reduction, and in analyzing the relative effect on the predominant varieties of NTD. For example, in the Liverpool-Bootle areas, after years of decline, the rate of NTD at birth had reached an "apparent" level of 2.4 per 1000 total births, but taking into consideration prenatal screening for NTD, which had begun in about 1974, the "real" incidence, i.e. including terminations, was about 3.2 per 1000; screening had thus led to a 25% reduction (Owens et al. 1981).

This soon improved. In 1980–2 in a relatively low-risk area of Scotland the total NTD prevalence (in aborted plus neonates) of 2.5 per 1000 pregnancies was reduced overall by 80% (90% and 72% for anencephalus and spina bifida respectively), to a prevalence at birth of 0.5 per 1000 pregnancies (Thom et al. 85). A study in Glasgow, a high-risk area, similarly found that the continuous decline in 1974–85 in the birth prevalence of NTD was due in great measure to prenatal screening, which enabled anencephalus to be reduced by 59% and spina bifida by 23%; the latter, as elsewhere, less effectively eliminated perhaps because of its lesser sensitivity to alphafetoprotein screening (Stone et al. 1988). Such reduced birth prevalence of NTD resulting from elective termination of affected fetuses has been widely and repeatedly reported; recent publications on the subject are cited below.

The question at this point in these pages is how much of the later secular decline was selective abortion responsible for. An early estimate, undoubtedly an underestimate, found that almost one-third of the 77% decline in NTD birth prevalence in 1964–85 in England and Wales could be attributed to prenatal detection and selective abortion (Cuckle and Wald 1987). This was obviously an average, Ferguson-Smith (1983) recognizing that "as the annual proportion of pregnancies screened. . .increased, so has the proportion of all NTD pregnancies which have been terminated."

This trend was occurring elsewhere as well. A survey of 20 European regions for 1980–6 found that to various extents all had experienced reduced birth prevalence of NTD as the proportion of pregnancies screened and the number of affected fetuses selectively eliminated increased (Eurocat Working Group 1991).

Such reports were late in coming from North America, where it seems prenatal screening was sporadic before the 1980s. A report from Manitoba noted a fall in the birth prevalence of NTD from 1.2 per 1000 total births in 1979–81 to 0.6 in 1988–90, a decrease of 50%, as prenatal screening and elective termination of pregnancies increased during those years (Evans et al. 1992). An even greater reduction occurred in Quebec, from 1.4 per 1000 total prevalence to 0.4 at birth, coincident with level of ultrasound screening (De Wals et al. 1999). The birth prevalence of anencephalus in a Boston hospital decreased as prenatal screening increased with time, so that by 1990 all affected pregnancies were detected prenatally and all were terminated (Limb and Holmes 1994). The quantitative repercussion of this practice on birth prevalence of the defect was not discussed.

A population-based study in metropolitan Atlanta in 1990–1 found that elective abortion reduced the NTD prevalence at birth from 1.1 per 1000 to 0.8, a reduction that was not as large as that found in the European and Canadian centers, perhaps because it dealt with a heterogeneous population. As was often true in other areas anencephalic fetuses were more successfully detected than those with spina bifida; and though not statistically significant the impact of termination was greater in whites than nonwhites (Roberts et al. 1995). A multistate population-based survey, extending the Atlanta findings, found that elective termination of NTD varied in six regions from 9% to 42%, with corresponding reductions in total prevalence (Cragan et al. 1995).

Other reports have told the same story: significant reduction of NTD birth prevalence due to screening of pregnancies and elective elimination of affected fetuses (Chan et al. 1993, Allen et al. 1996, Velie and Shaw 1996, Alembik et al. 1997, Forrester et al. 1998, Anon 1999, Palomaki et al. 1999, Forrester and Merz 2000). A recent example from British Columbia told of a reduction of 73%, predominantly due to pregnancy termination and relatively little ascribable to folic acid supplementation (Van Allen et al. 2006). Variances of one sort or another in such findings, noted in all such reports, newer and older, were due to variable success of the screening program in recruitment, difference in sensitivity of diagnostic method, variation in detectability of different types of NTD, etc.

Has the Secular Decline Continued?

An unsettled question remains. Discounting the effects of prenatal screening and pregnancy interruption, has the secular decline in NTD continued? An important contribution of the Glasgow study (Stone et al. 1988) was in addressing this question, and finding that the number of affected children born would have declined even in the absence of screening, i.e. that the NTD frequency in fetuses, the primary frequency, also participated in the secular decline. The latter was confirmed by data from Ireland, where selective termination was not officially permitted (Eurocat Working Group 1991). This was further emphasized by a study of births in East Ireland in 1980–94, especially so because of two reasons: folic acid had not entered the picture, and elective elimination of affected fetuses was all but nonexistent. Nevertheless in this interval the birth prevalence of NTD decreased fourfold, from 4.5 per 1000 to 0.99 (McDonnell et al. 1999). This was not always the case; e.g. in Manitoba, Australia, northeastern France, and Sweden, regions with disparate background rates, the total prevalence remained stable over a number of years (Evans et al. 1992, Chan et al. 1993, Alembik et al. 1997, Nikkilä et al. 2006). This question thus remains to be answered definitively.

We return to the question of whether folic acid supplementation and fortification have prevented NTD. As we have seen, it is transparently obvious that the matter of possible prevention cannot be answered definitively without simultaneously considering the roles of secular decline and prenatal diagnosis in the birth prevalence of these malformations. One group of authors gave a passing thought to these considerations (Cragan et al. 1995), but the entire problem has been glossed over.

A recent glance at this matter made some admissions, first that the decline in NTD frequency was independent of folic acid amount administered, and second that the decline had stalled (Heseker et al. 2009). The authors assumed, as had others previously, that a floor had been reached for the "preventable" ones, giving the usual 'reasons' for this – insufficient fortification, etc.; never giving thought to another possibility: one described in a recent publication (Kalter 2009): that the historic many decades long rise in the frequency of this defect followed by a historic decline had for now receded to a plateau.

Finale

In a brief discussion Nevin (1983) called attention to an important fact. He noted that fetuses with NTD occurred in only two of the 12 pregnancies he was aware of in which folic acid antimetabolites had failed their intended use as abortifacients. Since he wrote there were several more instances of such antimetabolites being used for this purpose, making a total at present, I believe, of at least 65, in which again only the original two, reported by Thiersch (1952), resulted in infants with NTD (Feldcamp and Carey 1993, Del Campo et al. 1999, Lloyd et al. 1999).

There is no doubt that folic acid antagonists are teratogens in humans and other animals (Schardein 2000). The typical malformations that have been induced by these agents however comprised a syndrome of skeletal defects, and did not include NTD (Warkany et al. 1959, Warkany 1978a). It is of interest to note that even in experimental settings large amounts of the potent folic acid antimetabolite aminopterin given during early pregnancy to monkeys (*Macaca mulatta* and *M. irus*) caused abortion but not malformation (Wilson 1969).

Thus one has every right to be puzzled by the fact that while powerful chemical teratogens have all but failed to induce these malformations in infants, "biochemical" folate deficiency is postulated to be capable of doing so (e.g. Wenstrom et al. 2000).

Addendum

The folic acid story has given rise to two main offshoots. One contended that folic acid, in addition to preventing NTD, prevents other congenital malformations, particularly oral clefts. A prefolic acid study of women with children with orofacial defects found that no protection was afforded by prenatal vitamin supplementation (Fraser and Warburton 1964). The first claim of protection, however qualified (Tolarova 1982, Tolarova and Harris 1995), was later refuted (Hayes et al. 1996, Shaw et al. 1998), but continued to be espoused (Czeizel et al. 1999, Hernández-Díaz et al. 2001, Itikala et al. 2001). Other claims of protection referred to certain congenital heart defects (Botto et al. 1996), imperforate anus (Myers et al. 2001), acute lymphoblastic leukemia in childhood (Thompson et al. 2001), and an assortment of abnormalities (Werler et al. 1999), upon some of which doubt has already been cast (Källén et al. 2002). Multiple malformations, defined as two or more congenital anomalies affecting more than one organ system or a major anomaly in combination with two minor anomalies, were not protected against (Shaw et al. 2000). Nor, if an old study with animals can get into the act for a moment, was cortisone-induced cleft palate in mice prevented by maternal administration of riboflavin or folic acid (Kalter 1959b).

Genes and NTD Risk

The other offshoot, based on the belief, considered to be a firmly proven and established fact, that folic acid prevents NTD, postulated that a genetically based abnormality in folate metabolism is a factor responsible for NTD, and was the basis of an exploration of this possibility (e.g. Wenstrom et al. 2000, Gelineau-van Waes and Finnell 2001). The resurrected cognizance of the multifactorial concept of the etiology of NTD has evoked the novel thought that included among the many predisposing variables there must also be genetic input; which has sparked a feverish campaign to explore the idea.

This has been done by seeking an association between NTD and folate-associated genetic variants. The evidence of this possibility has been mixed at best: a possible role of a folate-reductase allele, then refuted (Kirke et al. 1996, Mornet et al. 1997), no association of relevant mutations (Barber et al. 1998), a dubious role for a genetic factor (Shaw et al. 1999, 2002), no association of a variant and maternal folate level (Molloy et al. 1999), a possible moderate risk posed by a certain allele, perhaps depending on nutritional status (Botto and Yang 2000), a suggestion of a genotype being a NTD risk factor (Martinez de Villarreal et al. 2001), nonconfirmation of earlier findings that folic acid reduced the frequency of NTD in Splotch mice (Gefrides et al. 2002).

In summary, no evidence to date has conclusively shown "a significant association between a genetic polymorphism and increased NTD risk" (Gelineau-van Waes and Finnell 2001).

These results may mean that the wrong genes are now being looked at, and should not discourage the search for others that may be part of the multifactorial array associated with this risk. Facilitated by techniques not available heretofore, studies of the genes at work in the development of NTD in mice may point the way (Juriloff et al. 2001). Only the future, as usual, can tell.

Alcohol Consumption During Pregnancy

Alcohol has always aroused mixed feelings; it is both a blessing and a curse. There is no shortage of apropos quotations from the Bible, but these are the alpha and the omega, or in the Hebrew alphabet, the alef and the tav. Genesis says, may God give you of heaven's dew and of earth's richness – an abundance of grain and new wine. But Deuteronomy says, wine is the poison of dragons, and the cruel venom of asps.

In the twentieth century a novel accusation was leveled against alcohol: that it caused prenatal maldevelopment and mental retardation; accusations bolstered by the recalling of numerous historical instances of harm to infants due to parental alcohol consumption (Pratt 1982). And recently in the US an alarm pointing to this damage was sounded by an article that detailed a pattern of defects in children of chronically alcoholic women (Jones et al. 1973).

Early Seattle Study

First however came observations by Ulleland (1972). She noted that of 1594 consecutive births in an 18-month period at the Harborview Medical Center in Seattle 47 had a birth weight below the 10th percentile, i.e. were "undergrown for gestational age." Ten of them were children of relatively severe chronically alcoholic mothers, a larger proportion than of small babies born to nonalcoholic women.

In considering the reasons for this she noted that Harborview Hospital served an urban population of low socioeconomic status, "where a number of factors may combine to produce poor fetal development and infant performance" – sparse prenatal care, advanced maternal age, heavy smoking, various medical conditions, poor nutrition – an early realization in the study of the outcome of pregnancies of alcoholic women of the many poor environmental conditions that are associated with its detrimental neonatal consequences.

The Fetal Alcohol Syndrome

Eight of these ten small children were the subjects of a follow-up by Jones et al. (1973). Because the consequent report was essentially responsible for engendering

H. Kalter, *Teratology in the Twentieth Century Plus Ten*,
DOI 10.1007/978-90-481-8820-8_15, © Springer Science+Business Media B.V. 2010

a prolific area of medical endeavor [over 500 articles – Polygenis et al. 1998; an outpouring that has continued into the twenty-first century], it is necessary to note the salient facts regarding them and their parents.

First a statement from the report: "Four of these children were recognised as having a similar pattern of altered growth and morphogenesis. Thereafter, two other children were ascertained by the abnormal features identified in the first four patients, while the remaining two were ascertained because their mothers were chronically alcoholic."

In other words, the entire premise of what later came to be called the fetal alcohol syndrome was based on the condition of the four children with this "similar pattern." And only through them was it discovered that their mothers were alcoholic, that they all drank excessively throughout pregnancy, to the extent that they were in hospital with delirium tremens, one even in an alcoholic stupor at the time of delivery. The quantity of alcohol consumed, not specified, was assumed to satisfy "the criteria for alcoholism" published by the National Council on Alcoholism.

Mentioned only incidentally was the fact that some of the other children of the alcoholic women, who were also born while they were alcoholic, were partially affected (this was not explained) or normal. But these were "purposely" not considered. Also not emphasized was the fact that the four children were not seen at birth, but only weeks to years afterward; which was also true of many later studies.

The children were growth impaired and had a diverse and variable set of defects, comprising what was considered to be a "similar pattern of craniofacial, limb, and cardiovascular defects...." The similarity of the features in them and in a small number of additional ones with alcoholic mothers mentioned by Jones and Smith (1973) led to naming the condition the fetal alcohol syndrome (FAS).

The FAS Expanded

The first identified publication concerning children of chronically alcoholic women, aside from the fin de siècle article by Sullivan (1899), revisited below, was that of Lemoine et al. (1968), which described abnormal facial and other features and impaired growth in some of them (children of alcoholic fathers and mothers, but particularly of mothers). One set of children were observed during their first year of life only, in a second to 16 years of age. Regarding psychological difficulties in older children, the authors wrote that it is difficult to know whether the effect was due to alcoholism or to a number of traumatic conditions of life ("scènes de violence à la maison"), etc.

A smaller assortment of abnormalities was noted in Seattle, consisting of minor craniofacial defects, vague limb, mostly questionable cardiac conditions, and impaired prenatal growth (Jones et al. 1973). In the original enumeration the craniofacial defects consisted of microcephaly, short palpebral fissures, maxillary hypoplasia, and epicanthic folds; the limb defects, of altered palmar crease pattern and limited motion; and the cardiac anomalies, of ventricular septal defects. The last, as is so often the case, were transient, i.e. had closed postnatally, in three of the five children so affected.

Before long the FAS was expanded and modified, often with a loss of perspective (Jones et al. 1974, Clarren and Smith 1978). Added were a vast potpourri of morphological and neurobehavioral abnormalities: cleft palate, prominent lateral palatine ridges, hypoplastic philtrum, thinned upper vermilion, asymmetrical ptosis, posteriorly rotated ears, marked strabismus, anomalous external genitalia, labial hypoplasia, capillary hemangiomata, diminished adipose tissue, hypotonia, irritability in infancy, hyperactivity in childhood, fine-motor dysfunction, mild to moderate mental retardation, and many others seen occasionally. In fact, it has been claimed, rather ludicrously, that "nearly every human malformation has been reported in at least some FAS patients" (Clarren 1990), and that "scarcely an organ system exists which does not demonstrate effects of prenatal alcohol exposure" (Ginsburg et al. 1991)!

While many deviations, defects, anomalies, and malformations have been said to occur in children with the FAS the core of the condition is the triad of growth impairment, minor facial features, and neurodevelopmental aberrations; the first two, trivial and transient, seen in neonates, and the third mostly at older ages, thus properly in a separate class.

According to some accounts it was, in fact, mental retardation and not any neonatal physical attribute that was considered the most common, the "cardinal," effect of maternal alcoholism, even to the extent that it was named as the leading cause of mental retardation in the US (Abel and Sokol 1983, 1986), an assessment vigorously challenged (Livingstone and Lyall 1986). This judgment was made through examination of children of various ages, even in the absence of other signs of the syndrome. It is appropriate to note that Penrose in his clinical and genetic study of 1280 cases of mental defect in England in the 1930s did not mention alcohol once as a causative factor (Penrose 1938), nor will one find the word alcohol listed in the index in his monograph on the biology of mental defect (Penrose 1949).

FAS Specificity

I digress briefly to inquire into the distinctiveness of the physical features composing the FAS. Severe degrees of the dysmorphic craniofacial features of the syndrome were said to constitute a "unique constellation of anomalies" (Clarren and Smith 1978); or as later qualified – "essentially unique" (Clarren 1990) or "satisfactorily unique" (Sampson et al. 1997) – all contradicting it would seem its incomparability. Challenging such pronouncements is the fact that the several craniofacial features, which were said to serve as the primary means of identifying infants with the FAS, do not appear to be specific to maternal alcoholism, nor is the overall craniofacial pattern clearly pathognomonic. While the dogmatic assertions noted above were later weakened (Aase et al. 1995), the original formulation continued to be considered the standard; on the other hand, as noted, others on the contrary assert that it was not any neonatal physical attribute but mental retardation that was the hallmark of the maternal alcoholism effect.

Illustrating the diagnostic uncertainty, however, is the fact that children with numerous and diverse histories can have arrays of facial features similar if not identical to those found in the FAS. The ones presently recognized include a variety of etiologies (mentioned even by Jones and Smith 1973, Smith 1976, p. 477): antiepileptic drugs (Hill et al. 1974, Leavitt et al. 1992, Robert and Källén 1994, Moore et al. 2000), phenobarbitone (Seip 1976), maternal zinc deficiency (Flynn et al. 1981), phenylketonuria (Lipson et al. 1984), maternal marijuana use (Hingson et al. 1984), lymphotropic virus (Marion et al. 1986), benzodiazepine (Laegreid et al. 1989), toluene use (Arnold et al. 1994), heavy smoking (Rostand et al. 1990), cocaine (Fries et al. 1993, Robin and Zackai 1994). Whose abundance makes the assertion of uniqueness untenable.

Even short palpebral fissures, supposedly a cardinal feature of the FAS, lacks specificity. Jones et al. (1978) admitted that "short palpebral fissures are a nonspecific abnormality noted in a number of multiple abnormality syndromes," and Abel and Sokol (1991) went further and said that "palpebral fissure size. . .appears to be more a subjective than objective facial anomaly. . .." Regardless, it has been argued that the FAS can be distinguished from other malformation patterns that share its features (Mulvihill et al. 1976), that it is not "readily confused" with other malformation patterns (Clarren and Smith 1978). [Incredibly, this question of palpebral fissure length, as a sign of the FAS, is still with us 30 years on, and still it seems not clearly meaningful – Cranston et al. 2009, Jones et al. 2009]

After another 30 years, came the revelation, déjà vu all over again, to quote the inimitable message, that "Poor women of childbearing age who use crack, cocaine, marijuana, and heroin may be at risk for having an alcohol-exposed pregnancy because of concurrent alcohol use" (Sharpe and Velasquez 2008).

It seems most likely that what underlies these facial features of diverse etiology are nonspecific prenatal growth disturbances, hardly conferring upon them diagnostic trust and validity. Or, as another wording had it, essentially interpreting them similarly, the "many minor craniofacial anomalies represent persistence of fetal traits. . .thus they represent *dysmaturities*. . ." [author's emphasis] (Opitz 2000).

Diagnosing the FAS

Despite all this, it has been claimed that the requirements for the diagnosis of the FAS are well established and its diagnosis is easily made. On the contrary, even the optimal time for making the diagnosis is debated. "Birth is a convenient time to make the diagnosis. . .but only the most severely affected infants are likely to be identified then" (Sampson et al. 1997). Nevertheless these authors agreed that "the diagnosis of the FAS is particularly prone to unreliability," and requires "trained dysmorphologists." But even recognition training did not appear to be sufficient, since the minimal requirements for making the diagnosis include not only the physical features of the child but also association with "a characteristic history of chronic maternal alcoholism" (Mulvihill 1986), a patent admission that the features at birth may not be diagnostic. And for diagnosis at later ages, even more so, it was admitted that evidence of maternal alcoholism may be insufficient "since without the facial description [recognition is] quite difficult" (Clarren and Smith 1978).

Even more contradictory are statements such as "many diagnosticians believe that the FAS diagnosis can be made without knowledge of maternal alcohol exposure"; and "diagnosis rests on. . .neurobehavioral effects. . ." (Sampson et al. 1997). Thus neither the offspring state nor the maternal propensity seems to be sufficient or even necessary for diagnosis. Nor was the picture clarified by shifting requirements: "a woman who drinks heavily during pregnancy creates some risk of producing a child with FAS"; or conversely, "identifying a child with apparent FAS strongly implies that the child was damaged by alcohol in utero – even if a maternal history of alcohol consumption cannot be obtained" (Clarren et al. 1987b) – both, classics of ambiguity.

Fetal Alcohol Effects

Making recognition even more problematic was a widening of the alleged symptomatology: the effects of alcohol, it was declared, occur in a wide spectrum, the FAS does not only follow upon extreme levels of alcohol consumption; that "along the rest of the continuum toward normal are persons with every subcombination of fetal-alcohol-syndrome anomalies," and that each of the anomalies "can independently vary in severity and grade into the normal range" (Clarren and Smith 1978).

Thus was set forth the radical proposition that while heavy alcohol drinking carries a major fetal risk lesser alcohol abuse may lead to partial expression of the syndrome, termed "suspected fetal alcohol effects" – raising the specter of harm at any dose level. This was made concrete when it was asserted that "no absolutely safe level of ethanol consumption has yet been established," contravening every established teratological principle – and sparking a storm of concern (Kolata 1981, Anon 1983b,c, Raymond 1987).

This is a bold assertion, without validity of any sort; reiterated in these words: "Relationships between many neurobehavioral outcomes and measures of prenatal alcohol exposure are monotone without threshold down to the lowest nonzero levels of exposure. In sum, alcohol effects on the developing human brain appear to be a continuum without threshold" (Sampson et al. 2000). This is equivalent to equating alcohol effects with gene mutation dynamics, which is absurd.

Let me recall some words written above that are relevant here. The fundamental distinction, the characteristic that tells mutagenesis apart from teratogenesis, is the shape of the dose-response relation, which is the palpable expression of underlying mechanism. It is currently accepted that the response to mutagens, as to carcinogens, is linearly related to dose such that there is no dose that is without an effect.

The theoretical basis of this concept is that mutagens and carcinogens produce their consequences by hitting single targets, molecules in DNA without interaction with other entities or the outer environmental world, and thus that no matter how small the dose of the agent applied an effect, mutated genes or cancers, occurs.

In teratogenesis, on the contrary, the shape of the dose-response curve is not straight, but sigmoid, rising quite steeply from the no-effect dose and flattening at doses associated with maternal toxicity and total fetal death. What accounts for this unique pattern? Instead of the single unmediated, invariant target theorized of mutagenesis, teratogens deal with embryos, dynamic multicellular entities with reparative capacities and hereditary inputs, which in interacting with its enveloping membranes and the maternal body, create unimaginably versatile physiological, pharmacological, and metabolic efficiencies that continually divert and modify the effect of impinging agents. A totality that combines to produce unpredictability, and adds to the complexities of the drug tester's task.

It is therefore not possible to accept the notion that developmental phenomena are not subject to dose-response relationships; and thus the ideas expressed by Sampson are nonsensical and must be totally rejected.

To return to the subject at hand. The postulated partial or less severe manifestations of the syndrome made it necessary to devise new categories and new terminologies (the tacit admission that the original concepts were getting nowhere and needed sprucing up). Examples are the already mentioned "suspected fetal alcohol effects," then, less conditionally, "fetal alcohol effects" (FAE), and features "compatible with FAS" (Hanson et al. 1978), "alcohol-related birth defect" (Sokol and Clarren 1989), "fetal alcohol abuse syndrome" and "alcohol-abuse related birth defects" (Abel 1998), "alcohol-related neurodevelopmental disorder" (Sampson et al. 2000), and "prenatal exposure to alcohol" (Archibald et al. 2001); and in giving way to the irresistible urge to "acronyomania" (Sharp 1999), there came forth FAE, CFAS, ARBD, ARND, and PEA.

As may well have been foreseen, the vagueness of the numerous 'soft' signs of these fetal conditions soon led to indiscriminate assignment of FAE and the other terms to features resembling one or another of those of the fetal alcohol pattern, often unbutressed by evidence of anything more than a suspicion of maternal alcoholism. [See the syndrome of caudal regression above for another subversion of the integrity of a supposed syndrome.] This provoked even the proponents of the neologisms to protest at such abuses, and to recommend "a more conservative approach to diagnostic terminology," a difficult task, as they agreed, particularly when the subjects were single patients (Aase et al. 1995).

Reiterated by Aase et al. (1995) was the incantation that only the "combination with each other allows definition of the syndrome," i.e. the triad, newly elaborated, of growth deficiency, a distinctive pattern of mild facial dysmorphology, and mental deficit as a reflection of abnormal brain function – an added innovation.

In the face of the frequent transience of the facial features and growth deficits and the numerous and varied pre- and postnatal confounding influences on growth, the diagnosis largely depended on what has sometimes been claimed to be the commonest of the FAS characteristics – the neurodevelopmental deficits. These are expressed of course almost entirely in older childhood, making the diagnosis questionable and the passage of time increasing the difficulty of establishing a connection with maternal alcohol abuse during pregnancy.

The initial publications were followed by numerous reports of single cases or series of cases of maternal alcoholism associated with the FAS in children of different ages. The various inadequacies of method and objectivity they entailed were examined in detail (Neugut 1981, Roman et al. 1988), and supported the doubt that causation of alcoholism had been proved.

Nor to be ignored are the ubiquity and the workings of confounding cultural, social, and economic factors associated with maternal alcoholism on pre- and even more so postnatal development; as was already declared years ago, as noted above, by Lemoine et al. (1968); and reiterated many years later more generally, namely, "the postnatal environment has an overwhelming influence on cognitive function through to early adulthood..." (Jefferis et al. 2002). But these factors have hardly been explored in the FAS.

Even the very early examination of the effects of maternal alcoholism, a century ago, cogently noted, in the verbiage of the time, the truism that "it has to be borne in mind [the effect of alcoholism] is not solely the result of the direct influence of the intoxication on the organisms of the mother and child, but is also in part a consequence of the malign modification of the environment due to the parental vice" (Sullivan 1899).

Acknowledgment of this powerful influence has also come from pioneers in the alcoholism field. It was noted by Ulleland (1972) and given but glancing attention by Clarren and Smith (1978): "It is difficult to determine the extent to which the socioeconomic situation or factors related to continued maternal alcoholism may have adversely affected developmental progress." But unfortunately little study has been made of this difficult, but hardly to be neglected, area of concern.

The Epidemiological Process

In the context of malformations, epidemiology refers to determining whether and to what extent specific abnormalities may be associated with time, place, condition, and circumstance, for the implicit purpose of clarifying etiology. Epidemiological investigation of the alleged teratogenicity of maternal use or abuse of alcohol has been made both prospectively and retrospectively. The former begins with the putative cause: maternal alcohol consumption as teratogen; the latter with the supposed effect: the abnormalities in question.

Ideally, in order to strengthen causal relations, studies – prospective and retrospective – should be pursued by focusing on newborn children. But the difficulties of the neonatal diagnosis of FAS, as recounted above, forced the focus instead to be on conditions seen in older children (as with cardiovascular malformations) and the association with alcoholism to be examined retrospectively (Sokol and Clarren 1989). The following sections examine both FAS study methods.

Prospective Studies

It has been taken almost as axiomatic that intemporate drinking during pregnancy is harmful to fetuses, as exemplified by the following, "...children with the full fetal alcohol syndrome are born to mothers who are chronic alcoholics and not to light or moderate drinkers" (Streissguth et al. 1981). But, in time, as noted above, less than excessive – moderate – drinking also came to be held harmful.

Moderate Drinking

There has been little consensus however about what moderate or supposedly moderate consumption of alcohol consist of. Hanson et al. (1978) stated it to be $\geq$1.0 g/day; Kaminski et al. (1976) >1.6 g/day; Little (1977) $\sim$1.0 g/day; Ouellette et al. (1977) an average of 2.2 g/kg/day (i.e. about 5 g/day); Rosett and Weiner (1985) on average 4.6 oz/day; Tennes and Blackard (1980) a heroic 11–40 drinks/8 h; Davis et al. (1982) >20 mL/day; Marbury et al. (1983) 2–3 drinks/day, Gibson et al. (1983) $\leq$10 g/day, and Diav-Citrin and Ornoy (2000) at least 2 g/kg or 8 drinks/day; i.e. "moderate" can mean anywhere from 1 g/day to 40 drinks/day.

Most prospective studies of maternal consumption of such amounts examined the offspring of alcohol-using or -abusing women, despite the contradiction. Also despite the difficulty of substantiating or measuring maternal alcohol consumption with certainty, and the fact that most signs of the FAS are in dispute or difficult to recognize in newborn infants (Neugut 1981, Sokol and Clarren 1989).

Discussed first will be studies in neonates of effects on prenatal growth and physical features. Studies that confined attention to effects on growth alone will be little discussed here, confounded as such phenomena are by numerous factors beyond the focus of this work; the reader is directed to the many studies of this sort that have been conducted, with greatly varying findings (a selection of which is found below).

The NIH Study

Perhaps the earliest examination, of a possible prospective sort, of the fetal effects of maternal alcoholism was the NIH Collaborative Perinatal Project (Jones et al. 1974). Mining its records revealed that 69 of the over 55,000 women in the project had a history of alcoholism, a seemingly low proportion perhaps owing to failure of the project to focus on this problem. Twenty-three of the women, of predominantly lower social class standing, were considered, on the basis of "reasonably secure evidence," to have been chronic alcoholics before and during pregnancy. The records revealed that their offspring were growth deficient at birth and at 7 years of age; that six of 19 children surviving beyond infancy had physical features suggestive of the FAS more often than did those of matched nonalcoholic controls (who were not further mentioned); that six of 13 surviving 7-year-olds who lived with their mothers had lower mean IQs, but not significantly so, than those living in other households, and lower than they were at the age of 4 years – hints, it would seem, of the ramification of the postnatal environment; the last feature interpreted as constituting "deficient intellectual performance."

Little weight can be given to these findings, considering the nebulousness of the designation "chronic alcoholism," and the uncertainty of the children's condition. Despite these limitations, this study and others of the Seattle group became influential in calling forth and shaping studies in this area.

In contrast with this article, with its concentration on chronic alcoholism, almost all other prospective studies were concerned with far less drastic drinking practices, what has been characterized as 'moderate' drinking (Hanson et al. 1978). It is principally the findings of such studies, presented more or less chronologically, that will be described here.

(As an aside, a message sent by General George Washington to the Continental Congress in 1777 may be inserted here: "The benefits arising from moderate use of strong liquor have been experienced in all armies and are not to be disputed.")

Chronological Overview

An early study, not primarily designed to investigate the possible effects of alcohol, involved more than 9000 French-born women attending maternity hospitals in Paris (Kaminski et al. 1976). It was learned through interviews at the end of the first trimester that only a minority of the women, 5.5%, habitually drank more than 1.6 oz of absolute alcohol daily. These, the heavier drinkers, regardless of smoking practice, had babies of lighter birthweight but differing little in major malformation frequency from those of lighter drinkers. In contrast with several other studies, alcohol use by these women differed little before and during pregnancy.

In later Paris studies, although of women attending a hospital handling many complicated pregnancies, only 1.6% drank >1.6 oz/day (Kaminski et al. 1979, 1981). The malformation frequency approximated the usual one, and individually or in toto did not differ in the greater than the lesser drinking group, nor in the proportion of deaths due to malformations. Dysmorphic facial and other minor features were not specifically mentioned, perhaps because the studies were conducted before the FAS was discovered. But can that be the entire explanation?

Negative findings also marked a Seattle study of largely white, middle class women enrolled in a health maintenance organization (Little 1977). Almost half drank alcohol during pregnancy, with the greatest amount – on average ≥1 oz/day of absolute alcohol, considered to be "moderate" – consumed by 9.1%. No growth-retarded infant had the FAS; lesser FAS signs, other physical abnormalities, and major malformations were not mentioned. Excessive drinking in late pregnancy decreased birth weight far more than in earlier months, understandably since the later months is the preponderant period of fetal growth.

The findings in another Seattle study were said to be positive, but this is questionable (Hanson et al. 1978). It involved white middle class women enrolled in a health maintenance organization, as well as women of a range of economic classes seen at two Seattle hospitals. Judgment of the condition of the newborn children was based on a combination of abnormalities of growth and morphogenesis. With carefully chosen words the examinations were said to "provide a strong reason to suspect that moderate levels of alcohol consumption during early pregnancy can have an adverse effect on the fetus."

The data clearly showed otherwise. Analysis of the sample of 163 infants showed that the proportions with what were called "features of FAS" of mothers drinking ≤0.1, 0.11–0.9, and ≥1.0 oz/day were not statistically different from one another. Nevertheless it was asserted that moderate drinking was harmful to the fetus, the epithet 'moderate' apparently applied to the children of the heaviest drinking mothers.

A later account provided additional information regarding the sample (Streissguth et al. 1981). While repeating the misinterpretation that there was a significant relation between maternal alcohol use and FAS features, other information was imparted: children were selected for examination whose mothers reported alcohol use, but not all the children of the heavy users were examined; "head circumference at 8 months was not significantly related to maternal alcohol use"; and "women identified as 'heavy' drinkers on one [scoring procedure] may be entirely different women than those identified as 'heavy' drinkers on another. . .." Puzzling indeed.

Several studies were made in Boston hospitals. In a Boston City Hospital study, on questioning women about alcohol use at their first prenatal visit, it was learned that 13% were heavy drinkers – averaging 2.2 g/kg/day absolute alcohol; i.e. ~5 oz/day for a women of 140 lb (Ouellette et al. 1977). Their infants had an increased frequency of major and minor congenital anomalies and growth or neurological abnormalities (hypotonia, jitteriness). No defect however, aside from microcephaly, was named. That defect however may well have been associated with reduced infant size, and the weight impairment attributed to the drinker's heavy smoking, well known to be associated with fetal growth retardation (incidentally first brought to notice, as noted above, by Simpson (1957), and later substantiated e.g. by Ounsted and Scott (1982), Young 1983, Brooke et al. 1989, Peacock et al. 1991, Peacock et al. 1998).

A verbal miniduel followed the authors expressing muted disagreement with the concept of the FAS forming a specific pattern. The criticism leveled at this opinion (Jones and Smith 1978) – that it did not take into consideration the characteristic facial features of the syndrome – was replied to (Anon 1977) with the remark that the normal variability in one of these features, palpebral fissure size, precluded absolute judgment without controls, which previous studies had not included. Which dispute needs explication, as follows.

A Palpebral Fissure Parenthesis

Short palpebral fissures, said to be the commonest and perhaps the hallmark of the dysmorphic facial features of the FAS (Hanson et al. 1976), merit a clarifying word. To begin with, the mean palpebral fissure length in normal offspring at 38 weeks of gestation was 17.5 mm and 18.5 mm, in Seattle and Györ, Hungary, respectively (Jones et al. 1978, Méhes 1980); and found, among other things, to be affected by ethnicity, birthweight, and head size (Fuchs et al. 1980, Alpert et al. 1981, Iosub et al. 1985). With respect to short palpebral fissure in the FAS Fuchs et al. (1980) offered the opinion that it is related to overall growth retardation in offspring of alcohol abusers; which seemed to be concurred in by Opitz (2000), as quoted above.

To Continue

Adding further uncertainty to the findings of the Boston City Hospital study was the fact that the large percentage of heavy drinkers noted by it was not typical of the Boston population, and it seems of others. By comparison, in a Cleveland hospital study the frequency of relatively heavy drinkers was 1.7% (Sokol et al. 1980), in Denver 1.8% (Tennes and Blackard 1980), in a central English city 1.0% (Davis et al. 1982), in the Brigham and Women's Hospital in Boston 0.7% (Marbury et al. 1983), in northern California 0.5% (Mills and Graubard 1987), and in Dallas 0.7% (Little et al. 1990). A frequency close to that found in the Boston City Hospital, 7%, was found in an area in France said to be one of heavy alcohol consumption (Rostand et al. 1990); see further details below.

Additionally muddying the picture were the different results obtained by a study conducted some time later at the same Boston hospital (Hingson et al. 1984). The women studied were also a low-income group, with about 8% heavy or moderately heavy drinkers (consuming ≥2 drinks/day). But there was no significant difference in frequency of major malformations (though some of the latter were not major) in their children and those of nondrinkers; and this was also true of the so-called features compatible with the FAS and of infant growth measures. Only one instance of the "full FAS" was seen, the status of whose mother was not stated.

This study discovered that, when the confounding variables entailed in maternal consumption of alcohol were discounted, marijuana use during pregnancy was often associated with features compatible with the FAS. It was thus learned that such practices, especially as they affect maternal nutrition, can have fetal outcomes simulating those of alcohol per se.

Another report from the Boston City Hospital (Rosett et al. 1983), though noting no major malformations in the 469 infants examined, nevertheless still asserted that heavy drinking was associated with such defects; and that only one infant, also mentioned by Hingson et al. (1984), was diagnosed as having the FAS.

No association was found between the frequency of minor and major malformations (undescribed) and any level of maternal alcohol intake in a study of over 12,000 predominantly white women in the Brigham and Women's Hospital in Boston (Marbury et al. 1983). Major malformations occurred in 2.6% of offspring of abstainers, 3.1% in those of moderate drinkers, consuming 1–6 drinks/week, and 2.0% in the most frequent drinkers, consuming ≥7 drinks/week; in other words no intergroup difference. Once more there was a significant independent effect of smoking on birthweight.

Moving elsewhere, in Denver women were interviewed during early pregnancy at two public hospitals regarding alcohol consumption, use of illicit drugs, and other aspects of life style (Tennes and Blackard 1980). About 80% of the women abstained or rarely drank alcohol and the mean amount consumed by most of the others during the first trimester was about 0.7–2.0 oz/day, amounts considered to be moderate or socially prevalent. Level of alcohol consumption was not associated with significant differences in frequency of the minor facial and other anomalies associated with the FAS, i.e. they were no more common in infants exposed in utero to heavy or moderate amounts of alcohol than in nonexposed infants. This was also true of the fetal growth attributes weight, length, and head circumference. And as was noted in other studies, use of alcohol and other drugs was significantly reduced in late pregnancy.

At the Cleveland Metropolitan General Hospital alcohol abuse was identified in 1.7% of 12,127 women delivering consecutively, with 2.4% of their children (i.e. 5 of the 12127, or 0.41/1000) considered to have "enough stigmata to be probable examples of the fetal alcohol syndrome" (Sokol et al. 1980). Alcoholism was defined according to the supposedly rigid diagnostic criteria set forth by the Michigan Alcoholism Screening Test: attendance at an Alcoholic Anonymous meeting, seeking help for a drinking problem, hospitalization for drinking (Sokol et al. 1981). As usual, excess alcohol use was associated with reduced birthweight, intensified by maternal smoking. The frequency of other anomalies, mostly minor ones, was perhaps increased only in offspring of a subgroup of the abusing mothers, only one of whom however had the FAS.

Later findings of the Cleveland hospital study disclosed that the defects consisted of minor anomalies or morphological variants, "neither very unusual nor very worrisome" (Ernhart et al. 1985, 1987). Of these, only posteriorly rotated ears were commoner in infants of alcoholic than nonalcoholic women. Another feature, short palpebral fissures, was not significantly different in frequency in the two groups. Nevertheless, by 'tallying' a series of anomalies, a 'reasonably specific' pattern was formed that was significantly associated with a history of alcohol abuse; though it is difficult to imagine how this could be established. Attempts to relate anomaly tally to the first trimester as the critical period and to show a clear dose-response relation were not convincing, especially in light of the numerous confounding variables in play (Ernhart et al. 1987).

Such tallying was apparently disavowed afterward, when it was maintained once more that FAS is only diagnosable when the "patient has signs of abnormality in each of three categories": growth retardation, central nervous system involvement, and characteristic facial appearance, including among others short palpebral fissures (Sokol and Clarren 1989).

At the Kaiser-Permanente of Northern California over 32,000 women were questioned about alcohol use in the first trimester of pregnancy (Mills and Graubard 1987). About 50% were nondrinkers, most of the others consumed ≤ 2 and a very small minority (0.5%) ≥ 3 drinks daily. The overall frequency of minor and major congenital malformations was not different in children of women drinking heavily than in those drinking more moderately or not at all, and there was no consistent dose response. It was believed to be unlikely that even the relative rarity of heavy drinking would have obscured a maldevelopmental effect. The diagnosis of major malformations appears to have been inadequate however, since the overall frequency in children of drinkers was about half that usually found in neonates. A large number of abnormalities, including microcephaly, were considered individually, but none, including those thought to be associated with maternal alcohol abuse, was significantly increased in frequency.

The data were collected at a time before the FAS was widely recognized. This indeed may have made the women less reluctant to reveal their drinking practices, and the personnel evaluating the information more objective regarding the FAS. Nevertheless, as the authors reasoned, its rarity in this population was attested by the infrequency of microcephaly and prenatal growth retardation. Thus once again moderate consumption of alcohol was found to be without prenatal harm.

An account, presented in an abstract, and so far as I have able to ascertain not as yet otherwise reported, considered the interaction of alcohol consumption and maternal risk factors on infant outcome (Cherpitel and Room 1989). Preliminary findings indicated no association between drinking and growth parameters, neurobehavioral development, or major congenital malformations in infants of a group of high-risk women in Berkeley, California. The data suggested that early alcohol consumption did not adversely influence infant outcome, including the occurrence of minor abnormalities, beyond the effect of the maternal risk conditions.

Women of low socioeconomic status attending the outpatient prenatal clinic at the Pittsburgh Magee-Womens Hospital who drank during the first trimester were selected for study (Day et al. 1989). Twenty-four percent were heavy drinkers, taking $\geq$1 drink/day, and also used tobacco, marijuana, and other illicit drugs excessively. The others abstained or drank moderate or light amounts, i.e. <1 drink/day. As had been frequently noted, alcohol use was substantially reduced during early pregnancy stages.

Only 1% of infants had major anomalies, fewer than usual, but many more had minor defects, which apparently indicate that the attention of the study nurse clinicians examining the infants was inordinately directed at them. Diagnosis of anomalies was sketchily described, and those found, major, minor, and fetal alcohol effects, were not named. The defects were not significantly associated with drinking; nor was an instance of fullblown FAS seen. Low birthweight and reduced head circumference were associated with heavy drinking; the latter was also closely associated with the intertwined maternal features of low socioeconomic status, heavy smoking, and use of illicit drugs, all well known agents of fetal detriment. In the face of the negative outcomes and contradictions encountered, it is difficult to understand how the implicit conclusion was reached that any drinking in pregnancy is a fetal risk.

The records of 5602 women delivering at the Parkland Memorial Hospital in Dallas, Texas, a county facility serving a low income indigent population, indicated that FAS occurred only in children of the 0.27% drinking more than $\geq$6 drinks/day (Little et al. 1990). Major congenital malformations were increased, but the list of them, "reported exactly as they were recorded in the medical records," is confused and contradictory. As for growth, after maternal risk factors were considered, only head circumference appeared to be reduced. Heavy use of tobacco and illicit substances and "generally poor health and nutritional circumstances" probably explained some of the "features consistent with FAS/FAE."

The relation of maternal alcohol use and infant abnormalities was examined using blood alcohol markers in women attending obstetric clinics at five Boston area hospitals, over one-third in a substance-abuse clinic (Stoler et al. 1998). About 5% drank daily and 14% weekly (quantity not reported) in the month before questioning, which was reduced, to 2.7% and 4.2% respectively, after they knew they were pregnant.

The infants were judged to be affected if they possessed a minimum of FAS-associated dysmorphic facial features and to be growth retarded on the basis of size criteria. Of the four blood markers WBAA (whole blood-associated acetaldehyde) was more indicative of alcohol use than were the others, but did not seem to differentiate drinkers of different quantities with any certainty; even a large proportion of abstainers were positive for this marker. Mothers of affected infants had elevated WBBA compared to the other markers, but the latter did not generally appear to be reliable indicators of maternal alcohol consumption. Furthermore adding markers serially, as was done to strengthen the relation, was hardly a legitimate statistical exercise.

Only in the discussion were certain pertinent facts mentioned: that all the growth criteria that entered into the judgment of 'affectedness' were within the normal range; that not all the women with positive markers had affected infants and that not all affected infants had mothers with positive markers, a possible reason for the latter being that "many of the maternal blood samples were obtained late in pregnancy"; a revelation which, by itself, would appear to invalidate the entire effort.

An apparently belated realization has arrived on the scene to the effect that only a minority of instances of alcohol-drinking during pregnancy leads to detrimental offspring consequences, because there are maternal genetic and metabolic factors that are protective in this respect (Gemma et al. 2007, Green and Stoler 2007).

Studies of alcohol mediating enzymes in the FAS picture however have not always been illuminating. A particular dehydrogenase enzyme allele present in American blacks seems to make for reduced vulnerability to alcohol-related birth defects, as the summary of a report states (Scott and Taylor 2007), but how general this phenomenon is apparently not yet determined.

Studies from Europe and elsewhere were largely exonerating. A study in Germany of over 7500 women consuming moderate amounts of alcohol daily ("a little every day") found no effect on the frequency of major and minor congenital malformations, nor on growth, psychomotor characteristics, and postnatal development (Mau 1980). The outcomes for women who drank more or less than such amounts were not reported. Since the study was conducted before the FAS was known of, while the examining physicians may have been less attuned to the signs of the syndrome, their observations may also have been more objective.

Two English studies found little to support the FAS concept. In the Warneford Hospital, Leamington Spa, almost half the women taking a self-administered questionnaire said they abstained from alcohol during pregnancy, a bit more than a third that they consumed on average 1–10 mL/day, and the others >10 mL/day (Davis et al. 1982). But even in the offspring of the 2.9% who drank most, i.e. >20 mL/day, major and minor congenital malformations were not more common than in those drinking less. Although birthweight was also unaffected the authors claimed a trend toward smaller head circumference in the babies of the heavy drinkers; which was decisively refuted by an independent analysis (Anon 1983d).

The overall malformation frequency in this study was far lower however than usually occurs, throwing doubt on the quality of its diagnosis, despite being made by pediatricians. Last, bowing to political judiciousness, the authors felt compelled to repeat that "no 'safe level' of alcohol consumption was found..." What "safe" referred to is hard to imagine. Others were emphatic that "the study simply does not permit the conclusion [of any association] between alcohol consumption by pregnant women and major abnormality or reduced head size in their babies" (Campbell and Sullivan 1983).

Women attending three London maternity units reported that they drank "occasionally, every day, or not at all" (Grisso et al. 1984). The overall malformation frequency was 3.0%, approximating the usual one, and was not significantly different in the three groups. The authors were courageous in cautioning against attributing harmful pregnancy outcomes to alcohol, particularly at levels that could be defined as "social."

Possible examples of the fetal harmfulness of heavy drinking among socially deprived women were reported from Belfast. Pregnancy outcome was studied in 23 women, predominantly of the lower socioeconomic classes, whose drinking problems were known, who smelled of alcohol, and admitted heavy drinking, though exact alcohol intake was not estimated (Halliday et al. 1982). The great majority of their babies were small for dates and the survivors continued to be growth retarded at age 2 years; the separate effect on growth of their heavy smoking was not examined however. Almost half of the infants had facial features of the FAS (therefore the others did not). An addendum noted that in the 4.5-year period of the survey the incidence of what "could be considered true examples of the fetal alcohol syndrome" was 1.3 per 1000.

The book-length report of an ambitious investigation of the effects of maternal drinking undertaken at the Simpson Memorial Maternity Pavilion in Edinburgh was a disappointment (Plant 1985). After elaborate introductory and descriptive material concerning aims, design, methods, baseline and alcohol consumption data the chapter labeled 'pregnancy outcome,' arrived at with much anticipation, gave not a shred of concrete information regarding the physical condition of the offspring. For example, palpebral fissure length ("allegedly a feature of the fetal alcohol syndrome") was supposedly measured, but nowhere were the measurements recorded. Similarly, the babies' weight, length, and head circumference, full information regarding which was said to be found in an appendix, was not to be found there; only a list of what the babies were examined for. Almost 39% of surviving babies had "alcohol-related" abnormalities, but none was named, except that it is clear that many of them were trivial. All together, apart from an excellent historical overview and a valuable bibliography, a worthless effort.

Two Australian studies were negative. In the one from the Queen Elizabeth Hospital in Adelaide just over a quarter of about 7000 women interviewed prenatally were nondrinkers, about two-thirds moderate drinkers, and the remainder heavier drinkers (≤10, 11–30, and >30 g/day alcohol, respectively) (Gibson et al. 1983). Alcohol use and congenital anomalies were not significantly associated, the larger than usual frequency of major defects, 4.2%, possibly indicated a loosening of definition (Gibson et al. 1981). Multiple regression analysis, discounting the significant effect of smoking, showed no association of alcohol and prenatal growth retardation.

A report from the Royal Hobart Hospital described the drinking and smoking patterns in early pregnancy of almost 15,000 mothers of all births in Tasmania in 1981–2 (Lumley et al. 1985). The patterns differed markedly between socioeconomic groups, but moderate and especially heavier drinking (≥2 glasses/day) were very uncommon. Only children of the latter group had a frequency (unstated) of congenital malformations greater than usual. A later study was categorical in finding no suggestion of an increase in congenital malformations (Bell and Lumley 1989). But reduced birthweight, as usual, was associated both with drinking and smoking in excess.

Mother-infant pairs were studied in the public maternity hospital in Roubaix, France, an area characterized by "higher than average alcohol consumption" (Rostand et al. 1990). The mothers, all French-born, reported light, moderate, or heavy drinking (on average ≥3 glasses/day) during the first trimester, or were known or suspected alcoholics. Very few congenital malformations and only two apparent instances of FAS were found, neither in offspring of heavy drinkers. The infants were examined for 17 neonatal craniofacial traits "described in previous reports on fetal alcohol syndrome." A 'tally' analysis found that offspring of heavy-drinking mothers had a larger mean number of such traits and a larger percent of "features suggestive" of fetal alcohol effects than offspring of the light and moderate drinkers combined. Examined individually however the frequency of none of the principal features of the FAS – short palpebral fissures, hypoplastic filtrum, hypertelorism, micrognathia, etc., or height, weight, head circumference – was significantly increased. It appears that a tally gives spurious results, based as it is on statistical overkill. Also negating the tally, offspring of heavy drinkers who were heavy smokers had no increased occurrence of the craniofacial characteristics.

Almost all women attending a special outpatient clinic at a Central Hospital in Helsinki were heavy drinkers, consuming >20 g/day alcohol in early pregnancy or throughout (Autti-Rämö et al. 1992, Autti-Rämö 2000). The children of the consistent drinkers, when judged by "subjective impression" at older ages, had increased levels of minor physical and craniofacial features than controls. Certain nasty facts, however, contradicted these impressions. For example, short palpebral fissures (which, taking the party line, the authors held to be "the single facial feature mandatory for diagnosis of FAS"), when put to the objective test of being measured, proved not to be shorter in case children than controls. There were other such negations as well. The only features that were significantly different were those of growth impairment, and this was especially so, not surprisingly, in those exposed throughout pregnancy.

Evidence in this report suggested that the maternal alcohol abuse problems and associated detrimental practices did not stop at delivery; and since Finnish laws largely prevented children from being removed from the care of their parents, even those who were known alcoholics, it can be expected that distinguishing the baleful effects of social deprivation on postnatal growth and development from imputed ones of maternal alcohol abuse would be difficult. Thus the minor morphological effects claimed to be due to heavy maternal alcohol consumption were hardly proven.

A Danish study examined the fetal effects of social drinking in all pregnant women living in Odense in 1988–9 (Olsen and Tuntiseranaee 1995). Those when questioned at about gestation week 12 who reported an average consumption of ≥5 drinks/week in the first trimester were selected for study. Strangely, only one newborn trait, palpebral fissure length, was examined and this was found to be trivially reduced by binge drinking only (i.e. ≥5 drinks/occasion); but even this, when birth weight was included in the calculation, did not reach statistical significance. As the authors commented, "the facial characteristics. . .were indeed minor and will probably never be recognized."

Major Malformations

As the above survey demonstrated, prospective investigations conducted during the 1980s and 1990s found scant evidence that maternal use of moderate or even larger amounts of alcohol during pregnancy causes serious fetal maldevelopment. Since the predominant attention of the studies was on the minor abnormalities presumably due to maternal alcohol drinking, very few studies even mentioned major congenital malformations. In most of the latter there was little difference in frequency in infants of mothers abstaining or drinking little or much alcohol (Kaminski et al. 1976, Mau 1980, Davis et al. 1982, Hingson et al. 1984, Gibson et al. 1983, Marbury et al. 1983, Rosett et al. 1983, Grisso et al. 1984, Mills and Graubard 1987, Day et al. 1989, Rostand et al. 1990).

A few claimed an increase. Ouellette et al. (1977) stated that heavy drinkers had infants with an increased frequency of major malformations, but none was named. Lumley et al. (1985) found unnamed congenital malformations to be increased in neonates of the relatively small number of women consuming more than 2 drinks/day; whose reality however was doubted because of wide confidence limits. Little et al. (1990) said major congenital malformations were increased, but the charts recording them seemed faulty. That seems to sum up the record.

Minor Malformations

The question then turns to the minor craniofacial features thought not only to characterize but to identify the FAS. The pioneering study of Jones et al. (1973) started the ball rolling when it reported that certain physical abnormalities, outlined above, occurred more often in offspring of chronic alcoholic women than in those of nonalcoholics.

Continuing, Little (1977) found no FAS in infants of "moderate" drinkers; Hanson et al. (1978) no significant difference in frequency of FAS features in infants whose mothers drank various amounts of alcohol daily; Sokol et al. (1980) only one instance of FAS in children of alcohol abusing women, and a small minority with probable examples of the FAS; Tennes and Blackard (1980) no association of maternal alcohol consumption with the minor facial and other anomalies related to the FAS.

Davis et al. (1982) found minor congenital defects not more common in offspring of women drinking >20 mL/day than in those drinking less; Gibson et al. (1983) alcohol use and congenital anomalies not significantly associated; Grisso et al. (1984) the proportion of liveborn children with malformations not significantly different in groups drinking various amounts; Ernhart et al. (1985) the frequency of short palpebral fissures no different in children of drinkers and nondrinkers; Mills and Graubard (1987) the overall frequency of minor and major congenital malformations not different in children of women drinking heavily than in those drinking more moderately or not at all; Day et al. (1989) the association between one drink a day and minor anomalies hardly of significance; Rostand et al. (1990) only two instances of the FAS, neither in offspring of heavy drinkers, and short palpebral fissures, etc., not significantly more common in children of light, moderate, or heavy drinkers; and Autti-Rämö et al. (1992) that palpebral fissures were not shorter in case children than in controls.

Such unanimity forces the conclusion that drinking does not cause major congenital malformations, but also that minor facial malformations are seldom a consequence of what is called moderate drinking, as observed in neonates. A meta-analysis reached the same conclusion, namely that moderate alcohol consumption (defined as 2 drinks/day) in the first trimester of pregnancy does not increase the risk of major malformations (Polygenis et al. 1998). The opinion may be added that occasionally seen minor facial defects in children of alcohol-abusing women are in all probability secondary to the commonest outcome of intemporate consumption (and its frequent confounding socioeconomic and life style accompaniments): fetal growth retardation.

Long-Term Effects on Growth

A Seattle longitudinal series, discussed in full below, concentrated on behavior, giving meager attention to the effects of social drinking on growth characteristics in newborns and their persistence to older ages. It was early found that neonatal height, weight, and head circumference were all modestly inversely related to prenatal alcohol level (Streissguth et al. 1981, 1994a), and that the effects diminished with age, disappearing by 14 years of age if not sooner (Barr et al. 1984, Sampson et al. 1994). It may be remarked that the alleged absence of a threshold regarding these effects was contradicted by graphs which for one of them, weight, showed a cutoff at 1–2 oz/day, although less clearly so for head circumference (Barr et al. 1984).

Other authors dissented from the weak neonatal findings, and uniformly agreed that birth weight and other neonatal growth characteristics were little affected by moderate to light drinking in pregnancy (Wright et al. 1982, Mills et al. 1984, Little et al. 1986, O'Connor et al. 1986, Fried and O'Connell 1987, Kline et al. 1987, Kariniemi and Rosti 1988, Sulaiman et al. 1988, Day et al. 1990, Walpole et al. 1990, Lazzaroni et al. 1993). Roman et al. (1988) made the astute comment that "in general, studies finding no statistically significant association between alcohol consumption and intrauterine growth were those that controlled for the largest number of other potential risk factors."

Follow-up has been scanty, but with little disagreement. With regard to head circumference, it was trivially reduced in 6-year olds of very heavily drinking mothers in Buffalo (Russell et al. 1991); it had almost reached normality in Atlanta in school age children of socioeconomically deprived black women drinking throughout pregnancy (Coles et al. 1991); catch-up growth had been successful in Cleveland – contradicting a Pittsburgh study (Day et al. 1989, 1991) – and by 5 years no adverse effect of prenatal alcohol exposure on head circumference was seen (Greene et al. 1991).

As for the standard FAS-associated facial features, while their severity lessened at follow-up in school age children, they seemed still to be evident to trained personnel (Coles et al. 1991). The drawback in this study, as in others, is that it was not made clear whether the socioeconomically deprived mothers of these children continued drinking and engaging in other harmful practices after giving birth. A small number of the children were in the care of grandmothers and others, but the majority it seems continued to live with their mothers, and to be exposed to conditions potentially impairing their growth and development.

The evanescence of the growth impairment noted by Sampson et al. (1994) sparked them to reflect that, in the light of the many confounders impinging on it, the emphasis on the birth weight effect of maternal alcohol consumption is "unfortunate." This concession is extraordinary, in the light of the central role that was usually assigned to growth retardation in the spectrum of fetal alcohol effects. With the devaluation of this vital element and the unclear relation of alcohol exposure to the dysmorphic features said to be part of this FAE (see critique above of Hanson et al. 1978), the whole edifice of the FAE concept seems to fall to the ground. And was not repaired by the attempt to prove the persistence of dysmorphic features as a package (Graham et al. 1988).

Retrospective Studies

For all the reasons outlined above the emphasis turned in later studies to abnormal conditions discovered in older children, and their supposed association with maternal alcohol use during pregnancy or before its recognition. The exceptions to this predominant focus concerned the purported association of particular congenital malformations and exposure to alcohol, discussed below.

In an unusual exercise in Finland children, older and younger, were identified whose mothers had been *treated* with alcohol during pregnancy (Halmesmäki and Ylikorkala 1988). The treatment had consisted of administering alcohol in the first or second trimester for the purpose of arresting premature uterine contractions and averting threatened abortion. No harmful outcomes were seen, despite the deleterious effects said to be frequently caused by this procedure (Abel 1981). The women received 30–40 mL of alcohol or even more, orally and sometimes intravenously, 4–5 times a day for up to 22 days, for an average of 38 g/day and a total of 30–3630 g. In addition, after leaving hospital, many reported taking "drinks" at home to arrest contractions.

Periodic examination of the offspring from infancy to 7 years and in some cases even to 14 years of age showed no significantly different frequency of congenital anomalies in the alcohol and control groups. There was no evidence of the typical FAS or intrauterine growth retardation and no statistically significant effect on postnatal growth and psychomotor development, frequency of disease, and complications of childhood. The authors mainly offered two explanations for the lack of fetal damage: the treatment was short in duration, compared to the supposed usual abuse of alcohol, and the women were nutritionally well cared for. Still, taking no chances, they advocated abstinence during pregnancy.

Orofacial Defects

Several retrospective studies surveyed orofacial defects in newborns. Cleft lip, cleft palate, alone or together, are among the commonest of all congenital malformations, with frequencies of 1–2/1000 live births. Their nonsyndromic forms are known to be polygenically inherited, as revealed by various familial characteristics, especially that the incidence among close relatives of affected individuals is many times greater than that in the general population (Curtis et al. 1961).

The drinking practices were examined of mothers of children with cleft lip with or without cleft palate (usually abbreviated CL ± CP) being followed at the plastic surgery service of the Baltimore Children's Hospital (Niebyl et al. 1985). Comparison of historical data of case and control mothers revealed no difference in alcohol consumption during pregnancy.

An ongoing case-control surveillance program based in Boston compared the association with alcohol of malformations of two types, the first at least partly of cranial neural crest origin and the second not of such origin (Werler et al. 1991). The only defect associated with alcohol was CL ± CP, one of the first type, which was increased in children of mothers drinking heavily and frequently (i.e. >5 drinks/day). At the same time there was something unusual about another malformation in the case infants, ventricular septal defects: they were half as frequent as the oral defects (25.5% vs. 42.0%), while in white children generally – the subjects of this surveillance program (Mitchell et al. 1981) – this heart anomaly occurs five or more times as often as oral clefts (5/1000 vs. 1/1000) (Hoffman and Christianson 1978, Bear

1988). This evidence of apparent questionable ascertainment alone made the oral defect finding doubtful.

In a case-control study of Iowa residents mothers of children with orofacial clefts ascertained via multiple sources were questioned by telephone interview about alcohol drinking during pregnancy some years after birth of the children (Munger et al. 1996). Cases included a relatively small number of stillbirths and aborted fetuses while controls included live births only. Over 50% of case and control mothers were nondrinkers, and another third or more drank 1–3 drinks per month. No FAS or signs of the syndrome apparently occurred. The only significant association was between isolated CL ± CP and maternal consumption of more than 10 drinks a month; beyond all other considerations not a convincing relation.

A California population-based study ascertained affected offspring by reviewing medical records at hospitals and genetics centers (Shaw and Lammer 1999). Only heaviest drinking, ≥5 drinks weekly or more frequently, was associated with CL ± CP, isolated or not, as well as with syndromic clefts; but not with any sign of the minor dysmorphic features indicative of the FAS.

These orofacial-defect studies all contained similar disqualifying features: the cleft types were in some instances etiologically dissimilar (Fraser 1980b), impairing biological logic; the number of cases was relatively small, opening the door to spurious association; and the maternal drinking data, uncovered months and sometimes years after the births, were it cannot be doubted tainted by memory faults.

Limb Defects

Several studies examined the association of maternal alcohol drinking and limb defects. A population-based study of one class of such abnormalities, reduction limb defects (RLD), and maternal conditions and diseases was made in Helsinki (Aro 1983). The data were obtained from the Finnish Registry of Congenital Malformations, to which it is compulsory to report all malformations detected in the first year of life. Of the various maternal factors examined only influenza, smoking, and alcohol consumption were statistically associated with the defects. The registry, dividing mothers into alcohol users and nonusers, revealed an association of the former with isolated RLD only; smoking and many other confounding factors may have biased the outcome. The study probably suffered from a usual problem, maternal memory bias, but even more so from severe underreporting of the defects (Saxén 1983).

Part of an epidemiological study of limb defects in British Columbia was devoted to an analysis of alcohol consumption during pregnancy of mothers of affected children (Froster and Baird 1992). Of the 659 cases of various limb defects reported to the provincial health registry in 1952–84 there were ten whose mothers had severe documented maternal alcohol problems, two for whom there were non-alcohol explanations, and of the remainder, two with ulnar defects and six with a terminal transverse defect of the right arm or hand, for whom a maternal alcohol association was considered significant.

No more definite information was available about the drinking of the mothers of the case children than that it was severe. The registry apparently recorded severe instances only and no other degree of alcohol drinking. Thus comparing cases of defects exposed in utero to severe drinking with instances of defects for which there was no information about maternal drinking could prove nothing.

It must also be remembered that reduction defects are one of a large family of limb malformations that have been variably classified. They are known to be etiologically diverse, with many known or suspected modes of causation – genic, chromosomal, uterine factors, maternal infections, and external teratogens – but for which few clear risk factors have yet been identified (Calzolari et al. 1990). These studies did nothing to clarify the matter.

Conditions in Older Children

In a Göteborg study two groups of children of various ages with "fetal alcohol lesions," one whose mothers were discovered to be alcoholic retrospectively and the other of women ascertained prospectively on visiting an alcoholics clinic, were lumped together, making the study worthless (Olegård et al. 1979).

At a health center or teaching hospital in Vancouver children were diagnosed as having the FAS from birth to 18 years of age (Smith et al. 1981). The diagnosis was based on a history of heavy maternal alcohol consumption during pregnancy, poor growth, delayed development, and characteristic facial appearance, but the primary basis of the diagnosis was the maternal state. Almost all the children were North American Indian, at that time and place a guarantee of social and economic deprivation; and possibly of being in a high FAS risk group (Aase 1981). In another study of the FAS in American Indians, again, in addition to the usual growth retardation, etc., the diagnosis "required. . .documentation of alcohol abuse during pregnancy" (May et al. 1983). The obviously biased ascertainment in these studies can contribute little to clarifying questions of the relation of maternal drinking to child health and development.

In a study called retrospective the infants of women receiving inpatient care for alcoholism at the Karolinska Hospital in Stockholm not surprisingly were smaller, lighter, and had reduced head circumference at birth, but none the FAS (Hollstedt et al. 1983).

It is not clear whether in a study in Budapest of children of women who had been registered for treatment of chronic alcoholism, etc. it was the women or the children that were the ascertained group (Vitéz et al. 1984). The children, of various ages, examined in several locations, were divided into those most of whose mothers drank heavily during pregnancy ($\geq$30 mL/day), or whose mothers were otherwise alcoholic but abstained during pregnancy. The children were numerically graded according to 60 morphological, developmental, and neurological traits, and the two groups found to be quantitatively different. Because many of the traits were of the kinds affected prenatally by maternal life style and postnatally by environmental conditions, or were even without relevance to the FAS (e.g. undescended testes, vitamin D resistant rickets, etc.) it is not likely that the elaborate analysis and its outcome had anything to contribute to the question at hand.

Children of extended ages, selected from patient populations of various facilities in the then West Berlin, on the basis of developmental and psychiatric examinations, were diagnosed as having the classic FAS pattern (Steinhausen et al. 1982). It was somehow discovered that their mothers had consumed 140 $\pm$ 72 g of alcohol during the "critical" phase of pregnancy, and it was therefore concluded that their problems were "attributable primarily. . .to maternal alcohol abuse [and that] prenatal morphologic damage is the key determinant of development and psychopathology." At follow-up, 10 years later, even those designated as severely dysmorphic had caught up in weight but less so in head circumference, and while many of the major physical abnormalities had disappeared there was no significant improvement in intelligence (Spohr et al. 1993).

A similar report came from Glasgow (Beattie et al. 1983). With the aid of consultant pediatricans at the Royal Hospital for Sick Children and nearby pediatric units 40 children born in 1971–81 with facial features consistent with the FAS were identified. In addition all had low birthweight, and when examined at 6 weeks to 10 years of age continued to be undersized. Many also had cardiac abnormalities and were severely or moderately mentally retarded. Through the family physicians it became known that the mothers, who were of the lowest socioeconomic group, were heavy drinkers, the reported intake varying from six measures to a bottle of vodka per day, many had psychiatric problems related to alcohol abuse, etc., and many were heavy smokers. Can it be doubted that some large part of the physical and neurological impairment of the children was closely associated with the detrimental pre- and postnatal environmental conditions they had been exposed to. Furthermore, that ascertainment bias may have been widespread in this investigation came through startlingly from the casual statement that "[I]n several cases it was only when the affected child was diagnosed that the maternal alcohol problem became evident."

Children 3–18 years old in a native Indian community in British Columbia were examined and 14 of 116 found to have FAS or FAE (Robinson et al. 1987). The diagnosis was based on a history of maternal alcohol abuse or FAS in a sibling and characteristic signs of FAS/FAE; the abuse, while difficult to establish, apparently being the primary diagnostic factor. That alcohol abuse in such groups, and no doubt all the attendant societal detriments, is an old matter was shown by a study of medical problems in Northern Plains American Indian maternal grandmothers with grandchildren with FAS (Kvigne et al. 2008).

Young children diagnosed as having FAS, according to the usual array of indicators, but especially because of known prenatal exposure to heavy alcohol drinking, had brain hypoplasias, as analyzed by magnetic resonance imaging, while others also known to have been exposed prenatally to heavy maternal drinking (called PEA, prenatal exposure to alcohol) did not differ in these measurements from controls (Archibald et al. 2001).

Critique

Remarks about the retrospective studies can be brief. Principles that should govern true investigations of this type were flouted time and time again. Objective, unbiased, disinterested they were not. Children were selected for study not simply on the basis of possessing presumptive symptoms of the FAS or related traits, with the object of determining whether and to what extent they were associated with maternal consumption of alcohol during pregnancy, but with the latter usually taken as a given, that it was a matter of fact that alcohol consumption was not merely associated with but the cause of the symptoms studied. For the most part, it was as if the investigators had not the slightest understanding of or regard for the process by which association is examined for, not the slightest understanding of or respect for true epidemiological technique. It goes without saying that these studies are all but worthless as indicating the presence or absence of such an association.

Alcohol and Neurodevelopment

Longitudinal Studies

A long-ongoing longitudinal series examined the relation of maternal drinking during pregnancy to mental and behavioral characteristics of offspring through the first 14 years of life and longer. Its purpose was to examine whether lesser degrees of such characteristics were associated with so-called "social drinking." The term was not defined, but it can be construed as referring particularly to consumption of amounts less than those imbibed by explicitly alcoholic women, what was called "moderate" by many other authors. The design of the project was described in detail by Streissguth et al. (1981).

Over 1500 predominantly white middle class women consecutively attending two large Seattle hospitals over a 1-year period were interviewed in the 5th month of pregnancy regarding alcohol use, smoking, drugs, etc. in pregnancy and the month or so prior to pregnancy recognition. A web site described the project as follows:

> This longitudinal prospective non-clinical study, now in its 31st year, continues to evaluate the effects of alcohol and other prenatal factors on offspring development from birth through adulthood. Throughout those 31 years we have been studying one single sample of Seattle-born offspring, first "observed" via maternal interview in utero, who were examined or measured eight times in childhood and four times since then, and are now being interviewed again as they turn about 30.... demonstrating central nervous system effects of prenatal alcohol exposure independent of physical manifestations...

A selection of about 500 of the women, made prior to delivery to increase the proportion of heavier and reduce that of lighter drinkers, still included those with a great diversity of drinking patterns and practices, making it necessary, for the purpose of assessment, to devise a rough scheme ranging, contradictorily, from heavy drinking to abstinence. The specific amounts consumed were not clearly or consistently denoted from article to article in this series of papers. A further difficulty was met in classifying heavy drinkers, especially since women identified as such on one scale were entirely different from those identified on another. It is not surprising therefore that in relating alcohol consumption to pregnancy outcome uncertainties were sometimes encountered.

The study examined behavioral responses to various stimuli in 1- and 2-day-old infants, finding the frequency of aberrant responses related to degree of alcohol exposure (Streissguth et al. 1994a). The authors cautioned however that an apparent risk level sometimes found should not be "interpreted as a biological threshold," despite other behavioral outcomes tested in this series having supported a linear relation (Barr et al. 1990). Similar inconsistent claims were also made for other outcomes.

Some of the symptoms were possibly part of the syndrome of neonatal ethanol withdrawal, as found in alcohol-exposed children with or without the FAS (Pierog et al. 1977, Robe et al. 1981, Coles et al. 1984). This interpretation was held to be incorrect, since the withdrawal syndrome was said to be transient while the said central nervous system changes, on which the altered behavior was based, were persistent (Streissguth et al. 1994a). This was not entirely true, since elements related to the withdrawal syndrome were still present in 30-day-old alcohol-exposed children, and perhaps even in older ones (Coles et al. 1987).

At the 8-month follow-up offspring had mild but significant deficits in mental and motor development, only associated with a threshold of an average of $\geq$4–5 drinks/day in early pregnancy (Streissguth et al. 1980).

At 4 years of age the children's IQ was good, with a mean of 110.5 only 3.3% having scores below 85. A simple correlation indicated a trivial reduction in IQ associated with drinking >1.5 oz/day, when various confounding variables were considered. IQ however was also impaired in association with confounding factor – aspirin use, parental education, etc. (Streissguth et al. 1989).

Intellectual development and learning skills were tested at about 7.5 years (Streissguth et al. 1990). The overall IQ performance again was in general good, with a mean of 107.6, which was consistent with test scores at younger ages. IQ score was decreased minimally in children exposed to an average of $\geq$1 oz/day alcohol at midpregnancy, but again only when conjoined with other factors, in this case lower paternal education and large sibling number.

In a still later evaluation, made at age 14 years, a large battery of tests was administered, and aspects of scholastic and attention and memory performance were found to be associated with prenatal alcohol exposure, this time in a dose-dependent fashion, but especially involving binge drinking (Streissguth et al. 1994b,c).

Similar studies by others have been at odds with the Seattle neurobehavioral findings. Outcomes in Ottawa were inconsistent, but in essence "no unequivocal developmental consequences of prenatal maternal use of [moderate amounts of] alcohol" were found at 12 and 24 months of age (Fried and Watkinson 1988). Similarly, in Dundee no detectable adverse relations were found between mental and physical performance at 18 months and maternal alcohol consumption of about 1.5 oz absolute alcohol daily; with the remarkable additional finding that when confounding factors were allowed for, alcohol consumption was related to better performance (Forrest et al. 1991). Enhanced rather than lower Bayley scores were also noted in 18-month old children of women drinking 120 g/week or more, in a European Concerted Action summary report (Kaminski 1992). In a California study, while mental development at 1 year was related to prepregnancy drinking, it was not significantly related to drinking during pregnancy (O'Connor et al. 1986). But deficits in several aspects of intellectual functioning were seen in children about 6 years old of predominatly black women of low socioeconomic status who smoked heavily and drank a mean of 11.8 oz/week (Coles et al. 1991).

An examination of trends in alcohol use noted a substantial decrease in recent years in alcohol use during pregnancy in all demographic categories, but with the usual reluctance to admit any improvement in the dimension of the matter (Grant et al. 2009).

A recent overview (Anon 2009) of the question of the neurobehavioral consequences in childhood of moderate maternal alcohol drinking during pregnancy, as perhaps might have been expected of a governmental agency, was noncommital in concluding that –

> from a child perspective, it is safest if the mother abstains from alcohol consumption during pregnancy or when planning a pregnancy. As this literature review demonstrates, even small to moderate doses of alcohol intake might exert a negative affect on children.

This timidity was belied, however, by the discussion noting that:

> The effects found were relatively small. The effects of prenatal alcohol exposure seem to be more difficult to establish when the children are older. One explanation might be that. . .the effects of different kinds of exposure generally decline with time. The causes of observed behavioural deficits in children are often difficult to establish. In addition to prenatal alcohol exposure, their behaviour has other determinants. . .many studies fail to control for social factors that, in addition to prenatal alcohol exposure, cause postnatal effects on children.

The first decade of the new century has seen several contributions to the medical literature on various aspects of the FAS, particularly the effects on neurodevelopment, mostly however being overviews of the work of others. A general review of the effects on offspring of maternal drinking during pregnancy concerned what were called "underdiagnosed children," i.e. those lacking facial defects and growth deficiency, concluded uncritically that many neurobehavioral deficits were so linked (Anon 2000d). A "systematic review" discerned no evidence of adverse effects of prenatal binge drinking on many fetal characteristics, "except possibly on neurodevelopmental outcomes" (Henderson et al. 2007), exactly the aspect calling for independent judgment. Another noted the detrimental, perhaps additive, effects of prenatal and postnatal trauma (Henry et al. 2007). A lukewarm and mildly skeptical view (Holmgren 2009) engendered several rejoinders (Gray et al. 2009). Other such contributions, yielding no more critical insight into the question, need not be noted individually.

Finally, despite the vagueness of the diagnosis, the fetal alcohol disorders, having been taken for granted, led to programs for their intervention (Peadon et al.

2009). Intervention whose efficacy was demonstrated by their adverse outcomes being avoided by "being reared in good stable environments" (Streissguth et al. 2004); a twofold admission of the symptoms being of postnatal origin and being transitory.

Alcohol and Spontaneous Abortion

Drinking during pregnancy has been held responsible for causing abortion, an accusation necessitating a short detour into this fascinating subject. Spontaneous abortion (SAB) is defined as death of the conceptus before reaching viability, the time independent life becomes possible. Not too long ago this was thought to begin at 28 weeks of pregnancy after the first day of the last menstrual period (Hook and Porter 1980). Over time medical technology succeeded in keeping many young fetuses alive, so that the age of viability had to be redefined; till at present it is accepted that SAB is death of conceptuses before 20 weeks of pregnancy.

SAB, usually recognized only after there has been a missed menstrual period (Boué et al. 1976), is a common happening, its frequency being about 12–15%, the great majority in the first trimester. This however does not include the loss that occurs before pregnancy recognition. Assays of human beta-chorionic gonadotrophin, used to detect embryonic presence, have estimated the loss between implantation and pregnancy recognition to be about 60% (Regan and Rai 2000), and earlier an even larger loss, of 78% of fertilized eggs, the vast majority before clinical diagnosis of pregnancy (Roberts and Lowe 1975).

To discover the full extent of SAB pregnancies must therefore be monitored from as early as possible after conception. This goal has been poorly met in efforts to determine whether maternal alcohol consumption is related to SAB. It must first be understood that it was not excessive drinking and chronic alcoholism, but "moderate" or "social" drinking whose association with SAB was examined, as seen by the following studies.

Two early studies had confused findings. In a study at the Kaiser Foundation Health Plan of Northern California women were asked about alcohol use and other matters at their first antenatal visit, just over half said they were nondrinkers, most of the others that they drank <1 drink/day, and a small minority greater amounts. (Harlap and Shiono 1980).

In the large number of women there were only 4.7% SAB, meaning that the pregnancies were on average far advanced at the time they were first seen. The findings were inconsistent: drinking as much as >3 drinks/day was not associated with first-trimester SAB, while ≥1 drinks/day was associated with second-trimester SAB. It is not surprising that the authors, expressing caution, noted that the study had not been designed for SAB research.

In a retrospective case-control study in three Manhattan hospitals women with first-trimester SAB were compared with those who registered before 22 weeks for prenatal care and delivered at term (Kline et al. 1980a). As regards reported drinking, the women differed only with respect to drinking even as seldom as ≥2 drinks/week, namely 17% of cases and 8.1% of controls; moderate drinking thus apparently found to be a risk factor.

Later studies were no less ambiguous. A study in Leamington Spa found no relation between maternal alcohol consumption and SAB (Davis et al. 1982). But the SAB rate was a meager 2.3%, not surprising since only 15.7% of the women were surveyed in the first trimester; hence the study was inadequate to answer the question.

In a London study the frequency of SAB was unrelated to whether women did not drink or drank occasionally or every day, but again the SAB rate was a low 4.0%, probably because relatively few of the subjects were first seen early in pregnancy, and hence this effort was also inadequate to the task (Grisso et al. 1984).

The mean amount of alcohol consumed by a group of Finnish women who aborted was about one drink a week and by a matched control group half a drink a week. Moderate maternal alcohol use clearly was not associated with risk of abortion (Halmesmäki et al. 1989).

In a prospective study in Australia the total amount of alcohol consumed by women who aborted was not different than by those with completed pregnancies; the former however consumed significantly higher quantities of beer, which may have been related to maternal characteristics or variables such as smoking (Walpole et al. 1989).

A telephone interview study in Santa Clara County, California compared women spontaneously aborting and controls, ascertained respectively from hospital pathology reports and birth certificates (Windham et al. 1992). Comparing the responses about periconceptional alcohol consumption indicated that the SAB odds ratio was doubled in the 4.3% of the case women who drank an average of ≥1 drinks/day. The SAB rate in both groups was not disclosed. Two comments: The interviews took place many months after pregnancy was terminated, perhaps affecting alcohol consumption recall. The SAB were those for which there was a pathology specimen, implying that alcohol association was examined for only a sample of abortions.

In a prospective California Kaiser Permanente study alcohol consumption practices were discovered through telephone interviews conducted soon after the first prenatal appointment and pregnancy outcomes ascertained through hospital records (Windham et al. 1997). The 9.7% SAB rate was less than usual for SAB discovered early, indicating underascertainment. The rate was increased only in the 0.97% of the women who drank about ≥4 drinks/week, i.e. on average half a drink per day, a finding difficult to accept.

A study in Milan of first-trimester SAB and randomly selected controls found no evidence of an association between moderate alcohol consumption (≥1 drink/day) and the risk of SAB (Parazzini et al. 1994). Nor did a prospective study in the nearby city of Turin find that a larger but still low level (≥2 drinks/day) of alcohol consumption during pregnancy appear to be a significant risk for SAB (Cavallo et al. 1995).

A Danish study, defining SAB as fetal death before 28 completed weeks of pregnancy, despite this extended period, found the low SAB rate of 1.3% in women seen antenatally, and only in the 2.3% of women drinking ≥5 drinks/week an increased risk of first trimester SAB (Kesmodel et al. 2002); again fraught with inadeqacies.

Parenthetically, erroneously cited studies may mislead the casual reader. Thus, a study purporting to contribute to the matter at hand turned out to be without relevance, since what was examined was the relation of maternal drinking to SAB plus stillbirths, an incongrous combination (Anokute 1986); but even so it was erroneously cited, perhaps because of its misleading title (Abel 1997).

Confusing the unwary also was a study that combined SAB and stillbirth, but which redeemed itself by finding that this combination was not consistently related to light, moderate, and heavy drinking (Wilsnack et al. 1984). Finally, another study had to be read beyond its title and abstract to learn that it was actually concerned with SAB in prior pregnancies, not in current ones (Armstrong et al. 1992).

An overview of many articles noticed the beguiling fact that while most studies in the US and Canada found increases in SAB in mothers drinking moderate quantities of alcohol, most non-American ones did not; the explanation for which, not easy to interpret, was the socioeconomically greater homogeneity in the latter countries (Abel 1997).

The chromosome constitution of abortuses has rarely been considered in such studies. Many SAB being chromosomally aberrant (Carr 1963, 1983, Warburton et al. 1980) made it necessary to examine the relation of such abnormalities to maternal drinking. However a study in New York City hospitals only found an association with normal chromosomes, and only in women drinking ≥2 times/week, a tenuous finding indeed (Kline et al. 1980b).

Summary and Critique

Studies of the relation of moderate alcohol drinking and SAB may be judged by applying criteria for causal inference, the main ones being strength and consistency of association (Stein et al. 1984).

Strength is measured by relative risk, odds ratio, regression coefficient, and the like. In the above examples with positive findings prospective studies found risk ratios indicating that alcohol is a weak abortigen. Another aspect of the criterion of strength is proportionate agreement between dose and response. On this score there was little support for the alleged relation, since there was no evidence of constant increase in risk as alcohol quantity rose. Finally, as regards consistency – i.e. replicability – this too failed to uphold the proposition, since, not disregarding the varying validity of the studies, there were as many negative as positive findings.

Regarding plausibility, the rather small amount of alcohol imputed to cause early pregnancy loss defies credulity, especially since no study noted above linked excessive use of alcohol and SAB. Further challenging the likelihood that moderate alcohol intake can have this consequence is the failure to do so by actors like rubella, insulin-dependent diabetes, dioxin, massive doses of vitamin A, antiepileptic drugs, and others discussed throughout this work.

In the face of all this contrariety, it is strange that only Robertson et al. (1998) have pointed to the biological implausibility of the relation.

Prevalence of the FAS

It is ironic that 40 years after the first US article on the FAS fundamental matters regarding it, such as its prevalence, should still need to be clarified. A 2009 CDC statement noted that its reported prevalence rate varies widely depending on the population studied and the surveillance methods used, wide estimates that engender doubt as to the reality of the phenomenon or at least to its definition.

A statewide registry in New York noted a fairly low rate in 1995–8, 0.28/1000 children or 0.37 if all diagnoses under 2 years of age were included (Fox and Druschel 2003). Which continued to be low, even with a difference between two demographically similar counties with similar rates of binge drinking, 0.90/1000 births and 0.21 respectively, possibly explained by reporting age limits and differential clinician experience with the syndrome; but the persistent low estimates could not be denied (Druschel and Fox 2007). Another region found even lower rates, at most 0.01–0.03/1000 live births, in Victoria in Australia in 1995–2002 (Allen et al. 2007). For other low estimates see below. It is to be remembered, as explained above, that the diagnosis is difficult in neonates.

A publication of the National Institute of Alcohol Abuse and Alcoholism, going to the other extreme, says the literature of the 1980s–1990s points to 0.5–2.0/1000, which including alcohol related birth defects brings it to 10/1000 or 1% of all births (May and Gossage 2001). In Cleveland, in a survey of a large sample, the frequency at first was 0.08/1000 live births, and later 3/1000 (Sokol et al. 1986). The one possible explanation of the larger figure, though not explicitly given by the authors, was that it rested on a larger proportion of black subjects: "...black race may be a susceptibility factor..."

An even more extraordianary estimate made in nonwinegrowing communities in the Western Cape of South Africa arrived at one of over 40/1000 5–9 year olds, a society with innumerable medical and social problems, the high rate no doubt also reflected differences in local drinking patterns, alcohol availability, poverty, unemployment, health problems, and other risk factors (Viljoen et al. 2003). But also without doubt raises doubt of the definitions employed.

And in between were many attempted estimates. In Seattle in 1975 FAS was estimated as at least as 2.8/1000 (Sampson et al. 1997), etc., as outlined above and also below.

The continuing difficulty with accepting the reality of the FAS concept was tellingly voiced by a 1997 Report of the Subcommittee of the National Advisory Council on Alcohol Abuse and Alcoholism. It did so by listing the research that was still needed to accomplish this purpose. See www.csrincorporated.com/NIAAA-demo/extramural/grants/ FASfinal.htm. It included the following.

There is a need to develop a quantitative, objective method of classifying the entire spectrum of possible consequences of alcohol consumption during pregnancy.

Emphasis should be placed on characteristics that are measurable prenatally or at birth in order to increase the likelihood of early diagnosis and thereby enhance the possibility of early intervention. It is acknowledged that the ability to achieve this goal depends on the accuracy with which one can define/describe characteristics that are specific enough to be useful for diagnosis at these stages.

Many of the facial anomalies change with age. Racial and gender-specific norms need to be developed.

Accurate estimates of the prevalence of FAS and of partial presentations of FAS are needed. These estimates should include older children as well as adults, especially in populations where increased representation might be expected. Utilization of consensus criteria are necessary, including separate characterization of growth characteristics, facial anomalies, brain functioning, and extent of alcohol exposure.

In essence, the last desideratum cannot be achieved without successful accomplishment of the previous ones. To estimate how often the FAS occurs, whether in populations overall or in certain segments of the population, it is obvious that what the entity consists of must be established and standardized. But "a major difficulty in ascertainment. . .is that the facial features associated with the syndrome are not easily recognized and that one of the three salient features of FAS is central nervous system dysfunction, including mental retardation, which may not be identified until several years after birth" (Abel and Sokol 1987).

Several things must be cleared away. Prevalence at birth cannot be combined with findings of prevalence at later ages, because the signs of the syndrome in infants may not be objectively discernable or specific to the syndrome, while others may not persist to or be recognizable only at older ages. Prevalence can refer to frequency in children of alcohol-abusing women or to frequency in the overall population. Which of these is the more relevant depends on point of view. Obviously for either the meaningfulness of the finding depends on the definition of alcoholism or alcohol abusiveness.

Early studies, impaired by biased ascertainment and confused protocols, made unreliable estimates (Jones et al. 1974, Olegård et al. 1979). Prospective surveys made better stabs at it. The following are a selection of these. A Seattle study, with a comparatively limited number of births, based on disputed criteria (see Hanson et al. 1978 above), found an incidence of about 1.4/1000, or using different assumptions, 2.8/1000 live births (Streissguth et al. 1981, Sampson et al. 1997). Finally the results of two other prospective studies may be cited, 1.7/1000, determined by Halliday et al. (1982), and 1.2/1000, by Little et al. (1990).

Prevalence data were also derived from averages of aggregates of large numbers of studies. An early such report, of 19 prospective and retrospective studies combined, contained 164 cases in populations totaling 88,326 live births, giving an overall estimate of 1.9/1000, much higher however from prospective than retrospective ones (Abel and Sokol 1987). A later calculation, based solely on prospective studies, brought the estimate down to 0.33/1000, sharply lowered because of possible previous inclusion of false positives (Abel and Sokol 1991), and last a partial reversal to 0.97/1000 (Abel 1995).

But findings in different parts of the world have varied enormously – in the US the rate was 1.95, in other countries 0.08 – thus an overall average is deceptive. In fact the average of frequencies estimated by individual surveys rather than the average of the whole was much lower, considerably less than 0.5/1000, and the mode or median far less still (Abel 1995).

Why is this so? Perhaps because the diagnostic features of the syndrome are less clear in some ethnic or racial groups than others; or because of variable genetic propensities to alcohol teratogenesis; or because of population differences in drinking practices. As was seen above, the proportion of women in different studies who were alcohol abusers differed significantly, those more prone to abuse being those in straitened social and economic circumstances.

Examples were cited by Abel (1995). Those with low compared to middle/high socioeconomic status in US, 2.29 vs. 0.26/1000; locations with mothers mostly Indian and black (i.e. of low such status) compared with those primarily white, 0.48 vs. 0.29; British Columbia, native population 4.7, the rest 0.4. Similarly, the prevalence in Alaska natives in 1980–9 was 5.2/1000 live births and in non-Alaska natives 0.3 (Egeland et al. 1998).

Such wide differences between social strata prompted the question, as well as a host of others, of whether alcoholism has similar FAS consequences regardless of class (Bingol et al. 1987). To answer it, women were selected from the substance abuse units of an inner city and a suburban hospital, both groups proven chronic alcoholics, the first on public assistance and the second of upper middle/upper socioeconomic status, the former all black and Hispanic, the latter all Caucasian of north European origin. The amount of pure alcohol intake was equivalent in the two groups, though the first preferred beer and the second vodka.

The newborn infants proved to be very different in outcome: 45.1% of babies of the first group and 6.4% of those of the second were small for gestational age; 52% and 0.9% respectively had reduced head circumference; the general malformation rate (undoubtedly including minor defects, but unfortunately not enumerated) was 45.9% and 9.2% (in nonalcoholic controls it was 5%); full manifestations of the FAS (also unfortunately not specified) occurred in 40.5% in the first and only in a single instance in the second.

In older children the attention deficit disorder was present in 71.8% of group 1 children, 21.1% in group 2, compared to 5–10% in the general population; 39% of group 1 children had borderline intelligence or were mildly retarded, only one in group 2, who was institutionalized, was diagnosed as having mental retardation.

The Fundamental Problem

What is the reason for these profound differences? Bingol et al. (1987) listed many possibilities. The two groups of women, while alike in consuming alcohol excessively, differed in other respects important for fetal well being. Upper class women ate regularly and more balanced meals, supplemented with vitamins and minerals; lower class women ate irregularly, with little regard for nutritional value. Morbidity was increased in both, but cirrhosis was doubled in lower class women, which may reflect nutritional deficiency; the cumulative effects of chronic maternal alcoholism in successive generations (many alcoholic women had alcoholic mothers) may have fetal consequences (intergenerational effects, as postulated for general prenatal maldevelopment by Sever and Emanuel 1981 and Emanuel 1993).

If the principal determinant of the FAS – if not its sine qua non – is poverty and deprivation, as the lengthy enumeration by Abel (1995) of the contrast between the social classes gives great substance to, what might it mean so far as, if not the reality, at least the meaning of the FAS concept is concerned? One asks in the face of the theoretical preventability of this disease, not how are the women at risk to be identified – identifying them one by one would seem futile – but what is to be done for them generally, how is society to be reordered, for the 'disease' to be caused to vanish?

As Abel (2009) commented, "[N]umerous studies have noted that, even though they may supposedly drink the same, some women give birth to children with 'fetal alcohol spectrum disorder' and some do not. . .it affects mainly destitute women and is nearly always found among women who smoke. . . It is appropriate to note here a statement of a Teratology Society committee regarding fetal alcohol programs, which concurred in regarding social and economic deprivation as "underlying. . .bases of problem drinking. . ." (Adams et al. 2002).

Animal Alcohol Studies

A word should be said about what animal studies may contribute to amplify upon and clarify this relation. It must first be mentioned that the standard animals used in experimental studies – mice, rats, etc. – present no phenomenon comparable to abortion. Dead conceptuses are not expelled prior to term, but are retained in utero for their remains to be discovered later. Nonhuman primate species are an exception, abortuses being expelled before the usual term.

An extensive review of animal experiments on alcohol teratology (Blakley 1988) noted increased prenatal mortality in most species after prenatal alcohol exposure. Only a few studies, however, have been made with monkeys, considered most apt to reflect human handling of alcohol. In *Macaca fascicularis* the abortion rate was significantly increased by alcohol at 4–5 mg/kg/day (Scott and Fradkin 1984); and in *M. nemestrina* by about 2–4 mg/kg weekly (Clarren et al. 1987a), with treatment beginning in early gestation in both studies. As was commented, "these studies support the conclusion that very high blood levels [with a threshold of ~200 mg/dL]. . .are capable of producing spontaneous abortions" (Abel 1997). To achieve such blood levels – those expected in chronic alcoholics – Abel continued, "a 120 lb women would have to drink about eight drinks over a 3-h period." Any lesser drinking, in other words, would be unlikely to affect the rate of SAB.

This outlook was seconded by data from studies in Cleveland suggesting that the risk of SAB was perhaps increased only in the 2–4% of women who were heaviest drinkers, those at the extreme right end of the distribution of alcohol consumers

(Sokol 1980); and by an early judgment, regarding the wider question, that the full effects of maternal drinking occur only in the offspring of chronic alcoholic mothers (Fabro and Brown 1982).

Finale

Nevertheless the perceived developmental hazards of drinking, in the words of scholars of maternal alcoholism, became in the US a "social problem [that] escalated beyond the level warranted by the existing evidence. . .took on the status of a moral panic" (Armstrong and Abel 2000). A 'panic' intensified by exaggerated warnings on alcohol beverage labels, legislation found only in the US, and by irresponsible television portrayals during many evening news broadcasts linking alcohol and substance abuse during pregnancy generally (Golden 2000).

The puritanical streak in America seemingly was not expunged with the repeal of Amendment XVIII to the US Constitution. Regardless of what may have been its underlying motive legislation since 1985 has led to several hundred women being charged with fetal abuse and prosecuted for using illegal drugs or alcohol during pregnancy (Marshall 1999).

How can the years of work and the multitude of studies of the effects of maternal alcohol drinking on embryos, fetuses, and children up the present day in 2010 be judged? Are they right, are they wrong in the allegations, accusations they pose? Perhaps, as US Supreme Court Justice Breyer (2000) reminded his audience, at an address to a Whitehead Policy Symposium, of the reply of Wolfgang Pauli, when he was asked whether a certain scientific paper was wrong, and he said, "Certainly not. That paper is not good enough to be wrong."

B. Franklin on Wine

Benjamin Franklin in a letter to the Abbé Andre Morellet in about 1779 wrote "He made wine to gladden the heart of man," and – with unwitting teratological overtones – in a P.S.: "To confirm still more your piety and gratitude to Divine Providence, reflect upon the situation which it has given to the elbow. You see in animals who are intended to drink the waters that flow upon the earth, that if they have long legs, they also have a long neck, so that they can get at their drink without kneeling down. But man, who was destined to drink wine, must be able to raise the glass to his mouth. If the elbow had been placed nearer the hand, the part in advance would have been too short to bring the glass up to the mouth; and if it had been placed nearer the shoulder, that part would have been so long that it would have carried the wine far beyond the mouth. But by the actual situation, we are enabled to drink at our ease, the glass going exactly to our mouth. Let us, then, with glass in hand, adore this benevolent wisdom; – let us adore and drink!"

But I add, in moderation, and to your health!

The Accomplishment and the Expectation

What do we know and what can we hope to learn further about the environmental causes of congenital malformations in people? These pages have shown that the causes of major malformations of external origin that have been discovered and are now known are relatively few. And it may perhaps be that further discovery of environmental causes of congenital malformations with major significance has reached a plateau, perhaps a breathing spell. Should this be encouraging?

The human environmental teratogens, all discovered in the twentieth century, fall into an equally small number of categories. First there are the ones that have existed from as far back as humans have existed, and have undoubtedly been causing fetal damage during all this time: iodine deficiency, rubella and a handful of other infectious agents, and phenyketonuria – whose embryonic harmfulness was displayed when modern circumstances allowed its recognition. About the teratogenicity of another venerable condition, diabetes mellitus, there is in the author's opinion much doubt. A small number of other states, maternal diseases no doubt also of ancient status, fetal-virilizing tumors, dysthyroidisms, are potentially harmful, but merely for a vanishingly small percentage of babies. None other of this category is known.

Two remaining indubitable teratogens are in categories by themselves, the first one discovered, irradiation, medical and military, and the other, organic mercury, an environmental contaminant. All the others whose teratogenicity is incontrovertible are pharmaceutical chemicals, introduced, obviously, in the just past century, and just as obviously withdrawn or controlled after their danger was recognized – thalidomide, retinoids, certain antiepileptics, and folic acid antagonists.

You may ask, what about the others named in the pages above – excess vitamin A, Bendectin, blighted potatoes, hyperthermia, folic acid deficiency, diethylstilbestrol, female sex hormones, dioxin and other environmental contaminants, lithium, many antiepileptics, alcohol. The answer is that much study has found no consistent evidence that any of them causes major congenital malformations.

Lists of drugs that were once suspected of causing congenital malformations in human beings, but no longer considered teratogenic in customary use, and of environmental substances once suspected of causing such manifestations, but which have also been exonerated, were presented in an article almost 30 years old (Kalter and Warkany 1983). Little meriting credence, again in the author's opinion, has been added to them in the intervening period.

H. Kalter, *Teratology in the Twentieth Century Plus Ten*,
DOI 10.1007/978-90-481-8820-8_16, © Springer Science+Business Media B.V. 2010

Others are of a different opinion. In the 10th edition of the compendious Catalog of Teratogenic Agents there is included a prefatory section entitled "human teratogens: proven, possible, and unlikely," in which are listed a greater number as ostensibly proved than critical appraisal would accept, e.g. diethylstilbestrol, lithium, cigarette smoking (Shepard 2001).

But that is neither here nor there. Such differences will always exist in a subject where definition remains fluid. What is important is that the known environmental teratogens, according to whatever reckoning, are few in number. Of the roughly 3% of seriously malformed newborn children the causation of the great majority, whether exogenous, endogenous, or interactional, is still hardly known, after a 100 years of study and thought.

Will the proved column be lengthened in the future by discovery of congenital malformations due to aberrant genes, acting alone or in concert with others or with nongenetic factors? Probably. It is a given even now that environmental teratogens rarely act alone, but are almost invariably modified in their action by subsidiary genetic factors; whose presence today however is mostly undefined and whose properties are largely unidentified. A great deal more remains to be learned about these modifying genes; knowledge of which will surely promote preventive measures.

A good number of "developmental gene mutations as teratogenic agents," as Shepard labels them, are already listed in his Catalog. The question is, by the close of the present century how much of the large majority of malformations now without clear causation will have been revealed to have a mono- or polygenic genetic basis? And if this is a substantial one, the large question then will need to be addressed, of what to do about them. Will it be possible to short circuit them, so to speak, the way environmental teratogens have been dealt with?

But if at that time there still are some unknowns, the possibility that prenatal life, like the rest of the universe, is subject to accidents, meteors and the like, unforeseen and unpreventable, though repugnant to minds that refuse to accept unknowability, will have to be faced, along with unfathomabilities of the universe.

But this, it is to be hoped is a remote possibility for the most part; which is, for such as I am, an optimistic note on which to end this book.

References

Aase J. 1981. The fetal alcohol syndrome in American Indians: a high risk group. Neurobehav Toxicol Teratol 3:153–6.

Aase J, Jones KL, Clarren SK. 1995. Do we need the term "FAE"? Pediatrics 95:428–30.

Abate L, Basso P, Belloni A, et al. 1980. Mortality and birth defects from 1976 to 1979 in the population living in the TCDD polluted area of Seveso. In: Hutzinger O, Frei RW, Merian E, et al. eds. Chlorinated dioxins and related compounds: impact on the environment. Oxford, Pergamon, pp 571–87.

Abel EL. 1981. A critical evaluation of the obstetric use of alcohol in preterm labor. Drug Alcohol Depend 7:367–78.

Abel EL. 1995. An update on incidence of FAS: FAS is not an equal opportunity birth defect. Neurotoxicol Teratol 17:437–43.

Abel EL. 1997. Maternal alcohol consumption and spontaneous abortion. Alcohol Alcohol 32: 211–19.

Abel EL. 1998. Pevention of alcohol abuse-related birth defects. Alcohol Alcohol 33:346–52.

Abel EL. 2009. Fetal alcohol syndrome: same old, same old. Addiction 104:1474.

Abel EL, Sokol R. 1983. Fetal alcohol syndrome: how good is the criticism? Neurobehav Toxicol Teratol 5:491–2.

Abel EL, Sokol R. 1986. Fetal alcohol syndrome is now leading cause of mental retardation. Lancet 2:1222.

Abel EL, Sokol R. 1987. Incidence of fetal alcohol syndrome and economic impact of FAS-related anomalies. Drug Alcohol Depend 19:51–70.

Abel EL, Sokol R. 1991. A revised conservative estimate of the incidence of FAS and its economic impact. Alcohol Clin Exp Res 15:514–24.

Abramsky L, Botting B, Chapple J, et al. 1999. Has advice on periconceptional folate supplmentation reduced neural-tube defects? Lancet 35:998–9.

Abrishamchian AR, Khoury MJ, Calle EE. 1994. The contribution of maternal epilepsy and its treatment to the etiology of oral clefts: a population based case-control study. Genet Epidemiol 11:343–51.

Adams J, Bittner P, Buttar HS, et al. 2002. Statement of the Public Affairs Committee of the Teratology Society on the fetal alcohol syndrome. Teratology 66:344–7.

Agnish ND, Vane FM, Rusin G, et al. 1990. Teratogenicity of etretinate during early pregnancy in the rat and its correlation with maternal plasma concentrations of the drug. Teratology 42:25–33.

Alembik Y, Dott B, Roth MP, et al. 1997. Prevalence of neural tube defects in northeastern France, 1979–1994: impact of prenatal diagnosis. Ann Genet 40:69–71.

Alford CA, Pass RF, Stagno S. 1983. Chronic congenital infections: common environmental causes for severe and subtle birth defects. Birth Defects Orig Artic Ser 19:87–96.

Alkalay AL, Pomerance JJ, Rimoin DL. 1987. Fetal varicella syndrome. J Pediatr 111:320–3.

Allen JR, Marlar RJ, Chesney CF, et al. 1977. Teratogenicity studies on late blighted potatoes in nonhuman primates (*Macaca mulatta* and *Saguinus labiatus*). Teratology 15:17–24.

H. Kalter, *Teratology in the Twentieth Century Plus Ten*,

DOI 10.1007/978-90-481-8820-8,

Allen K, Riley M, Goldfeld S, et al. 2007. Estimating the prevalence of fetal alcohol syndrome in Victoria using routinely collected administrative data. Aust N Z J Public Health 31:62–6.
Allen WP, Stevenson RE, Thompson SJ, et al. 1996. The impact of prenatal diagnosis on NTD surveillance. Prenat Diag 6:531–5.
Alpert JL, Day N, Dooling E, et al. 1981. Maternal alcohol consumption and newborn assessment: methodology of the Boston City Hospital prospective study. Neurobehav Toxicol Teratol 3:195–201.
Altemus LA, Ferguson AD. 1965. Comparative incidence of birth defects in negro and white children. Pediatrics 36:56–61.
Alwan S, Reefhuis J, Rasmussen SA, et al. 2007. Use of selective serotonin-reuptake inhibitors in pregnancy and the risk of birth defects. N Engl J Med 356:2732–3.
Amin-Zaki L, Majeed MA, Elhassani SB, et al. 1979. Prenatal methylmercury poisoning: clinical observations over five years. Am J Dis Child 133:172–7.
Ances BM. 2002. New concerns about thalidomide. Obstet Gynecol 99:125–8.
Andermann E, Dansky L, Andermann F, et al. 1982. Minor congenital malformations and dermatoglyphic alterations in the offspring of epileptic women: a clinical investigation of the teratogenic effects of antiepileptic medication. In: Janz D ed. Epilepsy, pregnancy and the child. New York, Raven Press, pp 235–49.
Andersen DH. 1938. Cystic fibrosis of the pancreas and its relation to celiac disease: a clinical and pathological study. Am J Dis Child 56:344–99.
Andersen DH. 1941. Incidence of congenital diaphragmatic hernia in the young of rats bred on a diet deficient in vitamin A. Am J Dis Child 62:888–9.
Andersen DH. 1949. Effect of diet during pregnancy upon the incidence of congenital diaphragmatic hernia in the rat: failure to produce cystic fibrosis of the pancreas by maternal vitamin A deficiency. Am J Path 25:163–85.
Andersen DH. 1958. Cystic fibrosis of the pancreas. J Chron Dis 7:58–90.
Anderson H, Barr B, Wedenberg E. 1970. Genetic disposition – a prerequisite for maternal rubella deafness. Arch Otolaryngol 91:141–8.
Anderson SG. 1950. Epidemiological aspects of rubella. Med J Aust 2:389–90.
Anderson WJ, Baird D, Thomson AM. 1958. Epidemiology of stillbirths and infant deaths due to congenital malformations. Lancet 1:1304–6.
Annegers JF, Baumgartner KB, Hauser WA, et al. 1988. Epilepsy, antiepileptic drugs, and the risk of spontaneous abortion. Epilepsia 29:451–8.
Annegers JF, Elveback LR, Hauser WA, et al. 1974. Do antiepileptics have a teratogenic effect? Arch Neurol 31:364–73.
Annegers JF, Hauser WA, Elveback LR, et al. 1978. Congenital malformations and seizure disorders in the offspring of parents with epilepsy. Int J Epidemiol 7:241–7.
Anokute CC. 1986. Epidemiology of spontaneous abortions: the effects of alcohol consumption and cigarette smoking. J Nat Med Ass 78:771–5.
Anon. 1934. Vitamin A and normal eye development. In: Abt IA ed. Year book of pediatrics. Chicago, IL, Year Book Publishers, pp 72–4.
Anon. 1941. Congenital cataract following German measles in the mother. Med J Aust 2:651–2.
Anon. 1959. Congenital malformations: a conference on teratology. Pediatrics 23:193–251.
Anon. 1962. Report of the commission on drug safety. Washington, DC, Federation of American Societies for Experimental Biology.
Anon. 1963a. Funk and Wagnalls standard college dictionary. New York, Funk and Wagnalls.
Anon. 1963b. Conference on prenatal effects of drugs. Chicago, IL, Commission on Drug Safety.
Anon. 1964. Deformities caused by thalidomide. Reports on public health and medical subjects. No 112. London, Her Majesty's Stationery Office, pp 1–70.
Anon. 1965. Infant and perinatal mortality in the United States. Washington, DC, Nat Cent Health Stat, ser 3. no 4.
Anon. 1967. International comparison of perinatal and infant mortality: the United States and six West European countries. Washington, DC, Nat Cent Health Stat, ser 3, no 6.

Anon. 1969. The congenital malformations. In: Butler NR, Alberman ED eds. Perinatal problems: the second report of the 1958 British Perinatal Mortality Survey. Edinburgh, Livingstone, pp 283–320.
Anon. 1971. The compact edition of the Oxford English dictionary. Oxford, Oxford University Press.
Anon. 1973. The testing of chemicals for carcinogenicity, mutagenicity, and teratogenicity. Ottawa, ON, Health and Welfare Canada.
Anon. 1975. End of the potato avoidance hypothesis. BMJ 2:308–9.
Anon. 1977. Congenital deformities associated with Bendectin. CMAJ 117:721–2.
Anon. 1981. Fred Hale. J Anim Sci 51:512–3.
Anon. 1983a. Cytogenetic patterns in persons living near Love Canal – New York. MMWR Morb Mortal Wkly Rep 32:61–2.
Anon. 1983b. Alcohol and the fetus – is zero the only option? Lancet 1:682–3.
Anon. 1983c. Alcohol consumption in pregnancy. Arch Dis Child 58:474–6.
Anon. 1983d. Fetal effects of maternal alcohol use. JAMA 249:2517–21.
Anon. 1985. Vitamin A and teratogenesis. Lancet 1:319–20.
Anon. 1988. T. McKeown. BMJ 298:129.
Anon. 1989. Rubella and congenital rubella syndrome – United States, 1985–1988. MMWR 38:173–4.
Anon. 1990. Periconceptional use of multivitamins and the prevalence of neural-tube defects. N Engl J Med 322:1082–4.
Anon. 1991. Teratogenesis with carbamazepine. Lancet 337:316–7.
Anon. 1993. Congenital malformations surveillance. Teratology 48:545–709.
Anon. 1994. Stratification by size in the epidemiology of ventricular septal defects. EUROCAT 8(1):1–2.
Anon. 1995. Neural tube defects in the British Isles EUROCAT centres, 1980–1992. EUROCAT 9(2):1–2.
Anon. 1999. Wanted: a new order in protein nomenclature. Lancet 401:411.
Anon. 2000a. Neural tube defect surveillance and folic acid intervention – Texas-Mexico border, 1993–1998. MMWR Morb Mortal Wkly Rep 49:1–4.
Anon. 2000b. Accutane-exposed pregnancies – California, 1999. MMWR Morb Mortal Wkly Rep 49:28–31.
Anon. 2000c. Folate status in women of chlldbearing age – United States, 1999. MMWR Morb Mortal Wkly Rep 49:962–5.
Anon. 2000d. Prenatal exposure to alcohol. Alcohol Res Health 24:32–41.
Anon. 2002. Trends in spina bifida and anencephalus in the United States, 1991–2001. http://www.cdc.gov/nchs/products/pubs/pubd/ hestats/spine_anen.htm.
Anon. 2007. Folic acid effective in recurrence of NTDs. NTD Newsletter. Greenwood Genetic Center. No. 22. February 2007.
Anon. 2009. Racial/ethnic differences in the birth prevalence of spina bifida – United States, 1995–2995. MMWR Morb Mortal Wkly Rep 58:1409–13.
Aoki K. 2003. Long term follow-up of patients with inborn errors of metabolism detected by the newborn screening program in Japan. Southeast Asian J Trop Med Public Health 34: 19–23.
Archibald SL, Fennema-Notestine C, Riley EP, et al. 2001. Brain dysmorphology in individuals with severe prenatal alcohol exposure. Dev Med Child Neurol 43:48–54.
Ardinger HH, Atkin JF, Blackston RD, et al. 1988. Verification of the fetal valproate syndrome. Am J Med Genet 29:171–85.
Armstrong BG, McDonald AD, Sloan M. 1992. Cigarette, alcohol, and coffee consumption and spontaneous abortion. Am J Public Health 82:85–7.
Armstrong EM, Abel EL. 2000. Fetal alcohol syndrome: the origins of a moral panic. Alcohol Alcohol 35:276–82.
Arnold GI, Kirby RS, Lanendoerfer S, et al. 1994. Toluene embropathy: clinical delineation and developmental follow-up. Pediatrics 93:216–20.

Aro T. 1983. Maternal disease, alcohol consumption, and smoking during pregnancy associated with reduction limb defects. Early Hum Dev 9:49–57.

Arvin AM, Maldonado YA. 1995. Other viral infections of the fetus and newborn. In: Remington JS, Klein JO eds. Infectious diseases of the fetus and newborn infant. Philadelphia, PA, Saunders, pp 745–56.

Aschengrau A, Monson RR. 1989. Paternal military service in Vietnam and risk of spontaneous abortion. J Occup Med 7:618–23.

Aschengrau A, Monson RR. 1990. Paternal military service in Vietnam and the risk of late adverse pregnancy outcomes. Am J Public Health 80:1218–24.

Atanackovic G, Koren G. 1999. Fetal exposure to oral isotretinoin: failure to comply with the Pregnancy Prevention Program. CMAJ 60:1719–20.

Atreya CD, Mohan KV, Kulkarni S. 2004. Rubella virus and birth defects: molecular insights into the viral teratogenesis at the cellular level. Birth Defects Res A Clin Mol Teratol 70: 431–7.

Attenhofer Jost CH, Connolly HM, Edwards WD, et al. 2005. Ebstein's anomaly: review of a multifaceted congenital cardiac condition. Swiss Med Wkly 135:259–81.

Autret E, Radal M, Jonville-Béra AP, et al. 1997. Isotrétinoïne (Roaccutane®) chez la femme en âge de procréer: insuffisance de suivi des recommendations de prescription. Ann Dermat Venereol 124:518–22.

Autti-Rämö I. 2000. Twelve-year follow-up of children exposed to alcohol in utero. Dev Med Child Neurol 42:406–11.

Autti-Rämö I, Gaily E, Granström ML. 1992. Dysmorphic features in offspring of alcoholic mothers. Arch Dis Child 67:12–16.

Avigad S, Cohen BE, Bauer S, et al. 1990. A single origin of phenylketonuria in Yemenite Jews. Nature 344:168–70.

Axelrad DA, Bellinger DC, Ryan LM, et al. 2007. Dose-response relationship of prenatal mercury exposure and IQ: an integrative analysis of epidemiological data. Environ Health Perspect 115:609–15.

Baird CD, Nelson MM, Monie IW, et al. 1955. Congenital cardiovascular anomalies produced with the riboflavin antimetabolite, riboflavin, in the rat. Fed Proc 14:428 (abst).

Baird D. 1947. Social class and foetal mortality. Lancet 2:531–5.

Baird D. 1973. Diet and congenital defects. Lancet 1:291.

Bakir F, Damluji SF, Amin-Zaki L, et al. 1973. Methyl mercury poisoning in Iraq. Science 18:230–41.

Ballantyne JJ. 1904. Manual of antenatal patholgy and hygiene. The embryo. Edinburgh, William Green & Sons.

Banting FG, Best CH. 1923. The discovery and preparation of insulin. Univ Toronto Med J 1:94–8.

Bar-Oz B, Einarson T, Einarson A, et al. 2007. Paroxetine and congenital malformations: meta-analysis and consideration of potential confounding factors. Clin Ther 29:918–26.

Bar-Oz B, Levichek Z, Moretti ME, et al. 2005. Pregnancy outcome following rubella vaccination: a prospective controlled study. Am J Med Genet A 130:52–4.

Barber RC, Shaw GM, Lammer EJ, et al. 1998. Lack of association between mutations in the folate receptor-alpha gene and spina bifida. Am J Med Genet 76:310–17.

Barbero P, Lotersztein V, Bronberg R, et al. 2004. Acitretin embryopathy: a case report. Birth Defects Res A Clin Mol Teratol 70:831–3.

Barr HM, Streissguth AP, Darby BL, et al. 1990. Prenatal exposure to alcohol, caffeine, tobacco, and aspirin: effects on fine and gross motor performance in 4-year-old children. Dev Psychol 26:339–48.

Barr HM, Streissguth AP, Martin DC, et al. 1984. Infant size at 8 months of age: relationship to maternal use of alcohol, nicotine, and caffeine during pregnancy. Pediatrics 74: 336–41.

Barr M, Posznanski AK, Schmickel RD. 1973. Digital hypoplasia and antiepileptics during gestation: a teratogenic syndrome? J Pediatr 84:254–66.

Bass MH. 1952. Diseases of the pregnant woman affecting the offspring. Adv Int Med 5:15–58.

Battino D, Granata T, Binelli S, et al. 1992. Intrauterine growth in the offspring of epileptic mothers. Acta Neurol Scand 66:555–7.

Baur E, Fischer E, Lenz F. 1921. Grundriss der menschlichen Erblichkeitslehre und Rassenhygiene. Munich, Lehmanns.

Baur E, Fischer E, Lenz F. 1931. Human heredity. New York, Macmillan.

Baxter H, Fraser FC. 1950. The production of congenital defects in the offspring of female mice treated with cortisone: a preliminary report. McGill Med J 19:245–9.

Bear JC. 1988. Liability to cleft lip and palate: interpreting the human data. In: Kalter H ed. Issues and reviews in teratology, vol. 4. New York, Plenum Press, pp 163–204.

Beattie JO, Day RE, Cockburn F, et al. 1983. Alcohol and the fetus in the west of Scotland. BMJ 287:17–20.

Beck F, Lloyd JB. 1966. The teratogenic effects of azo dyes. In: Woollam DH ed. Advances in teratology, vol. 1. London, Logos Press, pp 131–93.

Beck-Mannagetta G, Drees B, Janz D. 1982. Malformations and minor anomalies in the offspring of epileptic parents: a retrospective study. In: Janz D, Bossi L, Dam M, et al. eds. Epilepsy, pregnancy, and the child. New York, Raven, pp 317–23.

Bell J. 1951. On brachydactyly and symphalangism. In: On hereditary digital anomalies. The treasury of human inheritance, vol. 5. Cambridge, Cambridge University Press, pp 1–31.

Bell KN, Oakley GP. 2008. Update on prevention of folic acid preventable spina bifida and anencephaly. Birth Defects Res A Clin Mol Teratol 85:102–7.

Bell R, Bailey K, Cresswell T, et al. 2008. Trends in prevalence and outcome of pregnancy in women with pre-existing type I and type II diabetes. Br J Obstet Gynecol 115:419–20.

Bell R, Lumley J. 1989. Alcohol consumption, cigarette smoking and fetal outcome in Victoria, 1985. Community Health Stud 13:484–91.

Benda CE. 1954. The Dandy-Walker syndrome or the so-called atresia of the foramen Magendie. J Neuropathol Exp Neurol 13:14–29.

Benirschke K. 1987. You need a sympathetic pathologist! the borderline of embryology and pathology revisited. Teratolgy 36:389–93.

Bennewitz HG. 1824. De diabete mellito, graviditatis symptomate. Inaugural dissertation in Medicine, Berlin.

Bérard A, Azoulay L, Koren G, et al. 2007. Isotretinoin, pregnancies, abortions and birth defects: a population-based perspective. Br J Clin Pharmacol 63:196–205.

Bercovich D, Elimelech A, Yardeni T, et al. 2008. A mutation analysis of the phenylalanine hydroxylase (PAH) gene in the Israel population. Ann Hum Genet 72:305–9.

Berry RJ, Li Z, Erickson JD, et al. 1999. Prevention of neural-tube defects with folic acid in China. N Engl J Med 341:485–90.

Bickel H, Bachmann C, Beckers R, et al. 1981. Neonatal mass screening for metabolic disorders: summary of recent sessions of the Committee of Experts to Study Inborn Metabolic Diseases, Public Health Committee, Council of Europe. Eur J Pediatr 37:133–9.

Bickel H, Gerrard J, Hickmans EM. 1953. Influence of phenylalanine intake on phenylketonuria. Lancet 2:812–3.

Biddle FG, Fraser FC. 1977. Maternal and cytoplasmic factors in experimental teratology. In: Wilson JG, Fraser FC eds. Handbook of teratology, vol. 3. New York, Plenum Press, pp 3–33.

Bierwolf D. 1956. Kleinhirnmisssbildungen durch hereditaren Hydrocephalus bei der Hausmaus. Wiss Z Univ Halle 5:1237–82.

Biggs GG, Freeman RK, Yaffe SJ. 1986. Drugs in pregnancy and lactation. Baltimore, MD, Wilkins & Wilkins.

Bingol N, Schuster C, Fuchs M, et al. 1987. The influence of socioeconomic factors on the occurrence of fetal alcohol syndrome. Adv Alcohol Subst Abuse 6:105–18.

Birnbaum R. 1912. A clinical manual of the malformations and congenital diseases of the fetus. Backer G (trans.). London, Churchill.

Bisanti L, Bonetti F, Caramaschi F, et al. 1980. Experiences from the accident of Seveso. Acta Morph Acad Sci Hung 28:139–57.

Bishop JB, Witt KL, Sloane RA. 1997. Genetic toxicities of human teratogens. Mutat Res 396:9–43.

Bjerkedal T. 1982. Outcome of pregnancy in women with epilepsy, Norway 1967 to 1978: congenital malformations. In: Janz D, Bossi L, Dam M eds. Epilepsy, pregnancy, and the child. New York, Raven Press, pp 289–95.

Bjerkedal T, Czeizel A, Goujard J, et al. 1982. Valproic acid and spina bifida. Lancet 2:1096.

Blakley PM. 1988. Experimental teratology of ethanol. In: Kalter H ed. Issues and reviews in teratology, vol. 4. New York, Plenum Press, pp 237–82.

Blatter BM, Hermens R, Bakker M, et al. 1997. Paternal occupational exposure around conception and spina bifida in offspring. Am J Ind Med 32:283–91.

Blencowe H, Cousens S, Modell B, et al. 2010. Folic acid to reduce neonatal mortality from neural tube disorders. Int J Epidemiol 39:110–21.

Blot WJ. 1975. Growth and development following prenatal and childhood exposure to atomic radiation. J Radiat Res Suppl 82:82–8.

Bodendorfer TW. 1978. Fetal effects of antiepileptic drugs. Drug Intell Clin Pharm 12:14–21.

Boneva RS, Moore CA, Botto L, et al. 1999. Nausea during pregnancy and congenital heart defects: a population-based case-control study. Am J Epidemiol 49:717–25.

Borgatta L, Burnhill MS, Tyson J, et al. 2001. Early medical abortion with methotrexate and misoprostol. Am J Obstet Gynecol 97:1–6.

Borman B. 2001. Policy for prevention of birth defects by periconceptional folic acid in New Zealand. Teratology 64:321–2.

Botto LD, Khoury MJ, Mulinare J, et al. 1996. Periconceptional multivitamin use and the occurrence of conotruncal heart defects: results from a population-based, case-control study. Pediatrics 98:911–17.

Botto LD, Yang Q. 2000. 5,10-Methylenetetrahydofolate reductase gene variants and congenital anomalies: a HuGE review. Am J Epidemiol 51:862–77.

Boué J, Philippe E, Giroud A, et al. 1976. Phenotypic expression of lethal chromosomal anomalies in human abortuses. Teratology 14:3–20.

Boulet SL, Yang Q, Mai C, et al. 2008. Trends in the postfortification prevalence of spinal bifida and anencephaly in the United States. Birth Defects Res A Clin Mol Teratol. 82:527–32.

Boulet SL, Gambrell D, Shin M, et al. 2009. Racial/ethnic differences in the birth prevalence of spinal bifida – United Ststes, 1995–2005. MMWR Morb Mortal Wkly Rep 57:1409–13.

Boulot P, Chabbert N, d'Ércole C, et al. 2003. French multicentric survey of outcome of pregnancy in women with pregestational diabetes. Diab Care 26:2990–3.

Bovier-Lapierre M, Saint-Dizier C, Freycon F, et al. 1974. Deux enfants nés de mère phénylcétonurique echec d'un régime pauvre en phénylalanine institué pendant la deuxiéme grossesse. Pédiatrie 29:51–72.

Bower C, D'Antoine HD, Stanley FJ. 2009. Neural tube defects in Australia: trends in encephaloceles and other neural tube defects before and after promotion of folic acid supplementation and voluntary food fortification. Birth Defects Res A Clin Mol Teratol 85:269–73.

Bower C, Stanley FJ. 1989. Dietary folate as a risk factor for neural-tube defects: evidence from a case-control study in Western Australia. Med J Aust 150:613–19.

Bower C, Stanley FJ. 1992. Periconceptional vitamin supplementation and neural-tube defects: evidence from a case-control study in Western Australia and a review of recent publications. J Epidemiol Community Health 46:57–61.

Brannigan VM, Bier VM, Berg C. 1992. Risk, statistical inference, and the law of evidence: the use of epidemiological data in toxic tort cases. Risk Anal 12:343–51.

Brent RL. 2000. Nongenital malformations and exposure to progestational drugs during pregnancy: the final chapter of an erroneous allegation. Teratology 61:449.

Breyer SG. 2000. Genetic advances and legal institutions. J Law Med Ethics 28:3–8.

Briggs GG, Freeman RK, Yaffe S. 1998. Drugs in pregnancy and lactation. Baltimore, MD, Williams & Wilkins.

Brock DJ, Sutcliffe RG. 1972. Alpha-fetoprotein in the antenatal diagnosis of anencephaly and spina bifida. Lancet 2:197–8.

Brockman DE, Kirby ML. 1984. Dependence of thymus development on derivatives of the neural crest. Science 223:498–500.

Brodie MJ. 2006. Major congenital malformations and epileptic drugs: prospective observations. J Neurol Neurosurg Psychiatry 77:145.

Brooke OG, Anderson HR, Bland MJ, et al. 1989. Effects on birth weight of smoking, alcohol, caffeine, socieconomic factors, psychosocial stress. BMJ 298:795–801.

Brown GC. 1966. Recent advances in the viral aetiology of congenital anomalies. In: Woollam DH ed. Advances in teratology, vol. 1. London, Logos, pp 55–80.

Brown NA. 1985. Are offspring at risk from their father's exposure to toxins? Nature 316:110.

Brown NA, Kao J, Fabro S. 1980. Teratogenic potential of valproic acid. Lancet 1:60–1.

Bruin KM, Buitendijk SE, Hirasing RA, et al. 2000. [Prevalence of neural tube defects in births before and after promotion of periconceptional folic acid supplementation.] In Dutch. Ned Tijdschr Geneeskd 144:1732–6.

Buccimazza SS, Molteno CD, Dunne TT, et al. 1994. Prevalence of neural tube defects in Cape Town, South Africa. Teratology 50:194–9.

Bunde CA, Bowles DM. 1963. A technique for controlled survey of case records. Curr Ther Res 5:245–8.

Bunde CA, Leyland HM. 1965. A controlled retrospective survey in evaluation of teratogenicity. J New Drugs 5:93–8.

Burgess MA. 1991. Gregg's rubella legacy 1941–1991. Med J Aust 155:355–7.

Burn J, Fisk NM. 1997. Terathanasia, folic acid, and birth defects. Lancet 350:322.

Burnet FM. 1985. Australia among the world leaders. Lantern Light 6:6–7.

Byrne J, Warburton D. 1986. Neural tube defects in spontaneous abortions. Am J Med Genet 25:327–33.

Byrne J, Warburton D. 1987. Male excess among anatomically normal fetuses in spontaneous abortions. Am J Med Genet 26:605–11.

Cade JF. 1949. Lithium salts in the treatment of psychotic excitement. Med J Aust 36:349–52.

Calzolari E, Manservigi D, Garani G, et al. 1990. Limb reduction defects in Emilia Romagna, Italy: epidemiological and genetic study in 173 109 consecutive births. J Med Genet 27:353–7.

Camera G, Pregliasco P. 1992. Ear malformation in baby born to mother using tretinoin cream. Lancet 339:687.

Campbell R, Sullivan O. 1983. Alcohol consumption in pregnancy. Arch Dis Child 58:474–5.

Campbell S, Pierce JM. 1983. Ultrasound visualization of congenital malformations. Br Med Bull 39:322–31.

Canfield MA, Annegers JF, Brender JD, et al. 1996a. Hispanic origin and neural tube defects in Houston/Harris County, Texas. I. Descriptive epidemiology. Am J Epidemiol 143:1–11.

Canfield MA, Annegers JF, Brender JD, et al. 1996b. Hispanic origin and neural tube defects in Houston/Harris County, Texas. II. Risk factors. Am J Epidemiol 143:12–24.

Canfield MA, Ramadhani TA, Shaw GM, et al. 2009. Anencephaly and spina bifida among Hispanics: maternal, sociodemographic, and acculturation factors in the National Birth Defects Prevention Study. Birth Defects Res A Clin Mol Teratol 85:637–46.

Canger R, Battino D, Canevini MP, et al. 1999. Malformations in offspring of women with epilepsy: a prospective study. Epilepsia 40:1231–6.

Cannon MD. 1940. Failure of vitamin A depletion to produce congenital anomalies in the young of rats. Proc Soc Exp Biol Med 44:29–32.

Cao J, Lu JC, Huang YF. 2008. [Advances in molecular biology of rubella virus structural proteins] In Chinese. Zhonghua Nan Ke Xue 14:645–9.

Carlin JB, Ryan LM, Harvey EA, et al. 2000. Anticonvulsant teratogenesis 4: inter-rater agreement in assessing minor physical features related to anticonvulsant therapy. Teratology 62: 406–12.

Carr DH. 1963. Chromosome studies in abortuses and stillborn infants. Lancet 2:603–6.
Carr DH. 1983. Cytogenetics of human reproductive wastage. In: Kalter H ed. Issues and reviews in teratology, vol. 1. New York, Plenum Press, pp 33–72.
Carter CO. 1961. The inheritance of pyloric stenosis. Br Med Bull 17:251–4.
Carter CO. 1973. Diet and congenital defects. Lancet 1:291.
Carter CO. 1977. Principles of polygenic inheritance. Birth Defects Orig Artic Ser 13:69–74.
Casson IF, Clarke CA, Howard CV, et al. 1997. Outcomes of pregnancy in insulin dependent diabetic women: results of a five year population study. BMJ 315:275–8.
Castronovo FP. 1999. Teratogen update: radiation and Chernobyl. Teratology 60:100–6.
Cavallo F, Russo R, Zotti C, et al. 1995. Moderate alcohol consumption and spontaneous abortion. Alcohol Alcohol 30:195–201.
CDC. 1987. Serum dioxin in Vietnam-era veterans: preliminary report. MMWR Morb Mortal Wkly Rep 36:470–5.
CDC. 1989. Health status of Vietnam veterans. III. Reproductive outcomes and child health. The Centers for Disease Control Vietnam Experience Study. JAMA 259:2715–9.
CDC. 1991. Use of folic acid for prevention of spina bifida and other neural tube defects – 1983–1991. MMWR Morb Mortal Wkly Rep 40:513–6.
CDC. 1992. Recommendations for the use of folic acid to reduce the number of cases of spina bifida and other neural tube defects. MMWR Morb Mortal Wkly Rep 41:1–5.
Centerwall SA, Centerwall WR. 2000. The discovery of phenylketonuria: the story of a young couple, two retarded children, and a scientist. Pediatrics 105:89–103.
Chambers CD, Johnson KA, Dick LM, et al. 1995. Birth outcomes in pregnant women taking fluoxetine. N Engl J Med 335:1010–5.
Chambers CD, Johnson KA, Dick LM, et al. 1998. Maternal fever and birth outcome: a prospective study. Teratology 58:251–7.
Chan A, Robertson EF, Haan EA, et al. 1993. Prevalence of neural tube defects in South Australia, 1961–91: effectiveness and impact of prenatal diagnosis. BMJ 307:703–6.
Chen BH, Carmichael SL, Selvin S, et al. 2008. NTD prevalences in central California before and after folic acid fortification. Birth Defects Res A Clin Mol Teratol 82:547–52.
Chen C, Jensen BK, Mistry G, et al. 1997. Negligible systemic absorption of topical isotretinoin cream: implications for teratogenicity. J Clin Pharmacol 37:279–84.
Chen G, Song X, Li Y, et al. 2008. Prevention of NTDs with periconceptional multivitamin supplementation containing folic acid in China. Birth Defects Res A Clin Mol Teratol. 82:592–6.
Cheng N, Bai Y, Hu X, et al. 2003. A base-line survey on birth defects in Gansu province, West China. Ann Trop Paediatr 23:25–9.
Cherpitel CJ, Room R. 1989. Infant outcomes associated with alcohol consumption and maternal risk. Alcohol Clin Exp Res 13:343.
Christianson AL, Chester N, Kromberg JG. 1994. Fetal valproate syndrome: clinical and neurodevelopmental features in two sibling pairs. Dev Med Child Neurol 36:357–69.
Christianson RE, van den Berg B, Milkovitch L, et al. 1981. Incidence of congenital anomalies among white and black births with long-term follow-up. Am J Public Health 71:1333–41.
Christiansen RL. 1975. Classification and nomenclature of morphological defects. Lancet 2:513.
Clark H, Pierro LJ. 1994. Walter Landauer – the man and his work. In: Kalter H ed. Issues and reviews in teratology, vol. 7. New York, Plenum Press, pp 217–83.
Clarke CA, McKendrick OM, Sheppard PM. 1973. Spina bifida and potatoes. BMJ 3:251–4.
Clarke CO. 1973. Diet and congenital defects. Lancet 1:290–1.
Clarren SK. 1990. Fetal alcohol syndrome: diagnosis, treatment, and mechanisms of teratogenesis. In: Transplacental disorders: perinatal detection, treatment, and management. New York, Liss, pp 37–55.
Clarren SK, Bowden DM, Astley SJ. 1987a. Pregnancy outcomes after weekly oral administration of ethanol during gestation in the pig-tailed macaque *(Macaca nemestrina)*. Teratology 35: 345–54.

Clarren SK, Sampson PD, Larsen J, et al. 1987b. Facial effects of fetal alcohol exposure: assessment by photographs and morphometric analysis. Am J Med Genet 26:651–66.
Clarren SK, Smith DW. 1978. The fetal alcohol syndrome. N Engl J Med 298:1063–7.
Coffey VP, Jessop WJ. 1955. Congenital abnormalities. Ir J Med Sci 6:30–48.
Coffey VP, Jessop WJ. 1959. Maternal influenza and congenital deformities: a prospective study. Lancet 2:935–8.
Cohen BE, Szeinberg A, Zarfin Y, et al. 1986. Maternal hyperphenylalaninemia in Israel. J Inher Metab Dis 9(Suppl 2):227–30.
Cohen LS, Friedman JM, Jefferson JW, et al. 1994. A reevaluation of risk of in utero exposure to lithium. JAMA 271:146–50.
Cohlan SQ. 1953. Excessive intake of vitamin A as a cause of congenital anomalies in the rat. Science 117:535–6.
Cohlan SQ. 1954. Congenital anomalies in the rat produced by excessive intake of vitamin A during pregnancy. Pediatrics 13:556–67.
Coles CD, Brown RT, Smith IE, et al. 1991. Effects of prenatal alcohol exposure at school age. I. Physical and cognitive development. Neurotoxicol Teratol 3:356–67.
Coles CD, Smith IE, Falek A. 1987. Prenatal alcohol exposure and infant behavior: immediate effects and implications for later development. Adv Alcohol Subst Abuse 6:87–104.
Coles CD, Smith IE, Fernhoff PM, et al. 1984. Neonatal ethanol withdrawal: characteristics in clinically normal, nonsyndromic neonates. J Pediatr 104:445–51.
Collins MD, Mao GE. 1999. Teratology of retinoids. Ann Rev Pharmacol Toxicol 39:399–430.
Collins MH. 1995. Founders of pediatric pathology: Dorothy H. Andersen, MD, Med, ScD. In: Dimmick JE, Singer DB eds. Forensic aspects in pediatric pathology: perspectives in pediatric pathology, vol. 19. Basel, Karger, pp 4–10.
Consonni D, Pesatori AC, Zocchetti C, et al. 2008. Mortality in a population exposed to dioxin after the Seveso, Italy, accident in 1975: 25 years of follow-up. Am J Epidemiol 167:847–58.
Constable JD, Hatch MC. 1985. Reproductive effects of herbicide exposure in Vietnam: recent studies by the Vietnamese and others. Teratog Carcinog Mutagen 5:231–50.
Coombs J. 1987. The interface of law and science: the work of the Agent Orange Royal Commission. Med Law 6:1–12.
Coombs J. 1988. The Agent Orange phenomenon: the report of the Australian Royal Commission. In: Young AL, Reggiani GM eds. Agent Orange and its associated dioxin: assessment of a controversy. Amsterdam, Elsevier.
Cooper LZ. 1968. Rubella: a preventable cause of birth defects. Birth Defects Orig Artic Ser 7:23–35.
Cooper LZ. 1975. Congenital rubella in the United States. In: Krugman S, Gershon AE eds. Infections of the fetus and the newborn infant. New York, Liss, pp 1–22.
Cooper LZ. 1985. The history and medical consequences of rubella. Rev Infect Dis 7(Suppl 1):S2–S10.
Cooper WO, Hernandez-Diaz S, Arbogast PG, et al. 2006. Major congenital malformations after first-trimester exposure to ACE inhibitors. N Engl J Med 354:2443–51.
Cornel MC, Erickson JD. 1997. Comparison of national policies on periconceptional use of folic acid to prevent spina bifida and anencephaly (SBA). Teratology 55:134–7.
Correa A, Gilboa SM, Besser LM, et al. 2008. Diabetes mellitus and birth defects. Am J Obstet Gynecol 199:237.e1–9.
Correa-Villaseñor A, Ferencz C, Neill CA, et al. 1994. Ebstein's malformation of the tricuspid valve: genetic and environmental factors. The Baltimore-Washington Infant Study Group. Teratology 50:137–47.
Cotlier E. 1972. Rubella in animals and experimental ocular aspects of congenital rubella. Int Ophthalmol Clin 12:137–46.
Cox C, Clarkson TW, Marsh DO, et al. 1989. Dose-response analysis of infants prenatally exposed to methylmercury: an application of a single compartment model to single-strand hair analysis. Environ Res 31:640–9.

Cragan JD, Roberts HE, Edmonds LD, et al. 1995. Surveillance for anencephaly and spina bifida and the impact of prenatal diagnosis – United States, 1985–1994. MMWR Morb Mortal Wkly Rep 44:1–13.

Cranston ME, Mhanni AA, Marles SL. 2009. Concordance of three methods for palpebral fissure length measurement in the assessment of fetal alcohol spectrum discorder. Can J Clin Pharmacol 16:e234–e41.

Cruz-Coke R. 1973. Anencephaly and potatoes in Chile. Lancet 1:269–70.

Cuckle H, Wald N. 1987. The impact of screening for open neural tube defects in England and Wales. Prenat Diag 7:91–9.

Cundy T, Gamble G, Neale L, et al. 2007. Differing causes of pregnancy loss in type 1 and type 2 diabetes. Diabetes Care 30:2603–7.

Cunnington M, Tennis P. 2005. Lamotrigine and the risk of malformation in pregnancy. Neurology 64:955–60.

Curtis E, Fraser FC, Warburton D. 1961. Congenital cleft lip and palate: risk figures for counselling. Am J Dis Child 102:853–7.

Cutting RT, Phuoc TH, Ballo JM, et al. 1970. Congenital malformations, hydatidiform moles, and stillbirths in the Republic of Vietnam 1960–1969. Washington, DC, US Government Printing Office, pp 1–29.

Czeizel A. 1993a. Epidemiological studies of congenital abnormalities in Hungary. In: Kalter H ed. Issues and reviews in teratology, vol. 6. New York, Plenum Press, pp 85–124.

Czeizel A. 1993b. Prevention of congenital abnormalities by periconceptional multivitamin supplementation. BMJ 306:645–9.

Czeizel A, Dudás I. 1992. Prevention of the first occurrence of neural-tube defects by periconceptional vitamin supplementation. N Engl J Med 327:832–5.

Czeizel A, Dudás I, Métneki J. 1994. Pregnancy outcomes in a randomised controlled trial of perconceptional multivitamin supplementation: final report. Arch Gynecol Obstet 255: 131–9.

Czeizel A, Fritz G. 1989. Ethics of a randomized trial of periconceptual vitamins. JAMA 262:1634.

Czeizel A, Karig G. 1985. Analysis of the changing birth prevalence of neural tube defects in Hungary. Acta Morphol 33:89–100.

Czeizel A, Keller S, Bod M. 1983. An aetiological evaluation of increased occurrence of congenital limb reduction abnormalities in Hungary, 1975–1978. Int J Epidemiol 12:445–9.

Czeizel AE, Puhó EH, Acs N, et al. 2008a. Use of specified critical periods of different congenital abnormalities instead of the first trimester concept. Birth Defects Res A Clin Mol Teratol 82:139–46.

Czeizel AE, Puhó EH, Acs N, et al. 2008b. Delineation of a multiple congenital anomaly syndrome in the offspring of pregnant women affected with high fever-related disorders: a population based study. Congenit Anom (Kyoto) 48:158–66.

Czeizel AE, Rockenbauer M. 1997. Population-based case-control study of teratogenic potential of corticosteroids. Teratology 56:335–40.

Czeizel AE, Tímár L, Sárközi A. 1999. Dose-dependent effect of folic acid on the prevention of orofacial defects. Pediatrics 104:e66.

Czeizel AE, Tóth M, Rockenbauer M. 1996. Population-based case control study of folic acid supplementation during pregnancy. Teratology 53:345–51.

Dai L, Zhu J, Zhou G, et al. 2002. [Dynamic monitoring of neural tube defects in China during 1996 to 2000] In Chinese. Zhonghua Yu Fang Yi Xue Za Zhi 36:402–5.

Dai WS, Hsu M-A, Itri LM. 1989. Safety of pregnancy after discontinuation of isotretinoin. Arch Dermatol 125:362–5.

Dai WS, LaBraico JM, Stern RS. 1992. Epidemiology of isotretinoin exposure during pregnancy. J Am Acad Dermatol 26:599–606.

Dally A. 1998. Thalidomide: was the tragedy preventable? Lancet 351:1197–9.

Daly LE, Kirke PN, Molloy A, et al. 1995. Folate levels and neural tube defects: implications for prevention. JAMA 274:1698–702.

Daly S, Mills JL, Molloy AM, et al. 1997. Minimum effective dose of folic acid for food fortification to prevent neural-tube defects. Lancet 350:1666–9.

D'Amato RJ, Loughnan MS, Flynn E, et al. 1994. Thalidomide is an inhibitor of angiogenesis. Proc Natl Acad Aci USA 91:4082–5.

Daniel WA, Mounger JR, Perkins JC. 1971. Obstetric and fetal complications in folate-deficient adolescent girls. Am J Obstet Gynecol 111:233–8.

Dansky L, Andermann E, Andermann F. 1982. Major congenital malformations in the offspring of epileptic parents: genetic and evironmental risk factors. In: Janz D, Bossi L, et al. eds. Epilepsy, pregnancy, and the child. New York, Raven, pp 223–34.

Dareste C. 1877. Recherches sur la production artificielle des monstruosités: ou essais de tératogénie expérimentale. Paris, Reinwald.

Daston GP. 1993. Do thresholds exist for developmental toxicants? A review of the theoretical and experimental evidence. In: Kalter H ed. Issues and reviews in teratology, vol. 6. New York, Plenum Press, pp 169–97.

Davidson PW, Myers GJ, Cox C, et al. 1998. Effects of prenatal and postnatal methylmercury exposure from fish consumption in neurodevelopment: outcomes at 66 months of age in the Seychelles Child Development Study. JAMA 280:701–7.

Davidson PW, Myers GJ, Weiss B, et al. 2006. Prenatal mercury exposure from fish consumption and child development: a review of evidence and perspectives from the Seychelles Child Development Study. Neurotoxicology 27:1106–9.

Davis AR, Miller L, Tamimi LH, et al. 1999. Methotrexate compared with mercaptopurine for early induced abortion. Obstet Gynecol 93:904–9.

Davis DL, Gottlieb MB, Stampnitzky JR. 1998. Reduced ratio of male to female births in several industrialized countries: a sentinel health indicator? JAMA 279:1018–23.

Davis PJ, Partridge JW, Storrs CN. 1982. Alcohol consumption in pregnancy: how much is safe? Arch Dis Child 57:940–3.

Davis RL, Rubanowice D, McPhillips H, et al. 2007. Risks of congenital malformations and perinatal events among infants exposed to antidepressant medications during pregnancy. Pharmacoepidemiol Drug Saf 16:1181–3.

Davison C. 1963. Definition and classification of human malformations. Dev Med Child Neurol 5:340–2.

Dawson EB, Evans DR, Harris WA, et al. 1999. Amniotic fluid B12, calcium, and lead levels associated with neural tube defects. Am J Perinatol 6:373–8.

Day NL, Jasperse D, Richardson G, et al. 1989. Prenatal exposure to alcohol: effect on infant growth and morphologic characteristics. Pediatrics 84:536–41.

Day NL, Richardson G, Robles N, et al. 1990. Effect of prenatal alcohol exposure on growth and morphology of offspring at 8 months of age. Pediatrics 85:748–52.

Day NL, Robles N, Richardson G, et al. 1991. The effects of prenatal alcohol use on the growth of children at three years of age. Alcohol Clin Exp Res 15:67–71.

de Die-Smulders CE, Sturkenboom MC, Veraart J, et al. 1995. Severe limb defects and craniofacial anomalies in a fetus conceived during acitretin therapy. Teratology 52:215–19.

De Wals P, Bloch D, Calabro A, et al. 1991. Association between holoprosencephaly and exposure to topical retinoids: results of the EUROCAT survey. Paediatr Perinat Epidemiol 5:445–7.

De Wals P, Tairou F, Van Allen MI. 2007. Reduction in neural-tube defects after folic acid fortification in Canada. N Engl J Med 357:135–42.

De Wals P, Trochet C, Pinsonneault L. 1999. Prevalence of neural tube defects in the Province of Quebec, 1992. Can J Public Health 90:237–9.

Dean JC, Hailey H, Moore SJ, et al. 2002. Long term health and neurodevelopment in children exposed to antiepileptic drugs before birth. J Med Genet 39:251–9.

Dekaban AS. 1968. Abnormalities in children exposed to x-radiation during various stages of gestation: tentative timetable of radiation injury to the human fetus, part I. J Nucl Med 9: 471–7.

Del Campo M, Kosaki K, Bennett F, et al. 1999. Developmental delay in fetal aminopterin/methotrexate syndrome. Teratology 60:10–12.
Dent CE. 1956. Discussion: etiological factors in mental retardation. Twenty-third Ross Pediatric Research Conference, Columbus, pp 32–3.
Department of Veterans Affairs. 2009. Presumptive service connection for disease associated with exposure to certain herbicide agents: AL amyloidosis. Final rule. Fed Regist 74:21258–60.
Dessens AB, Cohen-Kettenis PT, Mellenbergh GL, et al. 2001. Association of prenatal phenobarbital and phenytoin exposure with genital anomalies and menstrual disorders. Teratology 64:181–8.
Diav-Citrin O, Ornoy A. 2000. Adverse environment and prevention of early pregnancy disorders. Early Preg 4:5–18.
Diav-Citrin O, Shechtman S, Bar-Oz B, et al. 2008. Pregnancy outcome after in utero exposure to valproate: evidence of dose relationship in teratogenic effect. CNS Drugs 22:325–34.
Diav-Citrin O, Shechtman S, Weinbaum D, et al. 2008. Paroxetine and fluoxetine in pregnancy: a prospective, multicentre, controlled, observational study. Br J Clin Pharmacol 66:695–705.
Dietrich KN. 1999. Environmental chemicals and child development. J Pediatr 134:7–9.
Dieckmann WJ, Davis ME, Rynkiewicz LM, et al. 1953. Does the administration of diethylstilbestrol during pregnancy have therapeutic value? Am J Obstet Gynecol 166:1062–81.
DiGeorge AM. 1968. Congenital absence of the thymus and its immunological consequences: concurrence with congenital hypoparathyroidism. Birth Defects Orig Art Ser 4:116–21.
DiLiberti JH, Farndon PA, Dennis NR, et al. 1984. The fetal valproate syndrome. Am J Med Genet 19:473–81.
Dimich-Ward H, Hertzman C, Teschke K, et al. 1996. Reproductive effects of paternal exposure to chlorphenate wood preservatives in the sawmill industry. Scand J Work Environ Health 22:267–73.
Dinan TG. 2002. Lithium in bipolar mood disorder. BMJ 324:989–90.
Dobrovolskaïa-Zavadskaïa N. 1927. Sur la mortification spontanée de la queue chez la souris nouveau-née et sur l'existence d'un charactère (facteur) héréditaire "non-viable". C R Soc Biol 97:116–18.
Dodds EC, Goldberg L, Lawson W, et al. 1938. Oestrogenic activity of certain synthetic compounds. Nature 141:247–8.
Dods L. 1966. Obituary: Norman McAlister Gregg. Med J Aust 2:1166–9.
Dolk H. 2005. EUROCAT: 25 years of European surveillance of congenital anomalies. Arch Dis Child Fet Neonat Ed 90:F355–F58.
Donnai D, Harris R. 1978. Unusual fetal malformations after antiemetics in early pregnancy. BMJ 1:691–2.
Donovan JW, MacLennan R, Adena M. 1984. Vietnam service and the risk of congenital anomalies: a case-control study. Med J Aust 140:394–7.
Dorman C. 1991. Microcephaly and intelligence. Dev Med Child Neurol 33:267–72.
Dorman JS, Burke JP, McCarthy BJ, et al. 1999. Temporal trends in spontaneous abortion associated with type 1 diabetes. Diab Res Clin Pract 43:41–7.
Doull IJ. 2001. Recent advances in cystic fibrosis. Arch Dis Child 85:62–6.
Dowie M, Marshall C. 1980. The Bendectin cover-up. Mother Jones (November) 5:43–56.
Down JL. 1866. Observations on an ethnic classification of idiots. Lond Hosp Clin Lect Rep 3:259–62.
Dravet C, Julian C, Legras C, et al. 1992. Epilepsy, antiepileptic drugs, and malformations in children of women with epilepsy: a French prospective study. Neurology 42:75–82.
Driscoll SG. 1965. The pathology of pregnancy complicated by diabetes mellitus. Med Clin North Am 49:1053–67.
Driscoll SG, Benirschke K, Curtis GW. 1960. Neonatal deaths among infants of diabetic mothers: postmortem findings in ninety-five infants. Am J Dis Child 100:818–35.
Drogari E, Smith I, Beasley M, et al. 1987. Timing of strict diet in relation to fetal damage in maternal phenylketonuria: an international collaborative study by the MRC/DHSS Phenylketonuria Register. Lancet 2:927–30.

Druschel CM, Fox DJ. 2007. Issues in estimating the prevalence of fetal alcohol syndrome: examination of two counties in New York State. Pediatrics 119:e384–e90.

D'Souza SW, Robertson IG, Donnai D, et al. 1990. Fetal phenytoin exposure, hypoplastic nails, and jitteriness. Arch Dis Child 65:320–4.

Dudgeon JA. 1976. Infective causes of human malformations. Br Med Bull 32:77–83.

Duhamel B. 1961. From the mermaid to anal imperforation: the syndrome of caudal regression. Arch Dis Child 36:152–5.

Duncan S. 2007. Teratogenesis of sodium valproate. Curr Opin Neurol 20:175–80.

Dunn LC. 1965. A short history of genetics. New York, McGraw-Hill.

Edmonds LD, Layde PM, Ames LM, et al. 1981. Congenital malformation surveillance: two American systems. Int J Epidemiol 10:247–52.

Edmonds LD, Oakley GP. 1990. Ebstein's anomaly and maternal lithium exposure during pregnancy. Teratology 41:551–2.

Edwards JH. 1958. Congenital malformations of the central nervous system in Scotland. Br J Prev Med 12:115–30.

Edwards JH. 1973. Potato blight teratogen. Nature 243:475.

Edwards JH, Leck I, Record RG. 1964. A classification of malformations. Acta Genet 14:76–90.

Egeland GM, Perham-Hester KA, Gessner BD, et al. 1998. Fetal alcohol syndrome in Alaska, 1977 through 1992: an administrative prevalence derived from multiple sources. Am J Public Health 88:781–6.

Ehlers K, Stürje H, Merker HJ. 1992. Valproic acid-induced spina bifida: a mouse model. Teratology 45:145–54.

Ehrat R. 1948. Die Missbildungen der Neugeborenen an der Universitätsfrauenklinik Zurich 1921–1944. Thesis, Univ Zurich. Wädenswil, Villiger.

Einarson A, Choi J, Einarson TR, et al. 2009. Incidence of major malformations in infants following antidepressant exposure in pregnancy: results of a large prospective cohort study. Can J Psychiatry 54:242–6.

Einarson TR, Leeder JS, Koren G. 1988. A method for meta-analysis of epidemiological studies. Drug Intell Clin Pharmacol 22:813–24.

Ekelund H, Kullander S, Källén B. 1970. Major and minor malformations in newborns and infants up to one year of age. Acta Paediatr Scand 59:297–302.

Elizan TS, Fabiyi A. 1970. Congenital and neonatal anomalies linked with viral infections in experimental animals. Am J Obstet Gynecol 106:147–65.

Elizan TS, Fabiyi A, Sever JL. 1969. Study of rubella virus as a teratogen in experimental animals: a short review. J Mt Sinai Hosp 36:108–12.

Elwood JH. 1973. Epidemics of anencephalus and spina bifida in Ireland since 1900. Int J Epidemiol 2:171–5.

Elwood JH, MacKenzie G. 1973. Associations between the incidence of neurological malformations and potato blight outbreaks over 50 years in Ireland. Nature 243:476–7.

Elwood JM. 1973. Anencephaly and potato blight in eastern Canada. Lancet 1:769.

Elwood JM, Elwood JH. 1980. Epidemiology of anencephalus and spina bifida. Oxford, Oxford University Press.

Elwood M, Little J. 1992. Secular trends. In: Elwood JM, Little J, Elwood JH eds. Epidemiology and control of neural tube defects. Oxford, Oxford University Press, pp 168–94.

Ema M, Ise R, Kato H, et al. 2009. Fetal malformations and early embryonic gene expression response in cynomolgus monkeys maternally exposed to thalidomide. Reprod Toxicol Sep 12 [Epub ahead of print].

Emanuel I. 1972. Non-tuberous neural-tube defects. Lancet 2:879.

Emanuel I. 1983. Vitamin supplementation and neural tube defects. Lancet 2:39–40.

Emanuel I. 1993. Intergenerational factors in pregnancy outcome: implications for teratology. In: Kalter H ed. Issues and reviews in teratology, vol. 6. New York, Plenum Press, pp 47–84.

Emanuel I, Huang SW, Gutman LT, et al. 1972. The incidence of congenital malformations in a Chinese population: the Taipei collaborative study. Teratology 5:159–70.

Emanuel I, Sever LE. 1973. Questions concerning the possible association of potatoes and neural-tube defects, and an alternative hypothesis relating to maternal growth and development. Teratology 8:325–32.
Emery AE, Timson J, Watson-Williams EJ. 1969. Pathogenesis of spina bifida. Lancet 2:909–10.
Erickson JD. 1976. Racial variations in the incidence of congenital malformations. Ann Hum Genet 39:315–20.
Erickson JD, Mulinare J, McClain PW, et al. 1984. Vietnam veterans' risk for fathering babies with birth defects. JAMA 252:903–12.
Ernhart CB, Sokol R, Martier SS, et al. 1987. Alcohol teratogenicity in the human: a detailed assessment of specificity, critical period, and threshold. Am J Obstet Gynecol 56:33–9.
Ernhart CB, Wolf AW, Linn PL, et al. 1985. Alcohol-related birth defects: syndromal anomalies, intrauterine growth retardation, and neonatal behavioral assessment. Alcohol Clin Exp Res 9:447–53.
Eskanazi B, Mocarelli P, Warner M, et al. 2003. Maternal serum dioxin levels and birth outcomes in women of Seveso, Italy. Environ Health Perspect 111:947–53.
EUROCAT. 1991. Prevalence of neural tube defects in 20 regions of Europe and the impact of prenatal diagnosis, 1980–1986. J Epidemiol Community Health 45:52–8.
EUROCAT. 1993. Completed cluster investigations: a database of references. Eurocat Cluster Advisory Service, eurocat@ulster.ac.uk.
Evans BN, Chodirker BN, Harman C, et al. 1992. The impact of maternal serum alphafetoprotein screening on the birth prevalence of neural tube defects in Manitoba: 1979–1990. Am J Hum Genet 51:A267.
Evers IM, de Valk HW, Visser GH. 2004. Risk of complications of pregnancy in women with type 1 diabetes: nationwide prospective study in the Netherlands. BMJ 328:915. Epub 2004 Apr 5.
Fabro S. 1981. Biochemical basis of thalidomide teratogenicity. In: Juchau MR ed. The biochemical basis of chemical teratogenesis. New York, Elsevier/North Holland, pp 159–78.
Fabro S, Brown NA. 1982. In utero alcohol exposure: threshold for effects? Reprod Toxicol Med Lett 1:1–4.
Fabro S, Smith RL. 1965. The teratogenic activity of thalidomide in the rabbit. J Pathol 91:511–19.
Faich G, Rosa FW. 1989. When a uniquely effective drug is teratogenic: the case of isotretinoin. N Engl J Med 321:656–7.
Fairgrieve SD, Jackson M, Jonas P, et al. 2000. Population based, prospective study of the care of women with epilepsy in pregnancy. BMJ 321:674–5.
Fantel AG, Shepard TH, Newell-Morris LL, et al. 1977. Teratogenic effects of retinoic acid in pigtail monkeys (*Macaca nemestrina*). I. General features. Teratology 15:65–72.
Fara GM, Del Corno G. 1985. Pregnancy outcome in the Seveso area after TCDD contamination. Prog Clin Biol Res 163B:279–85.
Farabee C. 1905. Inheritance of digital malformations in man. Papers of the Peabody Museum of American Archæology and Ethnology, 3. Cambridge, Harvard University, 69–77.
FDA. 1993. Proposals for folic acid fortification and labeling of certain foods to reduce risk of neural tube defects. Fed Regist 58:53524–97.
FDA. 1996. Food standards: amendment of the standards of identity for enriched grain products to require addition of folic acid. Fed Regist 61:8781–97.
Fedrick J. 1973. Epilepsy and pregnancy: a report from the Oxford Record Linkage Study. BMJ 1:442–8.
Fedrick J, Adelstein P. 1976. Area differences in the incidence of neural tube defect and the rate of spontaneous abortion. Br J Prev Soc Med 30:32–5.
Feitosa MF, Kreiger H. 1992. Demography of the human sex ratio in some Latin American countries, 1967–1986. Human Biol 64:523–30.
Feldcamp M, Carey JC. 1993. Clinical teratology counseling and consultation case report: low dose methotrexate exposure in the early weeks of pregnancy. Teratology 47:533–9.
Felisati D. 1962. Thalidomide and congenital abnormalities. Lancet 2:724–5.

Ferencz C, Rubin JD, McCarter RJ, et al. 1987. Cardiac and noncardiac malformations: observations in a population-based study. Teratology 35:367–78.
Ferguson LR, Ford JH. 1997. Overlap between mutagens and teratogens. Mutat Res 396:1–8.
Ferguson-Smith MA. 1983. The reduction of anencephalic and spina bifida births by maternal serum alphafetoprotein screening. Br Med Bull 39:365–72.
Ferner RE, Smith JM. 1992. Lithium and pregnancy. Lancet 339:869.
Feuchtbaum LB, Currier RJ, Riggle S, et al. 1999. Neural tube defect prevalence in California (1990–1994): eliciting patterns by type of defect and maternal race/ethnicity. Genet Test 3:265–72.
Field B, Kerr C. 1973. Potato blight and neural-tube defects. Lancet 2:507–8.
Field B, Kerr C. 1988. Reproductive behaviour and consistent patterns of abnormality in offspring of Vietnam veterans. J Med Genet 25:819–26.
Fields H. 1972. Memoir of Douglas P. Murphy. Trans Stud Coll Physicians Phila 40:80–1.
Finnell RH, Chernoff GF. 1982. Mouse fetal hydantoin syndrome: effects of maternal seizures. Epilepsia 23:423–9.
Finnell RH, Dansky LV. 1991. Parental epilepsy, antiepileptic drugs, and reproductive outcome: epidemiologic and experimental findings spanning three decades; 1: animal studies. Reprod Toxicol 5:281–99.
Finnell RH, Nau H, Yerby MS. 1995. Teratogenicity of antiepileptic drugs. In: Levy RH, Mattson RH, Meldrum BS eds. Antiepileptic drugs, vol. 4. New York, Raven Press, pp 209–30.
Fisch RO, Walker WA, Anderson JA. 1996. Prenatal and postnatal developmental consequences of maternal phenylketonuria. Pediatrics 37:979–86.
Fisher RA. 1958. The genetical theory of natural selection. New York, Dover.
Fixler DE, Paster P, Chamberlain M, et al. 1989. Trends in congenital heart disease in Dallas County births. Circulation 81:137–42.
Fleming DM, Knox JD, Crombie DL. 1981. Debendox in early pregnancy and fetal malformation. BMJ 283:99–101.
Flynn A, Miller SI, Martier SS, et al. 1981. Zinc status of pregnant alcoholic women: a determinant of fetal outcome. Lancet 1:572–4.
Flynt JW. 1974. Trends in surveillance of congenital malformations. In: Janerich DT, Skalko RG, Porter IH eds. Congenital defects: new directions in research. New York, Academic Press, pp 119–28.
Fogh-Anderson P. 1942. Inheritance of harelip and cleft palate. Copenhagen, Busck.
Følling A. 1934. Utskillelse av fenylpyrodruesyre i urinen som stoffskifteanomali i forbindelse med imbecillitet. Norsk Med Tidskr 8:1054–9.
Følling I. 1994. The discovery of phenylketonuria. Acta Paediatr Suppl 407:4–10.
Forbes JA. 1969. Rubella: historical aspects. Am J Dis Child 118:5–11.
Forrest F, Florey CD, Taylor D, et al. 1991. Reported social alcohol consumption and infants' development at 18 months. BMJ 303:22–6.
Forrester MB, Merz RD. 2000. Prenatal diagnosis and elective termination of neural tube defects in Hawaii, 1986–1997. Fetal Diagn Ther 15:146–51.
Forrester MB, Merz RD, Yoon PW. 1998. Impact of prenatal diagnosis and elective termination on the prevalence of selected birth defects in Hawaii. Am J Epidemiol 148:1206–11.
Fox DJ, Druschel CM. 2003. Estimating prevalence of fetal alcohol syndrome (FAS). Birth Defects Res A Clin Mol Teratol 67:604–8.
Frank AL. 1998. Review of "A hazardous inquiry," by Alan Mazur. JAMA 280:1276.
Fraser FC. 1959. Antenatal factors in congenital defects: problems and pitfalls. N Y St J Med 59:597.
Fraser FC. 1965. Some teratological implications of quasi-continuous variation. Abstracts of the 5th Annual Meeting of the Teratology Society, San Francisco, May 26–28.
Fraser FC. 1971. The epidemiology of the common malformations as related to environmental monitoring. In: Hook EB, Janerich DT, Porter IH eds. Monitoring birth defects and environment: the problem of surveillance. New York, Academic Press, pp 85–96.
Fraser FC. 1976. The multifactorial/threshold concept – uses and misuses. Teratology 14:267–80.

Fraser FC. 1977. Relation of animal studies to the problem in man. In: Wilson JG, Fraser FC eds. Handbook of teratoilogy, vol. 1. New York, Plenum Press, pp 75–98.
Fraser FC. 1980a. The William Allan Memorial Award Address: evolution of a palatable multifactorial threshold model. Am J Hum Genet 32:796–813.
Fraser FC. 1980b. The genetics of cleft lip and palate: yet another look. In: Pratt RM, Christiansen RI eds. Current research trends in prenatal craniofacial development. New York, Elsevier, pp 357–66.
Fraser FC. 1990. Of mice and children: reminiscences of a teratogeneticist. In: Kalter H ed. Issues and reviews in teratology, vol. 5. New York, Plenum Press, pp 1–75.
Fraser FC. 1996. Liability, thresholds, malformations, and syndromes. Am J Med Genet 66:75–6.
Fraser FC. 1998. Some underlooked properties of the multifactorial/threshold model. Am J Hum Genet 62:1262–5.
Fraser FC, Fainstat TD. 1951. Production of congenital defects in the offspring of pregnant mice treated with cortisone: progress report. Pediatrics 8:27–33.
Fraser FC, Sajoo A. 1995. Teratogenic potential of corticosteroids in humans. Teratology 51:45–6.
Fraser FC, Warburton D. 1964. No association of emotional stress or vitamin supplement during pregnancy to cleft lip or palate in man. Plast Reconstr Surg 33:395–9.
Fraser GR, Calnan JS. 1961. Cleft lip and palate: seasonal incidence, birth weight, birth rank, sex, site, associated malformations and parental age. Arch Dis Child 36:420–3.
Fraser JL, Watt HJ. 1964. Megaloblastic anemia in pregnancy and the puerperium. Am J Obstet Gynecol 89:532–4.
Freinkel N, Gabbe SG, Hadden DR, et al. 1985. Summary and recommendations of the Second International Workshop-Conference on Gestational Diabetes Mellitus. Diabetes 34:123–6.
Frias ML, Castilla EE, Paz JE. 1977. Descriptive system for congenital limb defects. Teratology 15:63–70.
Fried PA, O'Connell CM. 1987. A comparison of the effects of prenatal exposure to tobacco, alcohol, cannibis and caffeine on birth size and subsequent growth. Neurotoxicol Teratol 9:78–85.
Fried PA, Watkinson B. 1988. 12- and 24-month neurobehavioral follow-up of children prenatally exposed to marihuana, cigarettes and alcohol. Neurtoxicol Teratol 10:305–13.
Friedman JM, Polifka JE. 2000. Teratogenic effects of drugs: a resource for clinicians (TERIS). Baltimore, MD, Johns Hopkins University Press.
Fries H. 1970. Lithium in pregnancy. Lancet 1:1233.
Fries MH, Kuller JA, Norton ME, et al. 1993. Facial features of infants prenatally exposed to cocaine. Teratology 48:413–20.
Friis ML. 1989. Facial clefts and congenital heart defects in children of parents with epilepsy: genetic and environmental etiologic factors. Act Neurol Scand 79:433–59.
Friis ML. 1990. Malformations in children of epileptic patients. In: Dam M, Gram L eds. Comprehensive epileptology. New York, Raven, pp 309–19.
Friis ML, Hauge M. 1985. Congenital heart defects in live-born children of epileptic parents. Arch Neurol 42:374–6.
Frith K. 1978. Fetal malformations after Debendox treatment in early pregnancy. BMJ 1:925.
Froster UG, Baird PA. 1992. Congenital defects of the limbs and alcohol exposure during pregnancy: data from a population based study. Am J Med Genet 44:782–5.
Fuchs M, Iosub S, Bingol N, et al. 1980. Palpebral fissure size revisited. J Pediatr 96:77–8.
Gaily E. 1990. Distal phalangeal hypoplasia in children with prenatal phenyoin exposure: results of a controlled anthropometric study. Am J Med Genet 35:574–8.
Gaily E, Granström ML. 1992. Minor anomalies in children of mothers with epilepsy. Neurology 42:128–31.
Gaily E, Granström ML, Hiilesmaa V, et al. 1988. Minor anomalies in offspring of epileptic mothers. J Pediatr 112:520–9.
Gal I, Kirman B, Stern J. 1967. Hormonal pregnancy tests and congenital malformation. Nature 216:83.

Gal I, Sharman IM, Pryse-Davies J. 1972. Vitamin A in relation to human congenital malformations. In: Woollam DH, ed. Advances in teratology, vol. 5. New York, Academic Press, pp 143–59.

Gal I, Sharman IM, Pryse-Davies J, et al. 1968. Vitamin A as a possible factor in human teratology. Proc Nutr Soc 28:9A.

Galindo A, Burguillo AG, Azriel S, et al. 2006. Outcome of fetuses in women with pregestational diabetes mellitus. J Perinat Med 34:323–31.

Galton F. 1889. Natural inheritance. London, Macmillan.

Garcia-Bournissen F, Tsur L, Goldstein LH, et al. 2008. Fetal exposure to isotretinoin – an international problem. Reprod Toxicol 25:124–8.

Gardiki-Kouidou P, Seller MJ. 1988. Amniotic fluid folate, vitamin B12 and transcobalamins in neural tube defects. Clin Genet 33:441–8.

Geelen JA. 1979. Hypervitaminosis A induced teratogenesis. CRC Crit Rev Toxicol 6:351–75.

Gefrides LA, Bennett GD, Finnell RH. 2002. Effects of folate supplementation on the risk of spontaneous and induced neural tube defects in Splotch mice. Teratology 65:63–9.

Geiger JM, Baudin M, Saurat JH. 1994. Teratogenic risk with etretinate and acitretin treatment. Dermatology 89:109–16.

Gelineau-van Waes J, Finnell RH. 2001. Importance of model organisms in understanding the biology and genetic basis of human nonsyndromic neural tube defects. Teratology 64:177–80.

Gemma S, Vichi S, Testai E. 2007. Metabolic and genetic factors contributing to alcohol induced effects and fetal alcohol syndrome. Neurosci Biobehav Rev 31:221–9.

Generoso WM, Shurbaji AG, Piergorsch WW, et al. 1991. Developmental response of zygotes exposed to similar mutagens. Mutat Res 250:439–46.

German J, Kowal A, Ehlers KH. 1970. Trimethadione and human teratogenesis. Teratology 3:349–62.

Gershon AA. 1995. Chickenpox, measles, and mumps. In: Remington JS, Klein JO eds. Infectious diseases of the fetus and newborn infant. Philadelphia, PA, Saunders, pp 565–618.

Ghosh A, Woo JS, Poon IM, et al. 1981. Neural-tube defects in Hong Kong Chinese. Lancet 2:468–9.

Giannoulis CH, Papathanasiou K, Tantanasis TH, et al. 2005. Isotretinoin (Ro-Accutane) teratogenesis. A case report. Clin Exp Obstet Gynecol 32:78.

Gibson GT, Baghurst PA, Colley DP. 1983. Maternal alcohol, tobacco, and cannabis consumption and the outcome of pregnancy. Aust N Z J Obstet Gynecol 23:15–19.

Gibson GT, Colley DP, McMichael AJ, et al. 1981. Congenital anomalies in relation to the use of doxylamine/dicyclomine and other antenatal factors: an ongoing prospective study. Med J Aust 1:410–14.

Gillman J, Gilbert C, Gillman T. 1948. A preliminary report on hydrocephalus, spina bifida and other congenital anomalies in the rat produced by trypan blue: the significance of these results in the interpretation of congenital malformations following rubella. S Afr J Med Sci 13:47–90.

Gillman J, Gilbert C, Spence I, et al. 1951. A further report on congenital anomalies in the rat produced by trypan blue. S Afr J Med Sci 16:125–35.

Gindler J, Li Z, Berry RJ, et al. 2001. Folic acid supplements during pregnancy and risk of miscarriage. Lancet 358:796–800.

Ginsburg KA, Blacker CM, Abel EL, et al. 1991. Fetal alcohol exposure and adverse pregnancy outcomes. Contrib Gynecol Obstet 18:115–29.

Giroud A, Martinet M. 1955. Malformations embryonnaires par hypervitaminose A. Arch Fr Pediatr 2:292–300.

Giroud A, Martinet M. 1956a. Tératogénèse par hautes doses de vitamine A en fonction des stades du développement. Arch Anat Micr 45:77–99.

Giroud A, Martinet M. 1956b. Malformations de la face et hypervitaminose A. Rev Stomatol 57:454–63.

Gittelsohn AM, Milham S. 1962. Declining incidence of central nervous system anomalies in New York State. Br J Prev Soc Med 16:153–8.

Gittelsohn AM, Milham S. 1965. Vital record incidence of congenital malformations in New York State. In: Neel JV, Shaw MW, Schull WJ eds. Genetics and epidemiology of chronic disease. Washington, DC, Department of Health, Education and Welfare, pp 305–20.

Giusti RM, Iwamoto K, Hatch EE. 1995. Diethylstilbestrol revisited: a review of the long-term health effects. Ann Int Med 122:778–88.

Gladstone DJ, Bologa M, Maguire C, et al. 1992. Course of pregnancy and fetal outcome following maternal exposure to carbamazepine and phenytoin: a prospective study. Reprod Toxicol 6:257–61.

Gold JA, Zaremski MJ, Lev ER, et al. 1993. *Daubert v Merrell Dow*; the Supreme Court tackles scientific evidence in the courtroom. Science 270:2964–7.

Golden J. 2000. "A tempest in a cocktail glass": mothers, alcohol, and television, 1977–1996. J Health Polit Policy Law 25:473–98.

Golding J. 1982. Evidence for an environmental aetiology for anencephalus. In: Perseaud TV ed. Advances in the study of birth defects, vol. 7. Lancaster, PA, MTP Press, pp 19–45.

Goldman LR, Paigen B, Magnant MM, et al. 1985. Low birth weight, prematurity and birth defects in children living near the hazardous waste site, Love Canal. Hazardous Waste Hazardous Materials 2:209–23.

Goldstein I, Wexler D. 1931. Rosette formation in the eyes of irradiated human embryos. Arch Ophthalmol 5:591–600.

Goldstein L. 1928. Ovarian irradiation: a review of the experimental literature. Am J Obstet Gynecol 6:747–62.

Goldstein L, Murphy DP. 1929. Microcephalic idiocy following radium therapy for uterine cancer during pregnancy. Am J Obstet Gynecol 18:89–95.

Gorman C. 1994. Thalidomide's back. Time (June 13), p 67.

Graham JM, Edwards MJ, Edwards MJ. 1998. Teratogen update: gestational effects of maternal hyperthermia due to febrile illnesses and resultant patterns of defects in humans. Teratology 58:209–21.

Graham JM, Hanson JW, Darby BL, et al. 1988. Independent dysmorphology evaluations at birth and 4 years of age for children exposed to varying amounts of alcohol in utero. Pediatrics 81:772–8.

Grant TM, Huggins JE, Sampson PD, et al. 2009. Alcohol use before and during pregnancy in Western Washington, 1989–2004: implications for the prevention of fetal alcohol spectrum disorders. Am J Obstet Gynecol 200:278.e1–8.

Gray R, Mukherjee RA, Rutter M. 2009. Alcohol consumption during pregnancy and its effects on neurodevelopment: what is known and what remains uncertain.

Green MD. 1996. Bendectin and birth defects: the challenge of mass toxic substances of abuse. Philadelphia, PA, University of Pennsylvania Press.

Green RF, Stoler JM. 2007. Alcohol dehydrogenase 1B genotype and fetal alcohol syndrome: a HuGe minireview. Am J Obstet Gynecol 197:12–25.

Greenberg G, Inman WHW, Weatherall JA, et al. 1975. Hormonal pregnancy tests and congenital malformations. BMJ 1:91–2.

Greenberg G, Inman WHW, Weatherall JA, et al. 1977. Maternal drug histories and congenital abnormalities. BMJ 2:853–6.

Greene MF. 2001. Diabetic embryopathy 2001: moving beyond the "diabetic milieu.". Teratology 63:16–18.

Greene T, Ernhart CB, Sokol RJ, et al. 1991. Prenatal alcohol exposure and preschool physical growth: a longitudinal analysis. Alcohol Clin Exp Res 5:905–13.

Greenwald P, Barlow JJ, Nasca PC, et al. 1971. Vaginal cancer after maternal treament with synthetic estrogens. N Engl J Med 285:390–2.

Greenwald P, Nasca PC. 1974. Stilbestrol exposure in utero: long term effect. In: Janerich DT, Skalko RG, Porter IH eds. Congenital defects: new directions in research. New York, Academic Press, pp 149–60.

Greenwood MR. 1985. Methylmercury poisoning in Irag: an epidemiological study of the 1971–1972 outbreak. J Appl Toxicol 5:148–59.

Gregg NM. 1941. Congenital cataract following German measles in the mother. Tr Ophthal Soc Aust 3:35–46.

Gregg NM. 1944. Further observations on congenital defects in infants following maternal rubella. Trans Ophthal Soc Aust 4:19–31.

Gregg NM, Beavis WR, Heseltine H, et al. 1945. The occurrence of congenital defects in children following maternal rubella during pregnancy. Med J Aust 2:22–6.

Griffith JP. 1887. Rubella (Rötheln: German measles): with a report of one hundred and fifty cases. Med Rec 32:11–17.

Grisso JA, Roman E, Inskip H, et al. 1984. Alcohol consumption and outcome of pregnancy. J Epidemiol Community Health 38:323–5.

Gruenwald P. 1947. Mechanisms of abnormal development. I. Causes of abnormal development in the embryo. Arch Pathol 44:398–436.

Grüneberg H. 1938. An analysis of the 'pleiotropic' effects of a new lethal mutation in the rat (*Mus norvegicus*). Proc Roy Soc B 125:123–44.

Grüneberg H. 1952. Genetical studies on the skeleton of the mouse. IV. quasi-continuous variations. J Genet 51:95–114.

Grüneberg H. 1963. The pathology of development: a study of inherited skeletal disorders in animals. New York, Wiley.

Gucciardi E, Pietrusiak MA, Reynolds DL, et al. 2002. Incidence of neural tube defects in Ontario, 1986–1999. CMAJ 167:237–40.

Guerin A, Nisenbaum R, Ray JG. 2007. Use of maternal GHb concentration to estimate the risk of congenital anomalies in the offspring of women with prepregnancy diabetes mellitus. Diabetes Care 30:1920–5.

Guthrie R, Susi A. 1963. A simple phenylalanine method for detecting phenylketonuria in large populations of newborn infants. Pediatrics 32:338–43.

Haeusler MC, Berghold A, Schoell W, et al. 1992. The influence of post-Chernobyl fallout on birth defects and abortion rates in Austria. Am J Obstet Gynecol 167:1025–31.

Hale F. 1933. Pigs born without eye balls. J Hered 24:105–6.

Hale F. 1935a. The relation of vitamin A to anophthalmos in pigs. Am J Ophthal 18:1087–93.

Hale F. 1935b. The relation of vitamin A to the eye development in the pig. J Anim Sci 1935:126–8.

Hale F. 1937. The relation of maternal vitamin deficiency to microphthalmia in pigs. Tex St J Med 33:228–32.

Hall J. 1997. Terathanasia, folic acid, and birth defects. Lancet 350:1322. author reply 1323–4.

Hall J. 2000. When is careless conception a form of child abuse? Lessons from maternal phenylketonuria. J Pediatr 136:12–3.

Hall MH. 1972. Folic acid deficiency and congenital malformations. J Obstet Gynecol Br Comm 79:159–6.

Hall MH. 1977. Folates and the fetus. Lancet 1:48–9.

Hall W, MacPhee D. 1985. The Agent Orange controversy in Australia: a contribution to the debate. Community Health Stud 9:109–19.

Halliday HL, Reid MM, McClure G. 1982. Results of heavy drinking in pregnancy. Br J Obstet Gynecol 89:892–5.

Halmesmäki E, Välimäki M, Roine R, et al. 1989. Maternal and paternal alcohol consumption and miscarriage. Br J Obstet Gynecol 96:188–91.

Halmesmäki E, Ylikorkala O. 1988. A retrospective study on the safety of prenatal ethanol treatment. Obstet Gynecol 72:545–9.

Hamkar R, Jalilvand S, Abdolbaghi MH, et al. 2006. Inadvertent rubella vaccination of pregnant women: evaluation of possible transplacental infection with rubella vaccine. Vaccine 24:3558–63.

Hanley WB, Platt LD, Bachman RP, et al. 1999. Undiagnosed maternal phenylketonuria: the need for prenatal selective screening or case finding. Am J Obstet Gynecol 180:986–94.

Hanson JW. 1986. Fetal hydantoin effects. Teratology 33:349–53.
Hanson JW, Buehler BA. 1982. Fetal hydantoin syndrome: current status. Pediatrics 101:816–8.
Hanson JW, Jones KL, Smith DW. 1976. Fetal alcohol syndrome. JAMA 235:1458–60.
Hanson JW, Smith DW. 1975. The fetal hydantoin syndrome. J Pediatr 87:285–90.
Hanson JW, Streissguth AP, Smith DW. 1978. The effects of moderate alcohol consumption on fetal growth and morphogenesis. J Pediatr. 92:457–60.
Hanson U, Persson B, Thunell S. 1990. Relationship between haemoglobin A1c in early type 1 (insulin-dependent) diabetic pregnancy and the occurrence of spontaneous abortion and fetal malformation in Sweden. Diabetologia 33:100–4.
Happle R, Traupe H, Bounameaux Y, et al. 1984. Teratogene Wirkung von Etretinat beim Menschen. Deut Med Woch 109:1476–80.
Harada M. 1978. Congenital Minamata disease: intrauterine methylmercury poisoning. Teratology 18:285–8.
Harada M. 1995. Minamata disease: methylmercury poisoning in Japan caused by environmental pollution. Crit Rev Toxicol 25:1–24.
Harada Y. 1968. Congenital (or fetal) Minamata disease. In: Study Group of Minamata/Disease. Kumamoto, Kumamoto University, pp 93–117.
Hardelid P, Cortina-Borja M, Munro A, et al. 2008. The birth prevalence of PKU in populations of European, South Asian, and sub-Saharan African ancestry living in South East England. Ann Hum Genet 72:65–71.
Harlap S, Shiono PH. 1980. Alcohol, smoking, and incidence of spontaneous abortions in the first and second trimester. Lancet 2:173–6.
Harlap S, Shiono PH, Ramcharan S. 1980. A life table of spontaneous abortions and the effects of age, parity, and other variables. In: Porter IH, Hook EB eds. Human embryonic and fetal death. New York, Academic Press, pp 145–58.
Harremoës P, Gee D, et al. eds. 2002. The precautionary principle in the 20th century. London, Earthscan, p 93.
Harris JA, Shaw GM. 1995. Neural tube defects – why are rates high among populations of Mexican descent? Environ Health Perspect 103(Suppl 6):163–4.
Harris LE, Steinberg AG. 1954. Abnormalities observed during the first six days of life in 8,716 live-born infants. Pediatrics 14:314–26.
Hartmann D, Bollag W. 1993. Historical aspects of the oral use of retinoids in acne. J Dermatol 20:674–8.
Hatch MC, Stein Z. 1986. Agent Orange and risks to reproduction: the limits of epidemiology. Teratag Carcinog Mutag 6:185–202.
Hay AW. 1977. Tetrachlorodibenzo-*p*-dioxin release at Seveso. Disasters 1:289–308.
Hayes C, Werler MM, Willett WC, et al. 1996. Case-control study of periconceptional folic acid supplementation and oral clefts. Am J Epidemiol 143:1229–34.
Heath CW. 1983. Field epidemiological studies of populations exposed to waste dumps. Environ Health Perspect 48:3–7.
Heinonen OP. 1973. Diethstilbestrol in pregnancy: frequency of exposure and usage patterns. Cancer 31:573–7.
Hellekson KL. 2001. NIH consensus statement on phenylketonuria. Am Fam Physician 63:1430–2.
Henderson IW. 1977. Congenital deformities associated with Bendectin. CMAJ 117:721–2.
Henderson J, Kesmodel U, Gray R. 2007. Systematic review of the fetal effects of parental binge-drinking. J Epidemiol Community Health 61:1069–73.
Hendricks KA, Simpson JS, Larsen RD. 1999. Neural tube defects along the Texas-Mexico border, 1993–1995. Am J Epidemiol 49:1119–27.
Hendrickx AG, Binkerd PE. 1990. Nonhuman primates and teratological research. J Med Primatol 19:81–108.
Hendrickx AG, Binkerd PE, Rowland JM. 1983. Developmental toxicity and nonhuman primates: interspecies comparisons. In: Kalter H ed. Issues and reviews in teratology, vol. 1. New York, Plenum Press, pp 149–80.

Hendrickx AG, Peterson P, Hartmann D, et al. 2000. Vitamin A teratogenicity and risk assessment in the macaque retinoid model. Reprod Toxicol 4:311–23.
Hendrickx AG, Silverman S, Pellegrini M, et al. 1980. Teratological and radiocephalometric analysis of craniofacial maformations induced with retinoic acid in rhesus monkeys (*Macaca mulatta*). Teratology 22:13–22.
Henningham J. 1975. Drug marketing rules criticised. The Australian (January 22), p 6.
Henry J, Sloane M, Black-Pond C. 2007. Neurobiology and neurodevelopmental impact of childhood traumatic stress and prenatal alcohol exposure. Lang Speech Hear Serv Sch 38:99–108.
Herbicide Assessment Commission. 1970. Summary of presentations. Chicago, IL, American Association for the Advancement of Science.
Herbst AL, Cole P, Colton T, et al. 1977. Age-incidence and risk of diethylstilbestrol-related clear cell adenocarcinoma of the vagina and cervix. Am J Obstet Gynecol 128:43–8.
Herbst AL, Cole P, Norusis MJ, et al. 1979. Epidemiologic aspects and factors related to survival in 384 Registry cases of clear cell adenocarcinoma of the vagina and cervix. Am J Obstet Gynecol 135:876–83.
Herbst AL, Kurman RJ, Scully RE. 1972a. Vaginal and cervical abnormalities after exposure to stilbestrol in utero. Obstet Gynecol 40:287–98.
Herbst AL, Kurman RJ, Scully RE, et al. 1972b. Clear-cell adenocarcinoma of the genital tract in young females: registry report. N Engl J Med 287:259–64.
Herbst AL, Scully RE. 1970. Adenocarcinoma of the vagina in adolescence: a report of 7 cases including 6 clear-cell carcinomas (so-called mesonephromas). Cancer 25:745–57.
Herbst AL, Ulfelder U, Poskanzer DC. 1971. Adenocarcinoma of the vagina: association of maternal stilbestrol therapy with tumor appearance in young women. N Engl J Med 284:878–81.
Hernández-Díaz S, Werler MM, Walker AM, et al. 2001. Neural tube defects in relation to use of folic acid antagonists during pregnancy. Am J Epidemiol 153:961–8.
Hernández-Herrera RJ, Alcalá-Galván LG, Flores-Santos R. 2008. [Neural tube defect in 248,352 consecutive newborns.] In Spanish. Rev Med Inst Mex Seguro Soc 46:201–4.
Hertig AT, Rock J, Adams EC, et al. 1959. Thirty-four fertilized human ova, good, bad, and indifferent, recovered from 210 women of known fertility: a study of biologic wastage in early human pregnancy. Pediatrics 23:202–11.
Heseker HB, Mason JB, Selhub J, et al. 2009. Not all cases of neural-tube defect can be prevented by increasing the intake of foic acid. Br J Nutr 102:173–80.
Hexter AC, Harris JA. 1991. Bias in congenital malformations information from the birth certificate. Teratology 44:177–80.
Hibbard BM, Hibbard ED, Jeffcoate TN. 1965. Folic acid and reproduction. Acta Obstet Gynecol Scand 44:375–400.
Hibbard ED, Smithells RW. 1965. Folic acid metabolism and human embryopathy. Lancet 1: 1254–5.
Hicks SP, D'Amato CJ. 1966. Effects of ionizing radiation on mammalian development. In: Woollam DH ed. Advances in teratology, vol. 1. New York, Academic Press, pp 196–250.
Higa K, Dan K, Manabe H. 1987. Varicella-zoster virus infections during pregnancy: hypothesis concerning the mechanisms of congenital malformations. Obstet Gynecol 69:214–22.
Hill RM, Verniaud VM, Horning MG, et al. 1974. Infants exposed in utero to antiepileptic drugs: a prospective study. Am J Dis Child 27:645–53.
Hingson R, Alpert JJ, Day N, et al. 1984. Effects of maternal drinking and marijuana use on fetal growth and development. Pediatrics 70:539–46.
Hippel E. 1907. Über experimentelle Erzeugung von angeborenen Star bei Kaninchen nebst Bemerkungen über gleichzeitig beobachteten Mikrophthalmus und Lidcolobom. Graefe Arch Ophthal 65:326–60.
Hobbs MS. 1969. Risk of anencephaly in migrant and non-migrant women in the Oxford area. Br J Prev Soc Med 23:174–8.

Hoffman JI. 1990. Congenital heart disease: incidence and inheritance. Paediatr Clin N Am 37:25–43.

Hoffman JI, Christianson R. 1978. Congenital heart disease in a cohort of 19,502 births with long-term follow-up. Am J Cardiol 42:641–7.

Hoffmann W. 2001. Fallout from the Chernobyl nuclear disaster and congenital malformations in Europe. Arch Environ Health 56:478–84.

Hofman KJ, Steel G, Kazazian HH, et al. 1991. Phenylketonuria in U.S. blacks: molecular analysis of the phenylhydroxylase gene. Am J Hum Genet 48:791–8.

Hofmann C, Eichele G. 1994. Retinoids in development. In: Sporn MB, Roberts AB, Goodman DS eds. The retinoids: biology, chemistry, and medicine. New York, Raven Press, pp 387–441.

Holden C. 1980. Love Canal residents under stress. Science 208:242–4.

Hollowell JG, Hannon WH. 1997. Teratogen update: iodine deficiency, a community teratogen. Teratology 55:389–405.

Hollstedt C, Dahlgren L, Rydberg U. 1983. Outcome of pregnancy in women treated at an alcohol clinic. Acta Psychiatr Scand 67:236–48.

Holmes LB. 1983. Teratogen update: bendectin. Teratology 27:277–81.

Holmes LB. 1989. Does taking vitamins at the time of conception prevent neural tube defects? JAMA 260:3181.

Holmes LB. 1992. Fetal environmental toxins. Pediatr Rev 13:364–9.

Holmes LB. 2002. The teratogenicity of antiepileptic drugs: a progress report. J Med Genet 39:245–7.

Holmes LB, Baldwin EJ, Smith CR, et al. 2009. Increased frequency of isolated cleft palate in infants exposed to lamotrigine during pregnancy. Neurology 70:2152–8.

Holmes LB, Cann C, Cook C. 1985. Examination of infants for both minor and major malformations to evaluate for possible teratogenic exposures. Prog Clin Biol Res 163B:59–63.

Holmes LB, Harvey EA, Coull BA, et al. 2001. The teratogenicity of antiepileptic drugs. N Engl J Med 344:1132–8.

Holmes LB, Rosenberger PB, Harvey EA, et al. 2000. Intelligence and physical features of children of women with epilepsy. Teratology 61:196–202.

Holmgren S. 2009. Low dose alcohol exposure during pregnancy – does it harm? A systematic literature survey. Östersund Stockholm, Swedish National Institute of Public Health.

Holzgreve W, Tercanli S, Pietrzik K. 1991. Vitamins to prevent neural tube defects. Lancet 338:639–40.

Honein MA, Paulozzi LJ, Erickson DM. 2001a. Continued occurrence of Accutane®-exposed pregnancies. Teratology 64:142–7.

Honein MA, Paulozzi LJ, Mathews TJ, et al. 2001b. Impact of folic acid fortification of the US food supply on the occurrence of neural tube defects. JAMA 285:2981–6.

Honoré LH. 1988. Male excesses among anatomically normal fetuses in spontaneous abortions. Am J Med Genet 30:843–4.

Hook EB. 1976. Changes in tobacco smoking and ingestion of alcohol and caffeinated beverages during early pregnancy: are these consequences, in part, of feto-protective mechanisms diminishing maternal exposure to embryotoxins? In: Kelly S, Hook EB, Janerich DT, et al. eds. Birth defects: risks and consequences. New York, Academic Press, pp 173–83.

Hook EB. 2001. Statistical and logical considerations in evaluating the association of prenatal folic-acid supplementation with pregnancy loss. Am J Med Genet 104:81–2.

Hook EB, Czeizel AE. 1997. Can terathansia explain the protective effect of folic-acid supplementation on birth defects? Lancet 350:513–15.

Hook EB, Porter IH. 1980. Terminological conventions, methodological considerations, temporal trends, specific genes, environmental hazards, and some other factors pertaining to embryonic and fetal death. In: Porter IH, Hook EB eds. Human embryonic and fetal death. New York, Academic Press, pp 1–18.

Hopf G, Mathias B. 1988. Teratogenicity of isotretinoin and etretinate. Lancet 2:1143.

Hopf NS, Ruder AM, Succop P. 2009. Background levels of polychlorinated biphenyls in the U.S. population. Sci Total Environ Sep 19 [Epub ahead of print].
Horne HW. 1985. Evidence of improved pregnancy outcome with diethylstilbestrol (DES) treatment of women with previous pregnancy failures: a retrospective analysis. J Chron Dis 38: 873–80.
Horowitz RI, McFarlane MJ, Brennan TA, et al. 1985. The role of susceptibility bias in epidemiologic research. Arch Int Med 145:909–12.
Horowitz RI, Viscoli CM, Merino M, et al. 1988. Clear cell carcinoma of the vagina and cervix: incidence, undetected disease, and diethylstilbestrol. J Clin Epidemiol 41:593–7.
Howkins J, Lawrie RS. 1939. Iniencephalus. J Obstet Gynecol Br Comm 46:25–31.
Hummler H, Korte R, Hendrickx AG. 1990. Induction of malformations in cynomolgus monkey with 13-cis retinoic acid. Teratology 42:263–72.
Hummler H, Schüpbach ME. 1981. Studies in reproductive toxicology and mutagenicity with Ro 10-9359. In: Orfanos CE, Braun-Falco O, et al. eds. Retinoids: advances in basic research and therapy. Berlin, Springer, pp 49–59.
Hutchings DE. 1993. A contemporary overview of behavioral teratology: a perspective from the field of substance abuse. In: Kalter H ed. Issues and reviews in teratology, vol. 6. New York, Plenum Press, pp 125–67.
Hvas CL, Henriksen TB, Østergaard JR, et al. 2000. Epilepsy and pregnancy: effect of antiepileptic drugs and lifestyle on birthweight. Br J Obstet Gynaecol 107:896–902.
Igata A. 1993. Epidemiological and clinical features of Minamata disease. Environ Res 63:157–69.
Iizuka T. 1973. High incidence of cleft lip and clft palate in the human embryos and early fetuses. Okajimas Fol Anat Jpn 50:259–71.
Ingalls TH. 1947. Etiology of mongolism: epidemiologic and teratologic implications. Am J Dis Child 74:147–65.
Ingalls TH. 1953. Preventive prenatal pediatrics. Ann Pediatr 6:3–62.
Ingalls TH, Avis FR, Curley FJ, et al. 1953. Genetic determinants of hypoxia-induced congenital anomalies. J Hered 44:185–9.
Ingalls TH, Curley FJ, Temin HM. 1950. Anoxia as a cause of fetal death and congenital defect in the mouse. Am J Dis Child 80:34–45.
Institute of Medicine (U.S.). 1994. Veterans and agent Orange: health effects of herbicides used in Vietnam. Washington, DC, National Academy Press.
Institute of Medicine (U.S.). 1998. Veterans and agent Orange: update. Washington, DC, National Academy Press.
Institute of Medicine (U.S.). 2006. Veterans and agent Orange: update. Washington, DC, National Academy Press.
Iosub S, Fuchs M, Bingol N, et al. 1985. Palpebral fissure length in black and Hispanic children: correlation with head circumference. Pediatrics 75:318–20.
Itikala PR, Watkins ML, Mulinare J, et al. 2001. Maternal multivitamin use and orofacial clefts in offspring. Teratology 63:79–86.
Jacobs JB, McNamara D. 1996. Vietnam veterans and the Agent Orange controversy. Armed Forces Soc 13:57–80.
Jacobson JL, Jacobson SW. 1997. Teratogen update: polychlorinated biphenyls. Teratology 55:338–47.
Jacobson SJ, Jones K, Johnson K, et al. 1992. Prospective multicentre study of pregnancy outcome after lithium exposure during first trimester. Lancet 339:530–3.
Jacques PF, Selhub J, Bostom AG, et al. 1999. The effect of folic acid fortification on plasma folate and total homocysteine concentrations. N Engl J Med 340:1449–54.
Jäger-Roman E, Deichi A, Jakob S, et al. 1986. Fetal growth, major malformations, and minor anomalies in infants born to women receiving valproic acid. J Pediatr 108:997–1004.
James WH. 1981. Recurrence rates for neural tube defects and vitamin supplementation. J Med Genet 18:249–51.
James WH. 1995. What stabilizes the sex ratio? Ann Hum Genet 59:243–9.

James WH. 1998. Declines in population sex ratios at birth. JAMA 280:1139.
James WH. 2000. Secular movements in sex ratios of adults and of births in populations during the past half century. Hum Reprod 15:1178–83.
James WH. 2002. Offspring sex ratios of people exposed to dioxin and dioxin-like compounds. Occup Environ Med 58:202.
James WH. 2008. Evidence that mammalian sex ratios at birth are partially controlled by parental hormone levels around the time of conception. J Endocrinol 198:3–15.
Janerich DT. 1973. Epidemic waves in the prevalence of anencephaly and spina bifida in New York State. Teratology 8:253–6.
Janerich DT, Burnett WS, Feck G, et al. 1981. Cancer incidence in the Love Canal area. Science 212:1404–7.
Janz D. 1982. On major malformations and minor anomalies in the offspring of parents with epilepsy: review of the literature. In: Janz D, Bossi L, Dam M, et al. eds. Epilepsy, pregnancy, and the child. New York, Raven Press, pp 211–22.
Janz D, Fuchs U. 1964. Sind antiepileptische Medicamente während der Schwangerschaft schädlich? Deut Med Woch 89:241–3.
Jarrett TJ. 1993. Gestational diabetes: a non-entity? BMJ 306:37–8.
Jarvin SV. 2009. Thryoid carcinoma in children and adolescents resulting from the Chernobyl accident: possible causes of the incidence increase overestimation. Cesk Patol 45:50–2.
Jefferis BJ, Power C, Hertzman C. 2002. Birth weight, childhood socioeconomic environment, and cognitive development in the. 1958. British birth cohort study. BMJ 325:305.
Jensen DM, Damm P, Mœlsted-Pedersen L, et al. 2004. Outcomes in type 1 diabetic pregnancies: a nationwide, population-based study. Diab Care 27:2819–23.
Jensen DM, Korsholm L, Ovesen P, et al. 2009. Peri-conceptional A1C and risk of serious adverse pregnancy outcome in 933 women with type 1 diabetes. Diab Care 32:1046–8.
Jewell D, Young G. 2000. Interventions for nausea and vomiting in early pregnancy. Cochrane Lib CD000145, pp 1–12.
Jick SS, Terris BZ, Jick H. 1993. First trimester topical tretinoin and congenital disorders. Lancet 341:1181–2.
Johns K, Olynik C, Kreisman S, et al. 2006. Gestational diabetes mellitus outcome in 394 patients. J Obstet Gynaecol Can 28:122–7.
Johnson KC, Rouleau J. 1997. Temporal trends in Canadian birth defects birth prevalences, 1979–1993. Can J Public Health 88:169–76.
Johnson LD, Driscoll SG, Hertig AT, et al. 1979. Vaginal adenosis in stillborns and neonates exposed to diethylstilbestrol and steroidal estrogens and progestins. Obstet Gynecol 53:671–9.
Jones KL, Hanson JW, Smith DW. 1978. Palpebral fissure size in newborn infants. J Pediatr 92:787.
Jones KL, Hoyme HE, Robinson LK, et al. 2009. Developmental pathogenesis of short palpebral fissure length in children with fetal alcohol syndrome. Birth Defects Res A Clin Mol Teratol Apr 6 [Epub ahead of print].
Jones KL, Lacro RJ, Johnson KA, et al. 1989. Pattern of malformations in the children of women treated with carbamazepine during pregnancy. N Engl J Med 320 1661–6.
Jones KL, Smith DW. 1973. Recognition of the fetal alcohol syndrome in early infancy. Lancet 2:999–1001.
Jones KL, Smith DW. 1978. Effects of alcohol on the fetus. N Engl J Med 298:55.
Jones KL, Smith DW, Streissguth AP, et al. 1974. Outcome of pregnancy in chronic alcoholic women. Lancet 1:1076–8.
Jones KL, Smith DW, Ulleland CN, et al. 1973. Pattern of malformation in offspring of chronic alcoholic mothers. Lancet 1:1267–71.
Juárez-Olguín H, Belmont-Gómez A, Flores-Pérez J, et al. 2008. Malformations in newborns associated to anticonvulsant consumption during pregnancy: experience in third level hospital of Mexico. Rev Invest Clin 60:15–20.
Juriloff DM, Gunn TM, Harris DG, et al. 2001. Multifactorial genetics of exencephaly in SELH/Bc mice. Teratology 64:189–200.

Kadir RA, Sabin C, Whitlow B, et al. 1999. Neural tube defects and periconceptional folic acid in England and Wales: retrospective study. BMJ 319:92–3.

Kajii T, Kida M, Takahashi K. 1972. The effects of thalidomide intake during 113 human pregnancies. Teratology 8:163–6.

Källén B. 1988. Comments on teratogen update: lithium. Teratology 38:597.

Källén B. 1994. Maternal carbamazepine and infant spina bifida. Reprod Toxicol 8:203–5.

Källén B. 2007. Congenital malformations in infants whose mothers reported use of folic acid in early pregnancy in Sweden: a prospective population study. Congenit Anom (Kyoto) 47: 119–24.

Källén B. 2008. Population surveillance of congenital malformations. Possibilities and limitations. Acta Paediatr 78:657–63.

Källén B, Hay S, Klingberg M. 1984. Birth defects monitoring systems: accomplishments and goals. In: Kalter H ed. Issues and reviews in teratology, vol. 2. New York, Plenum Press, pp 1–22.

Källén B, Löfkvist E. 1984. Time trends of spina bifida in Sweden 1947–81. J Epidemiol Community Health 38:103–7.

Källén B, Otterblad Olausson P. 2002. Use of folic acid and delivery outcome: a prospective registry study. Reprod Toxicol 16:327–33.

Källén B, Otterblad Olausson P. 2006. Antidepressant drugs during pregnancy and infant congenital heart defect. Reprod Toxicol 21:221–2.

Källén B, Robert E, Mastroiacovo P, et al. 1989. Anitconvulsant drugs and malformations: is there drug specificity? Eur J Epidemiol 5:31–6.

Källén B, Tandberg A. 1983. Lithium and pregnancy: a cohort study on manic-depressive women. Acta Psychiatr Scand 68:34–9.

Källén B, Winberg J. 1968. A Swedish registry of congenital malformations: experience with continuous registration during 2 years with special reference to multiple malformations. Pediatrics 41:765–76.

Kalter H. 1951. The incidence, inheritance, and significance of the taste reaction to phenylthiocarbamide (PTC). A thesis submitted to Faculty of Graduate Studies and Research, McGill University, in partial fulfillment of the requirements for the degree of Master of Science.

Kalter H. 1954. The inheritance of susceptibility to the teratogenic action of cortisone in mice. Genetics 39:185–96.

Kalter H. 1957. Factors influencing the frequency of cortisone-induced cleft palate in mice. J Exp Zool 134:449–67.

Kalter H. 1959a. Congenital malformations induced by riboflavin deficiency in strains of mice. Pediatrics 23:222–30.

Kalter H. 1959b. Attempts to modify the frequency of cortisone-induced cleft palate by vitamin, carbohydrate, and protein supplementation. Plastic Reconstr Surg 24:498–504.

Kalter H. 1960. The teratogenic effects of hypervitaminosis A upon the face and mouth of inbred mice. Ann N Y Acad Sci 85:42–55.

Kalter H. 1968. Teratology of the central nervous system: induced and spontaneous malformations of laboratory, agricultural, and domestic animals. Chicago, IL, University of Chicago.

Kalter H. 1971. Correlation between teratogenic and mutagenic effects of chemicals in mammals. In: Hollaender A ed. Chemical mutagens: principles and methods for their detection, vol. 1. New York, Plenum Press, pp 57–82.

Kalter H. 1978. The structure and uses of genetically homogeneous lines of animals. In: Wilson JG, Fraser FC eds. Handbook of teratology, vol. 4. New York, Plenum Press, pp 155–90.

Kalter H. 1980. The relation between congenital malformations and prenatal mortality in experimental animals. In: Porter IH, Hook EB eds. Human embryonic and fetal death. New York, Academic Press, pp 29–44.

Kalter H. 1987. Diabetes and spontaneous aborion: a historical review. Am J Obstet Gynecol 156:1243–53.

Kalter H. 1990. Perinatal mortality and congenital malformations in infants born to women with insulin-dependent diabetes mellitus – United States, Canada, and Europe, 1940–1988. MMWR Morb Mortal Wkly Rep 39:363–5.
Kalter H. 1991. Five-decade international trends in the relation of perinatal mortality and congenital malformations: stillbirth and neonatal death compared. Int J Epidemiol 20:173–9.
Kalter H. 1992. Cyclopia produced in a very early retinoic acid experiment. Teratology 46:207–8.
Kalter H. 1993. Case reports of malformations associated with maternal diabetes: history and critique. Clin Genet 43:174–9.
Kalter H. 1998. The non-teratogenicity of gestational diabetes. Paediatr Perinat Epidemiol 12: 456–8.
Kalter H. 2000. Of diabetic mothers and their babies: an examination of the impact of maternal diabetes on offspring development and survival. Amsterdam, Harwood Academic.
Kalter H. 2003. Teratology in the twentieth century: congenital malformations in humans and how their environmental causes were established. Amsterdam, Elsevier.
Kalter H. 2007. Mortality and maldevelopment. Part 1. Congenital cardiovascular malformations. The Netherlands, Springer.
Kalter H. 2009. Mortality and maldevelopment. Part 2. The saga of neural tube defects. The Netherlands, Springer.
Kalter H, Warkany J. 1957. Congenital malformations in inbred strains of mice induced by riboflavin-deficient, galactoflavin-containing diets. J Exp Zool 136:531–66.
Kalter H, Warkany J. 1959. Experimental production of congenital malformations by metabolic procedures. Physiol Rev 39:69–115.
Kalter H, Warkany J. 1961. Experimental production of congenital malformations in strains of inbred mice by maternal treatment with hypervitaminosis A. Am J Pathol 38:1–21.
Kalter H, Warkany J. 1983. Congenital malformations: etiologic factors and their role in prevention. N Engl J Med 308:424–31.
Kaminski M. 1992. Euromac: a European concerted action: maternal alcohol consumption and its relation to the outcome of pregnancy and child development at 18 months: relation to other studies. Int J Epidemiol 21:S79–S81.
Kaminski M, Franc M, Lebouvier M, et al. 1981. Moderate alcohol use and pregnancy outcome. Neurobehav Toxicol Teratol 3:73–81.
Kaminski M, Rumeau-Rouquette C, Schwartz D. 1976. Consommation d'alcool chez la femme enceinte et issue de la grossesse. Rev Epidemiol Sante Publ 24:27–40.
Kaminski M, Rumeau-Rouquette C, Schwartz D. 1979. Response to questions re: Consommation d'alcool chez la femme enceinte et issue de la grossesse, in Rev Epid Sante Publ 24:27–40, 1976. Alcohol Clin Exp Res 37:375–9.
Kaminski PF, Maier RC. 1983. Clear cell adenocarcinoma of the cervix unrelated to diethylstilbestrol exposure. Obstet Gynecol 62:720–7.
Kamm JJ. 1982. Toxicology, carcinogenicity, and teratogenicity of some orally administered retinoids. J Am Acad Dermatol 6:652–9.
Kaneko S, Otani K, Fukushima Y, et al. 1988. Teratogenicity of antiepileptic drugs: analysis of possible risk factors. Epilepsia 29:459–67.
Kang HK, Mahan CM, Lee KY, et al. 2000. Pregnancy outcomes among US women Vietnam veterans. Am J Ind Med 38:447–54.
Kariniemi V, Rosti J. 1988. Maternal smoking and alcohol consumption as determinants of birth weight in an unselected study population. J Perinatol Med 16:249–52.
Kaufman RH, Binder GC, Gray PM, et al. 1977. Upper genital tract changes associated with exposure in utero to diethylstilbestrol. Am J Obstet Gynecol 128:51–9.
Kaur I, Dogra S, Narang T, et al. 2009. Comparative efficacy of thalidomide and prednisolone in the treatment of moderate to severe erythema nodosum leprosum: a randomized staudy. Australas J Dermatol 50:181–5.
Kellokumpu-Lehtinen P, Pelliniemi LJ. 1984. Sex ratio of human conceptuses. Obstet Gynecol 64:220–2.

Kelly TE. 1984a. Teratogenicity of antiepileptic drugs. I: Review of the literature. Am J Med Genet 19:413–34.
Kelly TE. 1984b. Teratogenicity of antiepileptic drugs. III: Radiographic hand analysis of children exposed in utero to diphenylhydantoin. Am J Med Genet 19:445–50.
Kelly TE, Rein M, Edwards P. 1984. Teratogenicity of antiepileptic drugs. IV: The association of clefting and epilepsy. Am J Med Genet 19:451–8.
Kelsey FO. 1982. Regulatory aspects of teratology: role of the Food and Drug Administration. Teratology 25:193–9.
Kelsey FO. 1988. Thalidomide update: regulatory aspects. Teratology 38:221–6.
Kemper F. 1962. Thalidomide and congenital abnormalities. Lancet 2:836.
Kennedy WP. 1967. Epidemiological aspects of the problem of congenital malformations. Birth Defects Orig Art Ser 3:1–18.
Kesmodel U, Wisborg K, Olsen SF, et al. 2002. Moderate alcohol intake in pregnancy and the risk of spontaneous abortion. Alcohol Alcohol 37:87–92.
Khera KS. 1979. Teratogenic and genetic effects of mercury toxicity. In: Nriagu JO ed. The biogeochemistry of mercury in the environment. Amsterdam, Elsevier/North Holland, pp 503–18.
Khera KS. 1984. Adverse effects in humans and animals of prenatal exposure to selected therapeutic drugs and estimation of embryo-fetal sensitivity of animals for human risk assessment: a review. In: Kalter H ed. Issues and reviews in teratology, vol. 2. New York, Plenum Press, pp 399–507.
Khoury MJ, Holtzman NA. 1987. On the ability of birth defects monitoring to detect new teratogens. Am J Epidemiol 126:136–43.
Khoury MJ, Moore CA, Mulinare J. 1996. Vitamin A and birth defects. Lancet 347:322.
Kinlen L, Hewitt A. 1973. Potato blight and anencephalus in Scotland. Br J Prev Soc Med 27:208–13.
Kirby R, Petrini J, Alter C, et al. 2000. Collecting and interpreting birth defects surveillance data by Hispanic ethnicity: a comparative study. Teratology 61:21–7.
Kirke PN, Daly LE, Elwood JH. 1992. A randomised trial of low dose folic acid to prevent neural tube defects. Arch Dis Child 67:1442–6.
Kirke PN, Mills JL, Whitehead AS, et al. 1996. Methylenetetrahydrofolate reductase mutation and neural tube defects. Lancet 348:1037–8.
Kirke PN, Molloy AM, Daly LE, et al. 1993. Maternal plasma folate and vitamin B12 are independent risk factors for neural tube defects. Q J Med 86:703–8.
Kistler A, Hummler H. 1985. Teratogenesis and reproductive safety evaluation of the retinoid etretin (Ro 10-1670). Arch Toxicol 58:50–6.
Kitay DZ. 1969. Folic acid deficiency in pregnancy. On the recognition, pathogenesis, consequences, and therapy of the deficiency state in human reproduction. Am J Obstet Gynecol 104:1067–107.
Kline J, Shrout P, Stein Z, et al. 1980a. Drinking during pregnancy and spontaneous abortion. Lancet 2:176–80.
Kline J, Stein Z, Hutzler M. 1987. Cigarettes, alcohol and marijuana: varying associations with birthweight. Int J Epidemiol. 16:44–51.
Kline J, Stein Z, Susser M, et al. 1980b. Environmental influences on early reproductive loss in a current New York City study. In: Porter IH, Hook EB eds. Human embryonic and fetal death. New York, Academic Press, pp 225–40.
Knapp K, Lenz W. 1962. Die Formen der Thalidomid-Embrypathie. Roentgen-Eur 5:105–27.
Knightley P, Evans H, Potter E, et al. 1977. Suffer the children: the story of thalidomide. New York, Viking.
Knox EG, Lancashire RJ. 1991. Epidemiology of congenital malformations. London, Her Majesty's Stationery Office.
Koch R, Friedman E, Azen C, et al. 2000a. The international collaborative study of maternal phenylketonuria: status report 1998. Eur J Pediatr 159(suppl 2):S156–60.

Koch R, Hanley W, Levy H, et al. 2000b. Maternal phenylketonuria: an international study. Mol Genet Metabol 71:233–9.

Koch R, Hanley W, Levy H, et al. 2003. The maternal phenylketonuria international study: 1984–2002. Pediatrics 112:1523–9.

Koch R, Hartmann AM, Jäger-Roman E, et al. 1982. Major malformations in children of epileptic parents – due to epilepsy or its therapy? In: Janz D, Bossi L, Dam M, et al. eds. Epilepsy, pregnancy, and the child. New York, Raven Press, pp 313–5.

Koch R, Levy HL, Matalon R, et al. 1994. The international collaborative study of maternal phenylketonuria: status report 1994. Acta Paediatr Suppl 407:111–19.

Koch S, Lösche G, Jager-Romän E, et al. 1992. Major and minor malformations and epileptic drugs. Neurology 42:83–8.

Kochhar DM, Christian MS. 1997. Tretinoin: a review of the nonclinical developmental toxicology experience. J Am Acad Dermatol 36:S47–59.

Kolata GB. 1980a. How safe is Bendectin? Science 210:518–9.

Kolata GB. 1980b. Love Canal: false alarm caused by botched study. Science 208:239–42.

Kolata GB. 1981. Fetal alcohol advisory debated. Science 214:642–5.

Koren G, Pastuszak A, Ito S. 1998. Drugs in pregnancy. N Engl J Med 338:1128–37.

Koren G, Nava-Ocampo AA, Moretti ME, et al. 2006. Major malformations with valproic acid. Can Fam Physician 52:441–7.

Korones SB. 1986. Congenital rubella – an ecapsulated review. In: Sever JL, Brent RL eds. Teratogen update: environmentally induced birth defects. New York, Liss, pp 111–4.

Kozma C. 2001. Valproic acid embryopathy: report of two siblings with further expansion of the phenotypic abnormalities and a review of the literature. Am J Med Genet 98:168–75.

Kreipe U. 1967. Abnormalities of internal organs in thalidomide embryopathy. A contribution to the determination of the sensitivity phase in thalidomide administration during early pregnancy. Arch Kinderheilkd 176:33–61 (In German).

Kristensen P, Irgens LM, Andersen A, et al. 1997. Birth defects among offspring of Norwegian farmers, 1967–1991. Epidemiology 8:537–44.

Kučera J. 1971. Rate and type of congenital anomalies among offspring of diabetic women. J Reprod Med 7:61–70.

Kurent JE, Sever JL. 1977. Infectious diseases. In: Wilson JG, Fraser FC eds. Handbook of teratology, vol. 1. New York, Plenum Press, pp 225–59.

Kurman RJ, Scully RE. 1974. The incidence and histogenesis of vaginal adenosis: an autopsy study. Hum Pathol 5:265–76.

Kurtzke JF, Goldberg ID, Kurland LT. 1973. The disribution of deaths from congenital malformations of the nervous system. Neurology 23:483–96.

Kvigne VL, Leonardson GR, Borzelleca J, et al. 2008. Characteristics of grandmothers who have grandchildren with fetal alcohol syndrome or incomplete fetal alcohol syndrome. Matern Child Health 12:760–5.

Labarthe D, Adam E, Noller KL, et al. 1979. Design and preliminary observations of the National Cooperative Diethylstilbestrol Adenosis (DESAD) Project. Obstet Gynecol 51:453–8.

Lacroix R, Eason E, Melzack R. 2000. Nausea and vomiting during pregnancy: a prospective study of its frequency, intensity, and patterns of change. Am J Obstet Gynecol 182:931–7.

Laegreid L, Olegård R, Walström J, et al. 1989. Teratogenic effects of benzodiazapine use during pregnancy. J Pediatr 114:126–31.

Lafille E. 2006. [The revival of thalidomide: an old drug with new indications.] In French. Rev Prat 56:1977–83.

Lambert D, Dalac S, Escallier F, et al. 1988. Malformations foetales après etretinate. Nouv Dermatol 7:448–51.

Lammer EJ. 1988. Embryopathy of infant conceived one year after termination of maternal etretinate. Lancet 2:1080–1.

Lammer EJ, Chen DT, Hoar RM, et al. 1985. Retinoic acid embryopathy. N Engl J Med 313: 837–41.

Lammer EJ, Hayes AM, Schunior A, et al. 1988. Unusually high risk for adverse outcomes of pregnancy following fetal isotretinoin exposure. Am J Hum Genet 43:A58.

Lammer EJ, Opitz JM. 1986. The DiGeorge anomaly as a developmental field defect. Am J Med Genet Suppl 2:113–27.

Lammer EJ, Sever LE, Oakley GP. 1987. Teratogen update: valproic acid. Teratology 35:465–73.

Lancaster HO. 1954. The epidemiology of deafness due to maternal rubella. Acta Genet 5:12–24.

Landauer W. 1967. The hatchability of chicken eggs as influenced by environment and heredity. Storrs, CT, University of Connecticut.

Langmuir AG. 1971. New environmental factor in congenital disease. N Engl J Med 284:912–3.

Lanier AP, Noller KL, Decker DG, et al. 1973. Cancer and stilbestrol: a follow-up of 1,719 persons sexposed to estrogens in utero and born 1943–1959. Mayo Clin Proc 48:793–9.

Lasagna L. 1989. Congress, the FDA, and new drug development: before and after 1962. Perspect Biol Med 32:322–43.

Last JM. 1988. Dictionary of epidemiology. New York, Oxford University Press.

Lathrop GD, Wolfe EH, Albanese RA, et al. 1984. Air Force Health Study: an epidemiologic investigation of health effects in air force personnel following exposure to herbicides. Baseline morbidity study results. San Antonio, TX, USAF School of Areospace MedicineBrooks Air Force Base.

Lau SP, Fung KP. 1984. Ethnic variables in perinatal mortality rates. Lancet 2:402.

Laurence KM, Carter CO, David PA. 1967. Major central nervous system malformations in South Wales. I. Incidence, local variations, and geographical factors. Br J Prev Soc Med 21: 46–60.

Laurence KM, Carter CO, David PA. 1968. Major central nervous system malformations in South Wales. II. Pregnancy factors, seasonal variation, and social class effects. Br J Prev Soc Med 22:212–22.

Laurence KM, Janes N, Miller MH, et al. 1981. Double-blind randomised controlled trial of folate treatment before conception to prevent recurrence of neural tube defects. BMJ 282:509–11.

Laursen HB. 1980. Some epidemiological aspects of congenital heart disease in Denmark. Acta Paediatr Scand 69:619–24.

Lawrence JM, Pettiti DB, Watkins M, et al. 1999. Trends in serum folate after food fortification. Lancet 354:915–6.

Lawson CC, Schnorr TM, Whelan EA, et al. 2004. Paternal occupational exposure to 2.3.7.8-tetrachlorodibenzo-*p*-dioxin and birth outcomes of offspring; birth weight, preterm delivery, and birth defects. Environ Health Perspect 112:1403–8.

Lazzaroni F, Bonassi S, Magnani M, et al. 1993. Moderate maternal drinking and outcome of pregnancy. Eur J Epidemiol 9:599–606.

Leavitt AM, Yerby MS, Robinson N, et al. 1992. Epilepsy in pregnancy: developmental outcome of offspring at 12 months. Neurology 42:141–3.

Leck I. 1969. Ethnic differences in the incidence of malformations following migration. Br J Prev Soc Med 23:166–73.

Leck I. 1976. Descriptive epidemiology of common malformations (excluding central nervous system defects). Br Med Bull 32:45–52.

Leck I. 1977. Correlations of malformation frequency with environmental and genetic attributes in man. In: Wilson JG, Fraser FC eds. Handbook of teratology, vol. 3. New York, Plenum Press, pp 243–324.

Leck I. 1983. Spina bifida and anencephaly: fewer patients, more problems. BMJ 286:679–80.

Leck I. 1996. McKeown, Record, and the epidemiology of malformations. Pediatr Perinat Epidemiol 10:2–16.

Leck I, Lancashire RJ. 1995. Birth prevalence of malformations in members of different ethnic groups and in the offspring of matings between them, in Birmingham, England. J Epidemiol Community Health 49:171–9.

Leck I, Record RG, McKeown T, et al. 1968. The incidence of malformations in Birmingham, England, 1950–1959. Teratology 1:263–80.

Lee AL, Ananthakrishnan S, Pearce EN. 2009. Iodine deficiency. emed.medscape.com/article/122714-overview.
Lee JY, Bowden DS. 2000. Rubella virus replication and links to teratogenicity. Clin Microbiol Rev 13:571–87.
Lejeune J, Turpin R, Goutier M. 1959. Le mongolisme: premier example d'aberration autosomique humaine. Ann Genet 1:41–9.
Lemire RJ, Beckwith JB, Shepard TH. 1972. Iniencephaly and anencephaly with spinal retroflexion: a comparative study of eight human specimens. Teratology 6:27–36.
Lemire RJ, Beckwith JB, Warkany J. 1978. Anencephaly. New York, Raven Press.
Lemoine P, Harousseau M, Borteyru JP, et al. 1968. Les enfants de parents alcooliques. Anomalies observées. A propos de 127 cas. Ouest Med 21:476–82.
Lenke RR, Levy HL. 1980. Maternal phenylketonuria: an international survey of the outcome of untreated and treated pregnancies. N Engl J Med 303:1202–8.
Lenz W. 1961a. Diskussionsbemerkung zu dem Vontrag von R.A. Pfeiffer und K. Kosenow: Zur Frage der exogenen Entstehung schwerer Extremitätenmissbildung. Tagung der Rheinisch-Westfälischen Kinderärztevereinigung 18.11.
Lenz W. 1961b. Kindliche missbildungen nach Medikamenten-Einnahme während der Gravidität? Deut Med Woch 52:2555–6.
Lenz W. 1962. Thalidomide and congenital abnormalities. Lancet 1:45.
Lenz W. 1964. Chemicals and malformations in man. In: Fishbein M ed. Second international conference on congenital malformations. New York, International Medical Congress, pp 263–76.
Lenz W. 1965a. Epidemiologie von Missbildungen. Pädiatr Pädol 1:38–50.
Lenz W. 1965b. Epidemiology of congenital malformations. Ann N Y Acad Sci 123:228–36.
Lenz W. 1968. Ein Vergleich der sensiblen Phase für Thalidomid im Tierversuch und beim Menschen. Arch Kinderheilk 177:259–65.
Lenz W. 1979. Thalidomide: facts and inferences. In: Soda T ed. Drug-induced sufferings. Amsterdam, Excerpta Medica, pp 103–9.
Lenz W. 1985. Thalidomide embryopathy in Germany, 1959–1961. In: Marois M ed. Prevention of physical and mental congenital defects, Part C: basic and medical science. education and future strategies. New York, Liss, pp 77–83.
Lenz W. 1988. A short history of thalidomide embryopathy. Teratology 38:203–15.
Lenz W. 1990. Living history – biography: nature and nurture. Am J Med Genet 37:356–61.
Lenz W, Knapp K. 1962. Die Thalidomid-Embryopathie. Dtsch Med Wochenschr 87:1323–42.
Lenz W, Kučera J. 1967. L'étiologie de la régression caudale (agénésie du sacrum). Med Hyg 25:241–3.
Lenz W, Maier W. 1964. Congenital malformations and maternal diabetes. Lancet 2:1124–5.
Lenz W, Passarge E. 1965. Syndrome of caudal regression in infants of diabetic mothers. Abstracts of the 5th Annual Meeting of the Teratology Society, San Francisco, May 26–28, p 16.
Leppig KA, Werler MM, Cann CI, et al. 1987. Predictive value of minor anomalies. 1. Association with major malformations. J Pediatr 110:531–7.
Leridon H. 1976. Facts and artifacts in the study of intrauterine mortality: a reconsideration from pregnancy histories. Popul Stud 30:329–36.
Levy HL. 1987. Maternal phenylketonuria: review with emphasis on pathogenesis. Enzyme 38:312–20.
Levy HL. 1996. Reproductive effects of maternal metabolic disorders: implications for pediatrics and obstetrics. Turk J Pediatr 38:335–44.
Levy HL, Ghavami M. 1996. Maternal phenylketonuria: a metabolic teratogen. Teratology 53:176–84.
Levy HL, Guldberg P, Güttler F, et al. 2001. Congenital heart disease in maternal phenylketonuria: report from the Maternal PKU Collaborative Study. Pediatr Res 49:636–42.
Levy HL, Lenke RR, Crocker AC. 1982. Maternal PKU: proceedings of a conference. Rockville, MD, U.S. Department of Health and Human Services.

Levy HL, Waisbren SE. 1983. Effects of untreated maternal phenylketonuria and hyperphenylalaninemia on the fetus. N Engl J Med 309:269–74.
Levy HL, Waisbren SE, Güttler F, et al. 2003. Pregnancy experiences in the woman with mild hyperphenylalaninemia. Pediatrics 112:1548–52.
Levy HL, Waisbren SE, Lobbregt D, et al. 1994. Maternal mild hyperphenylalaninemia: an international survey of offspring outcome. Lancet 344:589–94.
Lewis K. 2009. Iodine deficiency disorders still prevalent in China. CMAJ 18:E (Epub 2009 Sep 14).
Li TN, Johansson A. 2001. Impact of chemical warfare with Agent Orange on women's reproductive lives in Vietnam: a pilot study. Reprod Health Matters 9:156–64.
Li Z, Ren A, Liu J, et al. 2007. Maternal flu or fever, medication use, and neural tube defects: a population based case-control study in Northern China. Birth Defects Res A Clin Mol Teratol 79:295–300.
Li Z, Ren A, Zhang L, et al. 2006. Extremely high prevalence of neural tube defects in a 4-county area in Shanxi Province, China. Birth Defects Res A Clin Mol Teratol 76:237–40.
Lian ZH, Yang HY, Li Z. 1987. Neural tube defects in Beijing-Tianjin area of China: urban-rural distribution and some epidemiological characteristics. J Epidemiol Community Health 41:259–62.
Limb CJ, Holmes L. 1994. Anencephaly: changes in prenatal detection and birth status, 1972 through 1990. Am J Obstet Gynecol 170:1333–8.
Lindhout D. 1994. Discussion, neural tube defects. Ciba foundation symposium. Chichester, Wiley, pp 154–5.
Lindhout D, Meinardi H, Meijer JW, et al. 1992a. Antiepileptic drugs and teratogenesis in two consecutive cohorts: changes in prescription policy paralleled by changes in pattern of malformations. Neurology 42:94–110.
Lindhout D, Omtzigt JG, Cornel MC. 1992b. Spectrum of neural-tube defects in 34 infants exposed prenatally to antiepileptic drugs. Neurology 42:111–18.
Lindhout D, Schmidt D. 1986. In-utero exposure to valproate and neural tube defects. Lancet 1:392–3.
Lipson A, Beuhler B, Bartley J, et al. 1984. Maternal hyperphenylalaninemia and fetal effects. J Pediatr 104:216–20.
Lipson AH. 1992. Thalidomide retrospective: what did the clinical teratologist learn? Teratology 46:411–13.
Lipson AH, Collins F, Webster WS. 1993. Multiple congenital defects associated with maternal use of topical tretinoin. Lancet 341:1352–3.
Little BB, Snell LM, Rosenfeld CR, et al. 1990. Failure to recognize fetal alcohol syndrome in newborn infants. Am J Dis Child 144:1142–6.
Little J, Elwood JM. 1991. Epidemiology of neural tube defects. In: Kiely M ed. Reproductive and perinatal epidemiology. Boca Raton, FL, CRC, pp 251–336.
Little J, Elwood M. 1992. Fetal loss. In: Elwood IM, Little J, Elwood JH eds. Epidemiology and control of neural tube defects. Monographs in epidemiology and biostatistics, vol. 20. Oxford, Oxford University Press, pp 324–34.
Little RE. 1977. Moderate alcohol use during pregnancy and reduced infant birth weight. Am J Public Health 67:1154–6.
Little RE, Asker FL, Sampson PD, et al. 1986. Fetal growth and moderate drinking in early pregnancy. Am J Epidemiol 123:270–9.
Liu J, Yang GZ, Zhou JL, et al. 2007. Prevalence of neural tube defects in economically and socially deprived area of China. Childs Nerv Syst 23:1119–24.
Livingstone J, Lyall H. 1986. Contribution of fetal alcohol syndrome to mental retardation. Lancet 2:1337–8.
Lloyd ME, Carr M, McElhatton P, et al. 1999. The effects of methotrexate on pregnancy, fertility and lactation. Q J Med 92:551–63.

Loffredo CA, Wilson PD, Ferencz C. 2003. Maternal diabetes: an independent risk factor for major cardiovascular malformations with increased mortality of affected infants. Teratology 64:98–106.

Lohnes D, Mark M, Mendelsohn C, et al. 1995. Developmental roles of the retinoic acid receptors. J Steroid Biochem Molec Biol 53:475–86.

Lorber J. 1974. The potato trial. Link 30:7.

Lorber J, Stewart CR, Ward AM. 1973. Alpha-fetoprotein in antenatal diagnosis of anencephaly and spina bifida. Lancet 1:1187.

Loughnan PM, Gold H, Vance JC. 1973. Phenytoin teratogenicity in man. Lancet 1:70–2.

Louik C, Lin AE, Werler MM, et al. 2007. First-trimester use of selective serotonin-reuptake inhibitors and the risk of birth defects. N Engl J Med 356:2675–83.

Loureiro KD, Kao KK, Jones KL, et al. 2005. Minor malformations characteristic of the retinoic acid embryopathy and other birth outcomes in children of women exposed to topical tretinoin during early pregnancy. Am J Med Genet A 136:117–21.

Lowry RB, Trimble BK. 1977. Incidence rates for cleft lip and palate in British Columbia 1952–71 for North American Indian, Japanese, Chinese and total populations: secular trends over twenty years. Teratology 16:277–83.

Lu MC, Sammel MD, Cleveleand RH, et al. 2000. Digit effects produced by prenatal exposure to antiepileptic drugs. Teratology 61:277–83.

Lumley J, Correy JF, Newman NM, et al. 1985. Cigarette smoking, alcohol consumption and fetal outcome in Tasmania 1981–82. Aust N Z J Obstet Gynecol 25:33–40.

Lummaa V, Merila J, Kause A. 1998. Adaptive sex ratio variation in pre-industrial human (*Homo sapiens*) populations. Proc R Soc Lond B 265:563–8.

Lundsberg L, Bracken MB, Saftlas AF. 1997. Low-to-moderate alcohol use and intrauterine growth retardation, low birthweight, and preterm delivery. Ann Epidemiol 7:498–508.

Lundström R. 1962. Rubella during pregnancy: a follow-up of children born after an epidemic of rubella in Sweden, 1951, with additional investigations on prophylaxis and treatment of maternal rubella. Acta Paediatr Suppl 133:1–110.

Lynberg MC, Khoury MJ, Lu X, et al. 1994. Maternal flu, fever, and the risk of neural tube defects: a population-based case-control study. Am J Epidemiol 140:244–55.

Mabry CC, Denniston JC, Nelson TI, et al. 1963. Maternal phenylketonuria: a cause of mental retardation in children without the metabolic defect. N Engl J Med 269:1404–8.

Macfarlane DW, Boyd RD, Codrill CB, et al. 1975. Intrauterine rubella, head size, and intellect. Pediatrics 55:797–801.

MacHenry JC, Nevin NC, Merrett JD. 1979. Comparison of central nervous system malformations in spontaneous abortions in Northern Ireland and south-east England. BMJ 1:1395–7.

Macintosh MC, Fleming KM, Bailey JA, et al. 2006. Perinatal mortality and congenital anomalies in babies of women with type 1 or type 2 diabetes in England, Wales, and Northern Ireland: population based study. BMJ 333:177.

Mackeprang M, Hay S, Lunde S. 1972. Completeness and accuracy of reporting of malformations on birth certificates. Public Health Rep 87:43–9.

MacMahon B. 1981. More on Bendectin. JAMA 246:371–2.

MacMahon B, Pugh TF. 1954. Sex ratio of white births in the United States during the Second World War. Am J Hum Genet 65:284–92.

MacMahon B, Yen S. 1971. Unrecognised epidemic of anencephaly and spina bifida. Lancet 1:31–3.

MacMahon B, Yen S, Rothman KJ. 1973. Potato blight and neural-tube defects. Lancet 1:598–9.

Maddox J. 1990. Sellafield makes news again. Lancet 343:690.

Maden M. 2000. The role of retinoic acid in embryonic and post-embryonic development. Proc Nutr Soc 59:65–73.

Majewski F, Steger M, Richter B, et al. 1981. The teratogenicity of hydantoins and barbiturates in humans, with considerations on the etiology of malformations and cerebral disturbances in the children of epileptic parents. Int J Biol Res Pregnancy 2(1):37–45.

Mall FP. 1908. A study of the causes underlying the origin of human monsters. J Morphol 19:1–367.
Malm H, Kajantie E, Kivirikkob S, et al. 2002. Valproate embryopathy in three sets of siblings: further proof of hereditary susceptibility. Neurology 59:630–3.
Malone FD, D'Alton ME. 1997. Drugs in pregnancy: antiepileptics. Semin Perinat 21:114–23.
Mantel N. 1985. Some problems with investigations and epidemiologic studies relating to pregnancy. Prog Clin Biol Res 163:45–8.
Maradit H, Geiger JM. 1999. Potential risk of birth defects after acitretin discontinuation. Dermatology 198:3–4.
Marbury MC, Linn S, Monson RR, et al. 1983. The association of alcohol consumption and outcome of pregnancy. Am J Public Health 73:1165–8.
Marden PM, Smith DW, McDonald J. 1964. Congenital anomalies in the newborn infant, including minor variations. J Pediatr 64:357–71.
Marion RW, Wiznia AA, Hutcheon RG, et al. 1986. Human T-cell lymphotropic virus type III (HTLV-III) embryopathy. Am J Dis Child 140:638–40.
Marsh DO, Clarkson TW, Cox C, et al. 1987. Fetal methylmercury poisoning: relationship between concentration in single strands of maternal hair and child effects. Arch Neurol 44:1017–22.
Marshall MF. 1999. Commentary: mal-intentioned illiteracy, willful ignorance, and fetal protection laws: is there a lexicologist in the house? J Law Med Ethics 27:343–6.
Martin E. 1880. Histoire des monstres. Depuis l'antiquité jusqu'a nos jours. Paris, Reinwald.
Martin GR, Perry LW, Ferencz C. 1989. Increased prevalence of ventricular septal defects: epidemic or improved diagnosis? Pediatrics 83:200–3.
Martin PJ, Millac PA. 1993. Pregnancy, epilepsy, management and outcome: a 10-year perspective. Seizure 2:277–80.
Martinez de Villarreal LE, Delgado-Enciso I, Valdez-Leal R, et al. 2001. Folate levels and N(5),N(10)-methylenetetrahydrofolate reductase. Arch Med Res 32:277–82.
Martínez-Frías ML. 1990. Clinical manifestations of prenatal exposure to valproic acid using case reports and epidemiologic information. Am J Med Genet 37:277–82.
Marwick C. 1998. Thalidomide back – under strict control. JAMA 278:1135–7.
Masterson JG, Frost C, Bourke GJ, et al. 1973. Anencephaly and potato blight in the Republic of Ireland. Br J Prev Soc Med 28:81–4.
Mastroiacovo P, Mazzone T, Addis A, et al. 1999. High vitamin A intake in early pregnancy and major malformations: a multicenter prospective controlled study. Teratology 59:7–11.
Mastroiacovo P, Spagnolo A, Marni E, et al. 1988. Birth defects in the Seveso area after TCDD contamination. JAMA 259:668–72.
Mathews TJ, Honein MA, Erickson JD. 2002. Spinal bifida and anencephaly prevalence – United States, 1991–2001. MMWR Morb Mortal Wkly Rep 51(RR13):9–11.
Mathews TJ, MacDorman MF. 2008. Infant mortality statistics from the 2005 period linked birth/infant death data set. Nat Vital Stat Rep 57:1–32.
Mau G. 1980. Moderate alcohol consumption during pregnancy and child development. Eur J Pediatr 133:233–7.
May PA, Gossage JP. 2001. Estimating the prevalence of fetal alcohol syndrome: a summary. Alcohol Res Health 25:159–67.
May PA, Hymbaugh KJ, Aase JM, et al. 1983. Epidemiology of fetal alcohol syndrome among American Indians of the southwest. Soc Biol 30:374–87.
Mazzotta P, Maltepe C, Navioz Y, et al. 2000. Attitudes, management and consequences of nausea and vomiting of pregnancy in the United States and Canada. Int J Gynaecol Obstet 70:359–65.
McBride WG. 1961a. Thalidomide and congenital abnormalities. Lancet 2:358.
McBride WG. 1961b. Congenital abnormalities and thalidomide. Med J Aust 2:1030.
McDonnell RJ, Johnson Z, Delaney V, et al. 1999. East Ireland 1980–1994: epidemiology of neural tube defects. J Epidemiol Community Health 53:782–8.
McElhatton PR. 2000. A review of the reproductive toxicity of methotrexate in human pregnancy. Reprod Toxic 14:549.

McFadyen RE. 1976. Thalidomide in America: a brush with tragedy. Clio Med 11:79–93.
McFarlane MJ, Feinstein AR, Horowitz RI. 1986. Diethylstilbestrol and clear cell vaginal carcinoma: reappraisal of the epidemiologic evidence. Am J Med 81:855–63.
McKeigue PM, Lamm SH, Linn S, et al. 1994. Bendectin and birth defects: a meta-analysis of the epidemiological studies. Teratology 50:27–37.
McKeown T. 1976. The modern rise of population. New York, Academic Press.
McKeown T. 1979. The role of medicine: dream, mirage, or nemesis? Oxford, Blackwell.
McKeown T. 1988. The origins of human disease. New York, Blackwell.
McKeown T, Lowe CR. 1951. The sex ratio of stillbirths related to cause and duration of gestation; an investigation of 7,066 stillbirths. Hum Biol 23:41–60.
McKeown T, Record RG. 1960. Malformations in a population observed for five years after birth. In: Wolstenholme GE, O'Connor CM eds. Ciba Foundation symposium on congenital malformations. Boston, MA, Little, Brown, pp 2–16.
McKeown T, Record RG. 1969. Sources of data for population study. In: Swinyard CA ed. Limb development and deformity: problems of evaluation and rehabilitation. Springfield, IL, Thomas, pp 212–20.
Meador KJ, Baker GA, Browning N, et al. 2009. Cognitive function at 3 years of age after fetal exposure to entiepileptic drugs. N Engl J Med 360:1597–605.
Meadow SR. 1968. antiepileptic drugs and congenital abnormalities. Lancet 2:1296.
Meadow SR. 1970. Congenital abnormalities and antiepileptic drugs. Proc Roy Soc Med 63: 48–9.
Meadow SR. 1991. Antiepileptics in pregnancy. Arch Dis Child 66:62–5.
Meberg A, Otterstad JE, Frøland G, et al. 1994. Increased incidence of ventricular septal defects caused by improved detection rate. Acta Paediatr 83:653–7.
Méhes K. 1980. Palpebral fissure length in newborn infant. Acta Paediatr Acad Sci Hung 21:55–6.
Méhes K. 1988. Informative morphogenetic variants in the newborn infant. Budapest, Akadémiai Kiadó.
Mellin GW. 1963. The frequency of birth defects. In: Fishbein M ed. Birth defects. Philadelphia, PA, Lippincott, pp 1–17.
Mellin GM, Katzenstein M. 1962. The saga of thalidomide: neuropathy to embryopathy, with case reports of congenital anomalies. N Engl J Med 267:1184–93.
Melnick M, Marazita ML. 1998. Neural tube defects, methylenetetrahydofolate reductase mutation, and north/south dietary differences in China. J Craniofac Genet Dev Biol 18:233–5.
Melnick S, Cole P, Anderson D, et al. 1987. Rates and risk of diethylstilbestrol-related clear-cell adenocarcinoma of the vagina and cervix: an update. N Engl J Med 316:514–16.
Mendelsohn C, Lohnes D, Décimo D, et al. 1994. Function of the retinoic acid receptors (RARs) during development. (II) Multiple abnormalities at various stages of organogenesis in RAR double mutants. Development 120:2749–71.
Merlob P. 1994. Mild errors of morphogenesis: one of the most controversial subjects in dysmorphology. In: Kalter H ed. Issues and reviews in teratology, vol. 7. New York, Plenum Press, pp 57–102.
Merlob P, Papier CM, Klingberg MA, et al. 1985. Incidence of congenital malformations in the newborn, particularly minor abnormalities. In: Marois M ed. Prevention of physical and mental congenital defects, Pt. C. New York, Liss, pp 51–6.
Meyer RE, Oakley GP. 1999. Folic acid fortification. Lancet 353:2168.
Michalek J, Rahe AJ, Boyle CA. 1998. Paternal dioxin and the sex of children fathered by veterans of Operation Ranch Hand. Epidemiology 9:474–5.
Michie CA. 1991. Neural tube defects in the 18th century. Lancet 337:504.
Milic AM. 1969. Racial factors in anencephaly. A comparison of cases among American Negroes and Caucasians. Am J Obstet Gynecol 104:34–9.
Miller E, Cradock-Watson JE, Pollack TM. 1982. Consequences of confirmed maternal rubella at successive stages of pregnancy. Lancet 2:781–4.
Miller JR, Lowry RB. 1978. Birth defect registries and surveillance. In: Wilson JG, Fraser FC eds. Handbook of teratology, vol. 3. New York, Plenum Press, pp 227–42.

Miller MT, Strömland K. 1999. Teratogen update: thalidomide: a review, with a focus on ocular findings and new potential uses. Teratoloogy 60:306–21.
Miller RW. 1956. Delayed effects occurring within the first decade after exposure of young individuals to the Hiroshima atomic bomb. Pediatrics 18:1–17.
Miller RW. 1968. Effects of ionizing radiation from the atomic bomb on Japanese children. Pediatrics 41:257–63.
Miller RW. 1971a. Studies in childhood cancer as a guide for monitoring congenital malformations. In: Hook EB, Janerich DT, Porter IH eds. Monitoring, birth defects and environment: the problem of surveillance. New York, Academic Press, pp 97–111.
Miller RW. 1971b. Cola-colored babies: chlorobiphenyl poisoning in Japan. Teratology 4:211–2.
Miller RW. 1985. Congenital PCB poisoning: a reevaluation. Environ Health Perspect 60:211–14.
Miller RW. 1993. My half-life in teratology: a key to cancer etiology. In: Kalter H ed. Issues and reviews in teratology, vol. 6. New York, Plenum Press, pp 1–46.
Miller RW. 1999. Severe mental retardation and cancer among atomic bomb survivors. Teratology 59:234–5.
Miller RW, Blot WJ. 1972. Small head size after in-utero exposure to atomic radiation. Lancet 2:784–7.
Miller RW, Mulvihill JJ. 1976. Small head size after atomic irradiation. Teratology 14:355–8.
Mills JL. 1982. Malformations in infants of diabetic mothers. Teratology 25:385–94.
Mills JL, Carter TC. 2009. Preventing neural tube defects and more via food fortification? Am J Epidemiol 169:18–21.
Mills JL, Graubard BI. 1987. Is moderate drinking during pregnancy associated with an increased risk for malformations? Pediatrics 80:309–14.
Mills JL, Graubard BI, Harley EE, et al. 1984. Maternal alcohol consumption and birth weight: how much drinking during pregnancy is safe? JAMA 252:1875–9.
Mills JL, Rhoads GG, Simpson JL, et al. 1989. The absence of a relation between the periconceptional use of vitamins and neural-tube defects. N Engl J Med 321:430–5.
Mills JL, Rhoads GG, Simpson JL, et al. 1990. Vitamins during pregnancy and neural tube defects. JAMA 263:2747–8.
Mills JL, Simpson JL, Cunningham GC, et al. 1997. Vitamin A and birth defects. Am J Obstet Gynecol 177:31–6.
Mills JL, Tuomilehto J, Yu KF, et al. 1992. Maternal vitamin levels during pregnancies producing infants with neural tube defects. J Pediatr 20:863–71.
Milunsky A, Jick H, Jick SS, et al. 1989. Multivitamin/folic acid supplementation in early pregnancy reduces the prevalence of neural tube defects. JAMA 262:2847–22.
Milunsky A, Jick H, Jick SS, et al. 1990. Vitamins during pregnancy and neural tube defects. JAMA 263:2748–9.
Mintz B ed. 1957. Environmental influences on prenatal development. Chicago, IL, University of Chicago Press.
Minussi L, Mohrdieck R, Bercini M, et al. 2008. Prospective evaluation of pregnant women vaccinated against rubella in southern Brazil. Reprod Toxicol 25:120–3.
Miskov S, Gjergia-Juraski R, Cvitanović-Sojat L, et al. 2009. Prospective surveillance of Croatian pregnant women on lamotrigine monotherapy–aspects of pre-pregnancy counseling and drug monitoring. Acta Clin Croat 48:271–81.
Mitchell AA, Rosenberg L, Shapiro S, et al. 1981. Birth defects related to Bendectin use in pregnancy. I. Oral clefts and cardiac defects. JAMA 245:2311–14.
Mittendorf R. 1995. Teratogen update: carcinogenesis and teratogenesis associated with exposure to diethylstilbestrol (DES) in utero. Teratology 51:435–45.
Mocarelli P, Brambilla P, Gerthoux PM, et al. 1996. Change in sex ratio with exposure to dioxin. Lancet 348:409.
Mocarelli P, Gerthoux PM, Ferrari E, et al. 2000. Paternal concentrations of dioxin and sex ratio of offspring. Lancet 355:1858–63.
Mole RH. 1979. Radiation effects on pre-natal development and their radiological significance. Br J Radiol 52:89–101.

Møller H. 1996. Change in male:female ratio among newborn infants in Denmark. Lancet 348:828–9.

Molloy AM, Kirke P, Hillary I, et al. 1985. Maternal serum folate and vitamin B12 concentrations in pregnancies associated with neural tube defects. Arch Dis Child 60:660–5.

Molloy AM, Mills JM, Kirke PN, et al. 1999. Folate status and neural tube defects. Biofactors 10:291–4.

Mølsted-Pedersen L, Tygstrup I, Pedersen J. 1964. Congenital malformations in newborn infants of diabetic women: correlation with maternal diabetic vascular complications. Lancet 1:1124–6.

Monif GR. 1970. Viral infections of the human fetus. London, Macmillan.

Moore CA, Li S, Li Z, et al. 1997. Elevated rates of severe neural tube defects in a high-prevalence area in northern China. Am J Med Genet 73:113–18.

Moore JA, IEHR Expert Scientific Committee. 1995. An assessment of lithium using the IEHR evaluative process for assessing human developmental and reproductive toxicity of agents. Reprod Toxicol 9:175–210.

Moore SJ, Turnpenny P, Quinn A, et al. 2000. A clinical study of 57 children with fetal antiepileptic syndromes. J Med Genet 37:489–97.

Mornet E, Muller F, Lenvoisé-Furet A, et al. 1997. Screening of the C677T mutation on the methylenetetrahydrofolate reductase gene in French patients with neural tube defects. Hum Genet 100:512–14.

Morrell MJ. 1998. Guidelines for the care of women with epilepsy. Neurology 51:S21–7.

Morris JK, Wald NJ. 1999. Quantifying the decline in the birth prevalence of neural tube defects in England and Wales. J Med Screen 6:82–5.

Morrison C, Rieder MJ. 1993. Practices of epilepsy during pregnancy: a survey of Canadian neurologists. Reprod Toxicol 7:55–9.

Morrow AW. 1972. Anencephaly and spina bifida: possible association with potato blight. Med J Aust 2:208.

Morrow JI, Hunt SJ, Russell AJ, et al. 2009. Folic acid use and major congenital malformations in offspring of women with epilepsy: a prospective study from the UK Epilepsy and Pregnancy Register. J Neurol Neurosurg Psychiatry 80:506–11.

Morrow JI, Russell A, Guthrie E, et al. 2006. Malformation risks of antiepileptic drugs in pregnancy: a prospective study from the UK Epilepsy and Pregnancy Register. J Neurol Neurosurg Psychiatry 77:193–8.

Mosley BS, Cleves MA, Siega-Riz AM, et al. 2009. Neural tube defects and maternal folate intake among pregnancies conceived after folic acid fortification in the United States. Am J Epidemiol 169:9–17.

MRC. 1991. Prevention of neural tube defects: results of the Medical Research Council Vitamin Study. MRC Vitamin Study Research Group. Lancet 338:131–7.

Mulinare J, Cordero JF, Erickson JD, et al. 1988. Periconceptional use of multivitamins and the occurrence of neural tube defects. JAMA 260:3141–5.

Müller-Küppers M. 1963. Zur frage der Fruchtschädigung in der Schwangerschaft durch Einnahme von Antieptileptica. Acta Paedopsychiat 30:401–5.

Mulvihill JJ. 1986. Fetal alcohol syndrome. In: Sever JL, Brent RL eds. Teratogen update: environmentally induced birth defect risks. New York, Liss, pp 13–8.

Mulvihill JJ, Klimas JT, Stokes DC, et al. 1976. Fetal alcohol syndrome: seven new cases. Am J Obstet Gynecol 125:937–41.

Munger RC, Romitti PA, Daack-Hirsch S, et al. 1996. Maternal alcohol use and risk of orofacial birth defects. Teratology 54:27–33.

Munro I. 1979. Thalidomide and the "Lancet". BMJ 1:1625.

Munro ND, Sheppard S, Smithells RW, et al. 1987. Temporal relations between maternal rubella and congenital defects. Lancet 2:201–4.

Murphy DP. 1928. Ovarian irradiation; its effect on the health of subsequent children: review of the literature, experimental and clinical, with a report of three-hundred and twenty human pregnancies. Surg Gynecol Obstet 47:201–5.

Murphy DP. 1929. Outcome of 625 pregnancies in women subjected to pelvic radium or roentgen irradiation. Am J Obstet Gynecol 18:79–87.

Murphy M, Seagroatt V, Hey K, et al. 1996. Neural tube defects 1974–94 – down but not out. Arch Dis Child 75:F.

Murphy ML. 1960. Teratogenic effects of tumour-inhibiting chemicals in the foetal rat. In: Wolstenholme GE, O'Connor CM eds. Ciba Foundation symposium on congenital malformations. Boston, Little, Brown, pp 78–107.

Myers GJ, Davidson PW. 1998. Prenatal methylmercury exposure and children: neurologic, developmental, and behavioral research. Environ Health Perspect 106(suppl 3):41–7.

Myers GJ, Davidson PW, Cox C, et al. 2003. Prenatal mercury exposure from ocean fish consumption in the Seycelles child development study. Lancet 361:1686–92.

Myers MF, Li S, Correa-Villaseñor A, et al. 2001. Folic acid supplementation and risk for imperforate anus in China. Am J Epidemiol 154:1051–6.

Myrianthopoulos NC, Melnick M. 1987. Studies in neural tube defects. I. Epidemiologic and and etiologic aspects. Am J Med Genet 26:783–96.

Naeye RL, Burt LS, Wright DL, et al. 1971. Neonatal mortality, the male advantage. Pediatrics 48:902–6.

Naggan L, MacMahon B. 1967. Ethnic differences in the prevalence of anencephaly and spina bifida in Boston, Massachusetts. N Engl J Med 277:1119–23.

Nakane Y. 1979. Congenital malformations among infants of epileptic mothers treated during pregnancy – the report of a collaborative study group in Japan. Fol Psychiat Neurol 33:363–9.

Nau H. 1986. Species differences in pharmacokinetics and drug teratogenesis. Environ Health Perspect 70:113–29.

Nau H. 1987. Species differences in pharmacokinetics and drug teratogenesis. Environ Health Perspect 70:113–20.

Nau H. 1993. Embryotoxicity and teratogenicity of topical retinoic acid. Skin Pharmacol 6:35–44.

Nau H. 2001. Teratogenicity of isotretinoin revisited: species variation and the role of all-trans-retinoic acid. J Am Acad Dermatol 45:S183–7.

Nau H, Hauck RS, Ehlers K. 1991. Valproic acid-induced neural tube defects in mouse and human: aspects of chirality, alternative drug mechanism, pharmacokinetics and possible mechanisms. Pharmacol Toxicol 69:310–21.

Neave C. 1967. Congenital malformations in offspring of diabetics, DPH Thesis, Harvard School of Public Health, Cambridge, MA.

Neave C. 1984. Congenital malformations in offspring of diabetics. Perspect Pediat Pathol 8:213–22.

Neel JV. 1958. A study of major congenital defects in Japanese infants. Am J Hum Genet 10:398–445.

Neel JV, Schull WJ. 1991. The children of atomic bomb survivors: a genetic study. Washington, DC, National Academy Press.

Neiger BL. 2000. The re-emergence of thalidomide: results of a scientific conference. Teratology 62:432–5.

Nelson KB, Ellenberg JH. 1982. Maternal seizure disorder, outcome of pregnancy, and neurologic abnormalities in the children. Neurology 32:1247–54.

Nelson MM. 1960. Teratogenic effects of pteroylglutamic acid deficiency in the rat. In: Wolstenholme GE, O'Connor CM eds. Ciba Foundation symposium on congenital malformations. Boston, MA, Little, Brown, pp 134–51.

Nelson MM, Baird CD, Wright HV, et al. 1956. Multiple congenital abnormalities in the rat resulting from riboflavin deficiency induced by the antimetabolite galactoflavin. J Nutr 58:25–34.

Nelson MM, Evans HM. 1949. Pteroylglutamic acid and reproduction in the rat. J Nutr 38:1–24.

Nelson MM, Forfar JO. 1971. Associations between drugs administered during pregnancy and congenital abnormalities of the fetus. BMJ 1:523–7.

Nelson N, Byerly TC, Kolbye AC, et al. 1971. Hazards of mercury: special report to the Secretary's Pesticide Advisory Committee, Department of Health, Education, and Welfare, November 1970. Environ Res 4:1–69.

Nelson RC, Warkany J. 1938. Changes in the ant. pituitary gland of rats with experimental goiter. Proc Soc Exp Biol Med 39:66–71.

Neubert R, Neubert D. 1997. Peculiarities and mode of action of thalidomide. In: Kavlock RJ, Daston GP eds. Drug toxicity in embryonic development II. New York, Springer, pp 41–119.

Neugut RH. 1981. Epidemiological appraisal of the literature on the fetal alcohol syndrome in humans. Early Hum Dev 5:411–29.

Nevin N. 1983. Discussion. In: Dobbing J ed. Prevention of spina bifida and other neural tube defects. New York, Academic Press, pp 114–5.

Nevin NC, Johnston WP, Merrett JD. 1981. Influence of social class on the risk of recurrence of anencephalus and spina bifida. Dev Med Child Neurol 23:155–9.

Nevin NC, Merrett JD. 1975. Potato avoidance during pregnancy in women with a previous infant with either anencephaly and/or spina bifida. Br J Prev Soc Med 29:111–15.

Newman TB. 1985. Etiology of ventricular septal defects: an epidemiologic approach. Pediatrics 76:741–9.

Ng SS, Gütschow M, Weiss M, et al. 2003. Antiangiogenic activity of N-substituted and tetrafluorinated thalidomide analogues. Cancer Res 63:3189–94.

Ngo AD, Taylor R, Roberts CL, et al. 2006. Association between Agent Orange and birth defects: sysyematic review and meta-analysis. Int J Epidemiol 35:1220–30.

Nguyen HT, Sharma V, McIntyre RS. 2009. Teratogenesis associated with antipolar agents. Adv Ther 26:281–94.

Niebyl JR, Blake DA, Rocco LE, et al. 1985. Lack of maternal metabolic, endocrine, and environmental influences in the etiology of cleft lip with or without cleft palate. Cleft Palate J 22:20–8.

Nielsen GL, Møller M, Sørensen HT. 2006. HbA1c in early diabetic pregnancy and pregnancy outcomes: a Danish population-based cohort study of 573 pregnancies in women with type 1 diabetes. Diabetes Care 29:2612–16.

Nightingale SL. 1998. Thalidomide approved for erythema nodosum leprosum. JAMA 280:872.

Nikkilä A, Rydhström H, Källén B. 2006. The incidence of spinal bifida in Sweden 1973–2003: the effect of prenatal diagnosis. Eur J Public Health 16:660–2.

Nishimura H, Takano K, Tanimura T, et al. 1968. Normal and abnormal development of human embryos: first report of the analysis of 1,213 intact embryos. Teratology 1:281–90.

Niswander KR, Gordon M. 1972. The women and their pregnancies. Philadelphia, PA, Saunders.

Noller KL, Fish CR. 1974. Diethylstilbestrol usage: its interesting past, important present, and questionalble future. Med Clin North Am 58:793–810.

Nora JJ, Nora AH, Toews WH. 1974. Lithium, Ebstein's anomaly, and other congenital heart defects. Lancet 2:594–5.

Nowack E. 1965. The sensitive phase in thalidomide embryopathy. Humangenetik 1:516–36 (In German).

Nulman I, Klaslo D, Koren G. 1999. Treatment of epilepsy in pregnancy. Drugs 57:535–44.

Nulman I, Rovet J, Barrera M, et al. 2009. Long-term neurodevelopment of children exposed to maternal nausea and vomiting of pregnancy and diclectin. J Pediatr 155:45–50.

Nulman I, Rovet J, Stewart DE, et al. 1997b. Neurodevelopment of children exposed in utero to antidepressant drugs. N Engl J Med 336:258–62.

Nulman I, Scolnik D, Chitayat D, et al. 1997a. Findings in children exposed in utero to phenytoin and carbamazepine monotherapy: independent effects of epilepsy and medications. Am J Med Genet 68:18–24.

Oakley GP. 1993. Folic acid-preventable spina bifida and anencephaly. JAMA 269:1292–3.

Oakley GP. 2002. Delaying folic acid fortification of flour: governments that do not ensure fortification are committing public health malpractice. BMJ 324:348–9.

Oakley GP, Adams MJ, James LM. 1983. Vitamins and neural tube defects. Lancet 2:798–9.

Oakley GP, Erickson JD, James LM, et al. 1994. Prevention of folic acid-preventable spina bifida and anencephaly. In: Bock G, Marsh J eds. Neural tube defects. Chichester, Wiley, pp 212–31.

O'Connell EJ, Feldt RH, Stickler GB. 1965. Head circumference, mental retardation, and growth failure. Pediatrics 6:62–6.
O'Connor MJ, Brill NJ, Sigman M. 1986. Alcohol use in primiparous women older than 30 years of age: relation to infant development. Pediatrics 78:444–50.
Oguni M, Dansky ML, Andermann E, et al. 1992. Improved pregnancy outcome in epileptic women in the last decade: relationship to maternal antiepileptic therapy. Brain Dev 14:371–80.
Oken E, Bellinger DC. 2008. Fish consumption, methylmercury and child neurodevelopment. Curr Opin Pediatr 20:178–83.
Olegård R, Sabel KG, Aronsson M, et al. 1979. Effects on the child of alcohol abuse during pregnancy: retrospective and prospective studies. Acta Paediatr Scand Suppl 275:112–20.
Olsen J, Tuntiseranaee P. 1995. Is moderate alcohol intake in pregnancy associated with the craniofacial features related to the fetal alcohol syndrome? Scand J Soc Med 23:156–61.
Olshan AF, Schnitzer PG. 1994. Paternal occupation and birth defects. In: Mattison DR, Olshan AF eds. Male-mediated developmental toxicity. New York, Plenum Press, pp 153–67.
Omtzigt JGC, Los FJ, Grobbee DE, et al. 1992. The risk of spina bifida aperta after first-trimester exposure to valproate in a prenatal cohort. Neurology 42:119–25.
Opitz JM. 2000. Heterogeneity and minor anomalies. Am J Med Genet 92:373–5.
Oppenheimer JH, Schwarz HL. 1997. Molecular basis of thyroid hormone-dependent brain development. Endoc Rev 18:462–75.
Orfanos CE, Ehlert R, Gollnick H. 1987. The retinoids: a review of their clinical pharmacology and therapeutic use. Drugs 34:459–503.
Ornoy A. 2009. Valproic acid in pregnancy: how much are we endangering the embryo and fetus? Reprod Toxicol 28:1–10.
Ornoy A, Cohen E. 1996. Outcome of children born to epileptic mothers treated with carbamazepine during pregnancy. Arch Dis Child 75:517–20.
Ornstein M, Einarson A, Koren G. 1995. Bendectin/Diclectin for morning sickness: a Canadian follow-up of an American tragedy. Reprod Toxicol 9:1–6.
O'Sullivan JB. 1961. Gestational diabetes: unsuspected, asymptomatic diabetes in pregnancy. N Engl J Med 264:1082–5.
Otake M, Schull WJ. 1984. *In utero* exposure to A-bomb radiation: a reassessment. Br J Radiol 57:409–14.
Otake M, Schull WJ, Lee S. 1996. Threshold for radiation-related severe mental retardation in prenatal exposed A-bomb surviors: a re-analysis. Int J Radiat Biol 70:755–63.
Otake M, Schull WJ, Yoshimaru H. 1991. Brain damage among the prenatally exposed. J Radiat Res 32:249–64.
Ouellette EM, Rosett HL, Rosman NP, et al. 1977. Adverse effects of maternal alcohol abuse during pregnancy. N Engl J Med 297:528–30.
Ounsted M, Scott A. 1982. Smoking during pregnancy: its association with other maternal factors and birthweight. Acta Obstet Gynecol Scand 61:367–71.
Owens JR, Harris F, McAllister E, et al. 1981. 19-year incidence of neural tube defects in an area under constant surveillance. Lancet 2:1032–5.
Oxford JS, Sutton RN. 1968. The effect of rubella and herpes virus hominis on the pre- and post-implantation stages of pregnancy in laboratory animnals. J Embryol Exp Morphol 20:285–94.
Pacheco SS, Braga C, Souza AI, et al. 2009 May 29. Efecto de la fortificaci6n alimentaria con ácido fólico en la prevelencia de defectos del tubo neural. Rev Saúde Pública (ahead of print Epub).
Paigen B. 1982. Controversy at Love Canal. Hastings Center Rep 12:29–37.
Palludan B. 1961. The teratogenic effect of vitamin A deficiency in pigs. Acta Vet Scand 2:32–59.
Palmer JR, Herbst AL, Noller KL, et al. 2009 Aug 18. Urogenital abnormalities in men exposed to diethylstilbestrol *in utero:* a cohort study. Environ Health 8:37.
Palomaki GE, Williams JR, Haddow JE. 1999. Prenatal screening for open neural-tube defects in Maine. N Engl J Med 340:1049–50.

Panagiotopoulos T, Antoniadou I, Valassi-Adam E. 1999. Increase in congenital rubella occurrence after immunisation in Greece: retrospective survey and systematic review. Lancet 319:1462–6.

Parazzini F, Tozzi L, Chatenoud L, et al. 1994. Alcohol and risk of spontaneous abortion. Hum Reprod 9:1950–3.

Park-Wyllie L, Mazzotta P, Pastuszak A, et al. 2000. Birth defects after maternal exposure to corticosteroids: prospective cohort study and meta-analysis of epidemiological studies. Teratology 62:385–92.

Passarge E, Lenz W. 1966. Syndrome of caudal regression in infants of diabetic mothers: observations of further cases. Pediatrics 37:672–5.

Pastel RH. 2002. Radiophobia: long-term psychological consequences of Chernobyl. Mil Med 167:134–6.

Paterson DC. 1969. Congenital deformities. CMAJ 101:175–6.

Paterson DC. 1977. Congenital deformities associated with Bendectin. CMAJ 116:348.

Patrias K, Gordner RL, Groft SC. 1997. Thalidomide: potential benefits and risks. Bethesda, MD, National Library of Medicine.

Patten BM. 1957. Varying developmental mechanisms in teratology. Pediatrics 19(part II):734–48.

Paul D. 2000. A double-edged sword. Nature 405:515.

Pauling L. 1986. Live longer and feel better. New York, Freeman.

Peacock JL, Bland JM, Anderson HR. 1991. Effects on birthweight of alcohol and caffeine consumption in smoking woman. J Epidemiol Community Health 45:59–63.

Peacock JL, Cook DG, Carey IM, et al. 1998. Maternal cotinine level during pregnancy and birthweight for gestational age. Int J Epidemiol 27:647–56.

Peadon E, Rhys-Jones B, Bower C, et al. 2009. Systematic review of interventions for children with Fetal Alcohol Spectrum Disorders. BMC Pediatr May 25;9:35.

Pearn JH. 1983. Teratogens and the male: an analysis with special reference to herbicide exposure. Med J Aust 2:16–20.

Pearson DW, Kernaghan D, Lee R, et al. 2007. The relationship between pre-pregnancy care and early pregnancy loss, major congenital anomaly or perinatal death in type 1 diabetes mellitus. Br J Obstet Gynecol 114:104–7.

Pearson K. ed. 1912. The treasury of human inheritance. London, Cambridge University Press.

Pedersen J. 1977. The pregnant diabetic and her mewborn: problems and management. Copenhagen, Munksgaard.

Pedersen LH, Hendriksen TB, Vertergaard M, et al. 2009. Selective serotonin reuptake inhibitors in pregnancy and congenital malformations: population based cohort study. BMJ 339:b3569.

Pei LJ, Li Z, Li S, et al. 2003. [The epidemiology of neural tube defects in high-prevalence and low-prevalence areas of China.] In Chinese. Zhonghua Liu Xing Bing Xue Za Zhi 24:465–70.

Pemberton M. 1992. Anencephaly alert in Texas. BMJ 304:732–3.

Penney GC, Mair G, Pearson DW, et al. 2003. Outcomes of pregnancies in women with type 1 diabetes in Scotland: a national population-based study. Br J Obstet Gynecol 110:315–18.

Penrose LS. 1938. A clinical and genetic study of 1280 cases of mental defect. London, His Majesty's Stationery Office.

Penrose LS. 1949. The biology of mental defect. London, Sidgwick & Jackson.

Penrose LS. 1951. Heredity and environment in the causation of foetal malformation. Practitioner 66:429–35.

Penrose LS. 1954. Mongolian idiocy (mongolism) and maternal age. Ann N Y Acad Sci 57:494–502.

Penrose LS. 1957. The genetics of anencephaly. J Ment Def Res 1:4–15.

Persad VL, Hof M, Dubé JM, et al. 2002. Incidence of open neural tube defects in Nova Scotia after folic acid fortification. CMAJ 167:241–5.

Persson B, Hanson U. 1998. Neonatal morbidities in gestational diabetes mellitus. Diabetes Care 21:879–84.

Persson M, Norman M, Hanson U. 2009. Aug 12 Obstetric and perinatal outcomes in type-1 diabetic pregnancies – a large, population-based study. Diabetes Care (Epub ahead of print).

Peticca A, Keely EJ, Walker MC, et al. 2009. Pregnancy outcomes in diabetes subtypes: how do they compare? A province-based study of Ontario, 2005–2006. J Obstet Gynaecol Can 31:487–96.

Pierog S, Chandavasu O, Wexler I. 1977. Withdrawal symptoms in infants with the fetal alcohol syndrome. J Pediatr 90:630–3.

Pinsky L. 1985. Informative morphogenetic variants: minor congenital anomalies revisited. In: Kalter H ed. Issues and reviews in teratology, vol. 3. New York, Plenum Press, pp 135–70.

Pittschieler S, Brezinka C, Jahn B, et al. 2008. Spontaneous abortion and the prophylactic effect of folic acid supplementation in epileptic women undergoing antiepileptic therapy. J Neurol 255:1926–31.

Plant M. 1985. Women, drinking, and pregnancy. London, Tavistock.

Platt LD, Koch R, Azen C, et al. 1992. Maternal phenylketonuria collaborative study, obstetric aspects and outcome: the first 6 years. Am J Obstet Gynecol 66:1150–62.

Platt LD, Koch R, Hanley WB, et al. 2000. The international study of pregnancy outcome in women with maternal phenylketonuria: report of a 12-year study. Am J Obstet Gynecol 182:326–33.

Plummer G. 1952. Anomalies occurring in children exposed *in utero* to the atomic bomb in Hiroshima. Pediatrics 10:687–93.

Pocchiari F, Silano V, Zampieri A. 1979. Human health effects from accidental release of tetrachlorodibenzo-*p*-dioxin (TCDD) at Seveso, Italy. Ann N Y Acad Sci 320:311–20.

Poland BJ, Miller JR, Harris H, et al. 1981. Spontaneous abortion: a study of 1961 women and their conceptuses. Acta Obstet Gynecol Scand Suppl 102:5–32.

Polifka JE, Rutledge JC, Kimmel JL, et al. 1996. Exposure to ethylene oxide during the early zygotic period induces skeletal anomalies in mouse fetuses. Teratology 53:1–9.

Polygenis D, Wharton S, Malmberg C, et al. 1998. Moderate alcohol consumption during pregnancy and the incidence of fetal malformations: a meta-analysis. Neurotoxicol Teratol 20:61–7.

Pomeranze J, Stone ML, King EJ. 1959. The obstetric importance of obesity and "benign" glycosuria in prediagnosis diabetes. Obstet Gynecol 13:81–4.

Poskanzer DC, Herbst AL. 1977. Epidemiology of vaginal adenosis and adenocarcinoma associated with exposure to stilbestrol *in utero*. Cancer 39:1892–5.

Poswillo DE, Hamilton WJ, Sopher D. 1972a. The marmoset as an animal model for teratological research. Nature 239:462–4.

Poswillo DE, Sopher D, Mitchell S. 1972b. Experimental induction of foetal malformation with "blighted" potato: a preliminary report. Nature 239:462–4.

Poswillo DE, Sopher D, Mitchell SJ, et al. 1973. Investigations into the teratogenic potential of imperfect potatoes. Teratology 8:339–48.

Potter EL. 1964. Classification and pathology of congenital anomalies. Am J Obstet Gynecol 90:985–93.

Powell RJ. 1999. Thalidomide: current uses. BioDrugs 11:409–26.

Pradat P, Robert-Gnansia E, Tanna GL, et al. 2003. First trimester exposure to corticosteroids and oral clefts. Birth Defects Res A Clin Mol Teratol 67:968–70.

Pratt OE. 1982. Alcohol and the developing fetus. Br Med Bull 38:48–52.

Pritchard JA, Scott DE, Whalley PJ, et al. 1970. Infants of mothers with megaloblastic anemia due to folate deficiency. JAMA 211:1982–4.

Pugach S, Clarkson T. 2009. Prenatal mercury exposure and postnatal outcome: clinical case report and analysis. Clin Toxicol (Phila) 47:366–70.

Quan A. 2006. Fetopathy associated with exposure to angiotensin converting enzyme inhibitors and angiotensisn receptor antagonists. Early Hum Dev 82:23–8.

Quinlan JD, Hill DA. 2003. Nausea and vomiting of pregnancy. Am Fam Physician 68:121–8.

Ramadhani T, Short V, Canfield MA, et al. 2009. Are birth defects among Hispanics related to maternal nativity or number of years lived in the United States? Birth Defects Res A Clin Mol Teratol 85:755–63.

Rankin J, Glinianaia S, Brown SR, et al. 2000. The changing prevalence of neural tube defects: a population-based study in the North of England, 1984–96. Paediatr Perinat Epidemiol 14:104–10.
Rating D, Jäger-Roman E, Koch R, et al. 1982. Minor anomalies in the offspring of epileptic parents. In: Janz D, Bossi L, Dam M, et al. eds. Epilepsy, pregnancy, and the child. New York, Raven Press, pp 283–8.
Ray JG, Vermeulen MJ, Koren G. 2007. Taking ACE inhibitors during pregnancy: is it safe? Can Fam Physician 53:1439–40.
Ray JG, Vermeulen MJ, Meier C, et al. 2004. Risk of congenital anomalies detected during antenatal serum screening in women with pregestational diabetes. Q J Med 97:651–3.
Raymond CA. 1987. Birth defects linked with specific level of maternal alcohol use, but abstinence is still the best policy. JAMA 258:177–8.
Record RG, McKeown T. 1949. Congenital malformations of the central nervous system. I. A survey of 930 cases. Br J Soc Med 4:183–219.
Record RG, McKeown T. 1950a. Congenital malformations of the central nervous system. II. Maternal reproductive history and familial incidence. Br J Soc Med 4:26–50.
Record RG, McKeown T. 1950b. Congenital malformations of the central nervous system. III. Risk of malformations in sibs of malformed individuals. Br J Soc Med 4:217–20.
Record RG, McKeown T. 1951. Congenital malformations of the central nervous system: data on 69 pairs of twins. Ann Eugen 15:285–92.
Regan L, Rai R. 2000. Epidemiology and the medical causes of miscarriage. Baillère Clin Obstet Gynecol 14:839–54.
Reggiani GM. 1980. Acute human exposure to TCDD in Seveso, Italy. J Toxicol Environ Health 6:27–43.
Reggiani GM. 1988. Historical overview of the controversy surrounding Agent Orange. In: Young Al, Reggiani GM eds. Agent Orange and its associated dioxin: assessment of a controversy. Amsterdam, Elsevier, pp 225–69.
Rehder H, Sanchioni L, Cefis F, et al. 1978. Pathologisch-embryologische Untersuchungen an Abortusfällen im Zusammenhang mit dem Seveso-Unglück. Schweiz Med Woch 108: 617–25.
Reiff-Eldridge R, Heffner CR, Ephross SA, et al. 2000. Monitoring pregnancy outcomes after prenatal drug exposure through prospective pregnancy registries: a pharmaceutical company commitment. Am J Obstet Gynecol 82:159–63.
Remington JS, McLeod R, Desmonts G. 1995. Toxoplasmosis. In: Remington JS, Klein JO eds. Infectious diseases of the fetus and newborn infant. Philadelphia, PA, Saunders, pp 140–267.
Renwick JH. 1972. Hypothesis: anencephaly and spina bifida are usually preventable by avoidance of a specific but unidentified substance present in certain potato tubers. Br J Prev Soc Med 26:67–88.
Renwick JH. 1973a. Potatoes and neural-tube defects. Lancet 2:562.
Renwick JH. 1973b. Prevention of anencephaly and spina bifida in man. Teratology 8:321–4.
Renwick JH, Possamai AP, Munday MR. 1974. Potatoes and spina bifida. Proc R Soc Med 67:110–14.
Reznik G, Gold J, Mandryk J. 1984. Vietnam service and congenital anomalies. Med J Aust 141:194–5.
Rivers TM. 1959. Introduction to the National Foundation conference on congenital malformations. J Chron Dis 10:83.
Rivers TM, Warkany J, Fraser FC, et al. 1959. Conference on congenital malformations: held under the auspices of The National Foundation. J Chron Dis 10:83–151.
Robboy SJ, Noller KL, O'Brien P, et al. 1984. Increased incidence of cervical and vaginal dysplasia in 3,980 diethylstilbestrol-exposed young women. JAMA 252:2979–83.
Robe LB, Gromisch DS, Iosub S. 1981. Symptoms of neonatal ethanol withdrawal. Curr Alcohol 8:485–93.

Robert E, Francannet C, Robert JM. 1988. [Registries of malformations in the Rhône-Alps/Auvergne region. Value and limits of monitoring teratogenesis. 11 years' experience (1976–1983).] In French. J Gynecol Obstet Biol Reprod (Paris) 17:601–7.
Robert E, Guibaud P. 1982. Maternal valproic acid and congenital neural tube defects. Lancet 2:937.
Robert E, Harris J, Källén B. 1997. The epidemiology of preaxial limb malformations. Reprod Toxicol 11:653–62.
Robert E, Källén B. 1994. In utero exposure to carbamazepine: effects on the fetus. In: Kalter H ed. Issues and reviews in teratology, vol. 7. New York, Plenum Press, pp 37–55.
Robert E, Källén B, Harris J. 1996. The epidemiology of orofacial clefts. I. Some general epidemiological characteristics. J Craniofac Dev Biol 16:234–41.
Roberts CJ, Lloyd S. 1973. Area differences in spontaneous abortion rates in South Wales and their relation to neural tube defect incidence. BMJ 4:20–2.
Roberts CJ, Lowe CR. 1975. Where have all the conceptions gone? Lancet 1:498–9.
Roberts CJ, Revington CJ, Lloyd S. 1973. Potato cultivation and storage in South Wales and its relation to neural tube malformation prevalence. Br J Prev Soc Med 27:214–16.
Roberts HE, Moore CA, Cragan JD, et al. 1995. Impact of prenatal diagnosis on the birth prevalence of neural tube defects, Atlanta, 1990–1991. Pediatrics 96:880–3.
Robertson J, Feldcamp M, Jennings J, et al. 1998. Maternal alcohol and caffeine consumption and spontaneous abortion. Epidemiology 9:583–4.
Robin NH, Zackai EH. 1994. Unusual craniofacial dysmorphia due to prenatal alcohol and cocaine exposure. Teratology 50:60–4.
Robinson GC, Conry JL, Conry RF. 1987. Clinical profile and prevalence of fetal alcohol syndrome in an isolated community in British Columbia. CMAJ 137:203–7.
Rodríguez-Pinilla E, Arroyo I, Fondevilla J, et al. 2000. Prenatal exposure to valproic acid during pregnancy and limb deficiencies: a case-control study. Am J Med Genet 90:376–81.
Rogan WJ. 1982. PCBs and cola-colored babies: Japan, 1968, and Taiwan, 1979. Teratology 26:259–61.
Rogan WJ, Gladen BC, Guo YL, et al. 1999. Sex ratio after exposure to dioxin-like chemicals in Taiwan. Lancet 353:206–7.
Rogan WJ, Gladen BC, Hung KL, et al. 1988. Congenital poisoning by polychlorinated biphenyls and the contaminants in Taiwan. Science 241:334–6.
Rogan WJ, Gladen BC, McKinney JD, et al. 1986. Neonatal effects of transplacental exposure to PCBs and DDE. J Pediatr 109:335–41.
Rogers JM, Daston GP, Ebron MT, et al. 1985. Studies on the mechanism of trypan blue teratogenicity in the rat developing in vivo and in vitro. Teratology 31:389–99.
Roland JM, Murphy HR, Ball V, et al. 2005. The pregnancies of women with type 2 diabetes: poor outcomes but opportunities for improvement. Diabet Med 22:1774–7.
Roman E, Beral V, Zuckerman B. 1988. The relation between alcohol consumption and pregnancy outcome in humans: a critique. In: Kalter H ed. Issues and reviews in teratology, vol. 4. New York, Plenum Press, pp 205–35.
Ronnenberg AG, Goldman MB, Chen DT, et al. 2002. Preconception folate and vitamin B6 status and clinical spontaneous abortion in Chinese women. Obstet Gynecol 100:107–13.
Rosa FW. 1983. Teratogenicity of isotretinoin. Lancet 2:513.
Rosa FW. 1991. Spina bifida in infants of women treated with carbamazepine during pregnancy. Lancet 324:674–7.
Rosa FW. 1993. Retinoid embryopathy in humans. In: Koren G ed. Retinoids in clinical practice. New York, Dekker, pp 77–109.
Rosa FW, Piazza-Hepp T, Goetsch R. 1994. Holoprosencephaly with 1st trimester topical tretinoin. Teratology 49:418–9.
Rosa FW, Wilk AL, Kelsey FO. 1986. Teratogen update: vitamin A congeners. Teratology 33:355–64.
Rosano A, Smithells D, Cacciani L, et al. 1999. Time trends in neural tube defects prevalence in relation to preventive strategies: an international study. J Epidemiol Community Health 53:630–5.

Rosenberg HS. 1987. Cardiovascular effects of congenital infections. Am J Cardiovasc Pathol 1:147–56.
Rosett HL, Weiner L. 1985. Alcohol and pregnancy: a clinical perspective. Annu Rev Med 36:73–80.
Rosett HL, Weiner L, Lee A, et al. 1983. Patterns of alcohol consumption and fetal development. Obstet Gynecol 61:539–46.
Rostand A, Kaminski M, Lelong N, et al. 1990. Alcohol use in pregnancy, craniofacial features, and fetal growth. J Epidemiol Community Health 44:302–6.
Rothman KJ, Fyler DC. 1976. Sex, birth order, and maternal age characteristics of infants with congenital heart defects. Am J Epidemiol 104:527–34.
Rothman KJ, Flyer DC, Goldblatt A, et al. 1979. Exogenous hormones and other drug exposures of children with congenital heart disease. Am J Epidemiol 109:433–9.
Rothman KJ, Moore MR, Singer MR, et al. 1995. Teratogenicity of high vitamin A intake. N Engl J Med 333:369–73.
Rouse B, Azen C. 2004. Effect of high maternal blood phenylalanine on offspring congenital anomalies and developmental outcome at ages 4 and 6 years: the importance of strict dietary control preconception and throughout pregnancy. J Pediatr 144:235–9.
Rouse B, Azen C, Koch R, et al. 1997. Maternal Phenylketonuria Collaborative Study (MPKUCS) offspring: facial anomalies, malformations, and early neurological sequelae. Am J Med Genet 69:89–95.
Rouse B, Lockhart L, Matalon R, et al. 1990. Maternal phenylketonuria pregnancy outcome: preliminary report of facial dysmorphology and major malformations. J Inher Metab Dis 13:289–91.
Rouse B, Matalon R, Koch R, et al. 2000. Maternal phenylketonuria syndrome: congenital heart defects, microcephaly, and developmental outcomes. J Pediatr 136:57–61.
Rubin A, Murphy DP. 1958. Studies in human reproduction. III. The frequency of congenital malformations in the offspring of nondiabetic and diabetic individuals. J Pediatr 53:579–85.
Rugh R, Wolff J. 1955. Reparation of the fetal eye following radiation insult. Arch Ophthal 54: 351–9.
Rush D, Rosenberg IH. 1992. Folate supplements and neural tube defects. Nutr Rev 50:25–8.
Rushton DI. 1985. The nature and causes of spontaneous abortions with normal karyotypes. In: Kalter H ed. Issues and reviews in teratology, vol. 3. New York, Plenum Press, pp 21–63.
Russell LB. 1954. The effects of radiation on mammalian prenatal development. In: Hollaender A ed. Radiation biology, vol. 1, pt 2. New York, McGraw-Hill, pp 861–918.
Russell M, Czarnecki DM, Cowan R, et al. 1991. Measures of maternal alcohol use as predictors of development in early childhood. Alcohol Clin Exp Res 15:991–1000.
Rutledge JC. 1997. Developmental toxicity induced during early stages of mammalian embryogenesis. Mutat Res 396:113–27.
Rutledge JC, Generoso WM, Shurbaji A, et al. 1992. Developmental anomalies derived from exposure of zygotes and first-cleavage embryos to mutagens. Mutat Res 296:167–77.
Sackett DL. 1979. Bias in analytic research. J Chron Dis 32:51–63.
Saint-Hilaire IG. 1832–1837 Histoire générale et particulière des anomalies de l'organisation chez l'homme et les animaux, ouvrage comprenant des recherches sur les caractères, la classification, l'influence physiologique et pathologique, les rapport généraux, les lois et les causes des monstruosités, des varietés et vices de conformation, ou traité de teratologie. 3 vols and atlas. Paris, Baillière.
Sakamoto M, Nakano A, Akagi H. 2001. Declining Minamata male birth ratio associated with increased male fetal death due to heavy methylmercury pollution. Environ Res 87:92–8.
Sampson PD, Bookstein FL, Barr HM, et al. 1994. Prenatal alcohol exposure, birthweight, and measures of child size from birth to age 14 years. Am J Public Health 84:1421–8.

Sampson PD, Streissguth AP, Bookstein FL, et al. 1997. Incidence of fetal alcohol syndrome and prevalence of alcohol-related neurodevelopmental disorder. Teratology 56:317–26.

Sampson PD, Streissguth AP, Bookstein FL, et al. 2000. On categorizations in analyses of alcohol teratogenesis. Environ Health Perspect 108(Suppl 3):421–8.

Samrén EB, Van Duijn CM, Christiaens GC, et al. 1999. Antiepileptic drug regimens and major congenital abnormalities in the offspring. Ann Neurol 46:739–46.

Samrén EB, Van Duijn CM, Koch R, et al. 1997. Maternal use of antiepileptic drugs and risk of major congenital malformations: a joint European prospective study of human teratogenesis associated with maternal epilepsy. Epilepsia 38:981–90.

Sanders J. 1993. From science to evidence: the testimony on causation in the Bendectin cases. Stanford Law Rev 46:1–86.

Sauer H, Günther J, Hescheleer J, et al. 2000. Thalidomide inhibits angiogenesis in embryoid bodies by the generation of hydroxyl radicals. Am J Pathol 156:151–8.

Sawin PB, Crary D, Fox RR, et al. 1965. Thalidomide malformations and genetic background in the rabbit. Experientia 21:672–6.

Saxén L. 1983. Twenty years of study of the etiology of congenital malformations in Finland. In: Kalter H ed. Issues and reviews in teratology, vol. 1. New York, Plenum Press, pp 73–110.

Saxén L, Holmberg PC, Kurppa K, et al. 1990. Influenza epidemics and anencephaly. Am J Public Health 80:473–5.

Schaefer C, Meister R, Weber-Schoendorfer C. 2009. May 15 Isotretinoin exposure and pregnancy outcome: an observational study of the Berlin Institute for Clinical Teratology and Drug Risk Assessment in Pregnancy. Arch Gynecol Obstet (Epub ahesd of print).

Schardein JL. 1980. Congenital abnormalities and hormones during pregnancy: a clinical review. Teratology 22:251–70.

Schardein JL. 1983. Teratogenic risk assessment: past, present, and future. In: Kalter H ed. Issues and reviews in teratology, vol. 1. New York, Plenum Press, pp 181–214.

Schardein JL. 1988. Teratology testing: status and issues after two decades of evolution. Rev Environ Contamin Toxic 102:1–78.

Schardein JL. 2000. Chemically induced birth defects. New York, Marcel Dekker.

Schnorr TM, Lawson CC, Whelan EA, et al. 2001. Spontaneous abortion, sex ratio, and paternal occupational exposure to 2,3,7,8-tetrachlorodibenzo-*p*-dioxin. Environ Health Perspect 109:1127–32.

Schorah CJ, Smithells RW, Seller MJ. 1997. Terathanasia, folic acid, and birth defects. Lancet 350:1323.

Schorah CJ, Wild J, Hartley R, et al. 1983. The effect of periconceptional supplementation on blood folate vitamin concentration in women at risk for neural tube defect. Br J Nutr 49:203–11.

Schou M. 1976. What happened later to the lithium babies? Acta Psychiatr Scand 54:193–7.

Schou M, Amdisen A. 1970. Lithium in pregnancy. Lancet 1:1391.

Schou M, Amdisen A. 1971. Lithium teratogenicity. Lancet 1:1132.

Schou M, Goldfield MD, Weinstein MR, et al. 1973. Lithium and pregnancy – I, report from the Register of Lithium Babies. BMJ 2:135–6.

Schuett VE, Brown ES, Michals K. 1985. Reinstitution of diet therapy in PKU patients from twenty-two US clinics. Am J Public Health 75:39–42.

Schull WJ, Otake M. 1999. Cognitive function and prenatal exposure to ionizing radiation. Int J Radiat Biol 74:59–71.

Schumacher H, Smith RL, Williams RT. 1965. The metabolism of thalidomide: the fate of thalidomide and some of its hydrolysis products in various species. Brit J Pharmacol 25:338–51.

Schupf N, Ottman R. 1997. Reproduction among individuals with idiopathic/cryptogenic epilepsy: risk factors for spontaneous abortion. Epilepsia 38:824–9.

Schwalbe E. ed. 1906–1937 Die Morphologie der Missbildungen des Menschen und der Tiere. Jena, Fischer.

Scott DE, Whalley PJ, Pritchard JA. 1970. Maternal folate deficiency and pregnancy wastage. II. Fetal malformation. Obstet Gynecol 36:26–8.
Scott DM, Taylor RE. 2007. Health-related effects of genetic variations of alcohol-mediating enzymes in African Americans. Alcohol Res Health 30:18–21.
Scott WJ, Fradkin R. 1984. The effects of prenatal ethanol in cynomolgus mankeys. Teratology 29:49–56.
Scott WJ, Fradkin R, Wilson JG. 1977. Non-confirmation of thalidomide induced teratogenesis in rats and mice. Teratology 6:333–5.
Scriver CR. 2007. The PAH gene, phenylketonuria, and a paradigm shift. Hum Mutat 28:831–45.
Scriver CR, Beaudet AL, Sly WS, et al. 1995. The metabolic and molecular bases of inherited disease. New York, McGraw-Hill.
Seip M. 1976. Growth retardation, dysmorphic facies and minor malformations following massive exposure to phenobarbitone in utero. Acta Paediatr Scand 65:617–21.
Selcen D, Seidman S, Nigro MA. 2000. Otocerebral anomales associated with topical retinoin use. Brain Dev 22:218–20.
Seller MJ, Nevin NC. 1984. Periconceptional vitamin supplementation and the prevention of neural tube defects in south-east England and Northern Ireland. J Med Genet 21:325–30.
Seller MJ, Nevin NC. 1990. Vitamins during pregnancy and neural tube defects. JAMA 263:2749.
Sentrakul P, Potter EL. 1966. Pathological diagnosis on 2,681 abortions at the Chicago Lying-in Hospital, 1957–1965. Am J Public Health 56:2083–92.
Sever LE, Arbuckle TE, Sweeney A. 1997. Reproductive and developmental effects of occupational pesticide exposure: the epidemiologic evidence. Occup Med 12:305–25.
Sever LE, Emanuel I. 1981. Intergenerational factors in the etiology of anencephalus and spina bifida. Dev Med Child Neurol 23:151–4.
Shapiro L, Pastuszak A, Curto G, et al. 1997. Safety of first-trimester exposure to topical tretinoin: prospective cohort study. Lancet 350:1143–4.
Shapiro S. 1990. Vitamins during pregnancy and neural tube defects. JAMA 263:2748.
Shapiro S, Heinonen OP, Siskind V, et al. 1977. Antenatal exposure to doxylamine succinate and dicyclomine hydrochloride (Bendectin) in relation to congenital malformations, perinatal mortality rate, and intelligence quotient score. Am J Obstet Gynecol 28:480–5.
Shapiro S, Slone D. 1979. The effects of exogenous female hormones on the fetus. Epidemiol Rev 1:110–23.
Shapiro S, Slone D, Hartz SC, et al. 1976. Antiepileptics and parental epilepsy in the development of birth defects. Lancet 1:272–5.
Sharp D. 1999. Acronymania. Lancet 353:1728.
Sharp GB, Cole P. 1990. Vaginal bleeding and diethylstilbestrol exposure during pregnancy: relationship to genital tract clear cell adenocarcinoma and vaginal adenosis in daughters. Am J Obstet Gynecol 162:994–1001.
Sharpe PB, Chan A, Haan EA, et al. 2005. Maternal diabetes and congenital anomalies in South Australia 1986–2000: a population-based cohort study. Birth Defects Res A Clin Mol Teratol 73:605–11.
Sharpe TT, Velasquez MM. 2008. Risk of alcohol-exposed pregnancies among low-income, illicit drug-using women. J Womens Health (Larchmt) 17:1339–44.
Shaw GM, Croen LA, Todoroff K, et al. 2000. Periconceptional intake of vitamin supplements and risk of multiple congenital anomalies. Am J Med Genet 93:188–93.
Shaw GM, Lammer EJ. 1999. Maternal periconceptional alcohol consumption and risk for orofacial clefts. J Pediatr 134:298–303.
Shaw GM, Lammer EJ, Zhu H, et al. 2002. Maternal periconceptional vitamin use, genetic variation of infant reduced folate carrier (A80G), and risk of spina bifida. Am J Med Genet 108:1–6.
Shaw GM, Rozen R, Finnell RH, et al. 1998. Infant C677T mutation in MTHFR, maternal periconceptional vitamin use, and cleft lip. Am J Med Genet 80:96–8.
Shaw GM, Rozen R, Finnell RH, et al. 1999. Maternal vitamin use, genetic variation of infant methylenetetrahydrofolate reductase and risk for spina bifida. Am J Epidemiol 150:323–4.

Shaw GM, Schaffer D, Velie EM, et al. 1995. Periconceptional vitamin use, dietary folate, and the occurrence of neural tube defects. Epidemiology 6:219–26.

Shaw GM, Velie EM, Wasserman CR. 1997. Risk for neural tube defect-affected pregnancies among women of Mexican descent and white women in California. Am J Public Health 87: 1467–71.

Shaw GM, Wasserman CR, Block G, et al. 1996. High maternal vitamin A intake and risk of anomalies of structures with a cranial neural cell contribution. Lancet 347:899–900.

Shepard TH. 1992. Catalog of teratogenic agents, 7th ed. Baltimore, MD, Johns Hopkins University Press.

Shepard TH. 2001. Catalog of teratogenic agents, 10th ed. Baltimore, MD, Johns Hopkins University Press.

Shepard TH, Fantel AG, Mirkes PE. 1986. Megadoses of vitamin A. Teratology 34:366.

Shepard TH, Fantel AG, Mirkes PE. 1988. Collection and use of human embryonic and fetal material: 25 years of experience. In: Kalter H ed. Issues and reviews in teratology, vol. 4. New York, Plenum Press, pp 129–62.

Sheridan MD. 1964. Final report of a prospective study of children whose mothers had rubella in early pregnancy. BMJ 2:536–9.

Shi L, Tu N, Patterson PH. 2005. Maternal influenza infection is likely to alter fetal brain development indirectly: the virus is not detected in the fetus. Int J Dev Neurosci 23:299–305.

Shiota K. 1993. Teratothanasia: prenatal loss of abnormal conceptuses and the prevalence of various malformations during human gestation. Birth Defects Orig Artic Ser 29:189–99.

Sievers G. 1964. Klinisch-statistische Studien zu aktuellen Missbildungsproblemen: Eine Ubersicht. Arzneimittelforschung 14:605–55.

Silverman JA, Winters RW, Strande C. 1971. Lithium carbonate therapy during pregnancy: apparent lack of effect upon the fetus. Am J Obstet Gynecol 109:934–6.

Simpson JL. 1978. Genetics of diabetes mellitus (DM) and anomalies in offspring of diabetic mothers. Sem Perinat 2:383–94.

Simpson WJ. 1957. A preliminary report on cigarette smoking and the incidence of prematurity. Am J Obstet Gynecol 73:808–15.

Skolnick A. 1990. Key witness against morning sickness drug faces scientific fraud charges. JAMA 263:1468–9, 1473.

Smith AH, Fisher DO, Pearce N, et al. 1982. Congenital defects and miscarriages among New Zealand 2,4,5-T sprayers. Arch Environ Health 37:197–200.

Smith C, Watt M, Boyd AE, et al. 1973. Anencephaly, spina bifida, and potato blight in the Edinburgh area. Lancet 1:269.

Smith DF, Sandor GG, MacLeod PM, et al. 1981. Intrinsic defects in the fetal alcohol syndrome: studies on 76 cases from British Columbia and the Yukon Territory. Neurobehav Toxicol Teratol 3:145–52.

Smith DW. 1971. Minor malformations: their relevance and significance. In: Hook EB, Janerich DT, Porter IH eds. Monitoring, birth defects and environment: the problem of surveillance. New York, Academic Press, pp 169–75.

Smith DW. 1975. Classification, nomenclature, and naming of morphologic defects. J Pediatr 87: 62–4.

Smith DW. 1976. Recognizable patterns of human malformations. Philadelphia, PA, Saunders.

Smith I, Glossop J, Beasley M. 1990. Fetal damage due to maternal phenylketonuria: effects of dietary treatment and maternal phenylalanine concentration around the time of conception (an interim report from the U.K. phenylketonuria register). J Inher Met Dis 13:651–7.

Smith OW, Smith GV. 1946. Use of diethylstilbestrol to prevent fetal loss from complications of late pregnancy. N Engl J Med 241:562–8.

Smithberg M, Dixit PK. 1982. Teratogenic effects of lithium in mice. Teratology 26:239–46.

Smithells RW. 1962. Thalidomide and malformations in Liverpool. Lancet 1:1270–3.

Smithells RW. 1994. Teratoserendipity. In: Kalter H ed. Issues and reviews in teratology, vol. 7. New York, Plenum Press, pp 1–36.

Smithells RW, Nevin NC, Seller MJ, et al. 1983a. Further experience of vitamin supplementation for prevention of neural tube defect recurrences. Lancet 1:1027–31.
Smithells RW, Newman CG. 1992. Recognition of thalidomide defects. J Med Genet 29:716–23.
Smithells RW, Seller MJ, Harris R, et al. 1983b. Vitamin supplementation and neural tube defects. Lancet 2:40.
Smithells RW, Sheppard S, Schorah CJ. 1976. Vitamin deficiencies and neural tube defects. Arch Dis Child 51:944–50.
Smithells RW, Sheppard S, Schorah CJ, et al. 1980. Possible prevention of neural-tube defects by periconceptional vitamin supplementation. Lancet 1:339–40.
Smithells RW, Sheppard S, Schorah CJ, et al. 1981. Apparent prevent of neural tube defects by periconceptional vitamin supplementation. Arch Dis Child 56:911–18.
Smithells RW, Sheppard S, Schorah CJ, et al. 1991. Folic acid to prevent neural tube defects. Lancet 338:379–80.
Smithells RW, Sheppard S, Wild J, et al. 1989. Prevention of neural tube defect recurrences in Yorkshire: final report. Lancet 2:498–9.
Sokol RJ. 1980. Alcohol and spontaneous abortion. Lancet 2:1079.
Sokol RJ, Ager J, Martier S, et al. 1986. Significant determinants of susceptibility to alcohol teratogenesis. Ann N Y Acad Sci 477:87–102.
Sokol RJ, Clarren SK. 1989. Guidelines for use of terminology describing the impact of prenatal alcohol on the offspring. Alcohol Clin Exp Res 13:597–8.
Sokol RJ, Miller SI, Debanne S, et al. 1981. The Cleveland NIAAA prospective alcohol-in-pregnancy study: the first year. Neurobehav Toxicol Teratol 3:203–9.
Sokol RJ, Miller SI, Reed G. 1980. Alcohol abuse during pregnancy: an epidemiologic study. Alcohol Clin Exp Res 4:135–45.
Somers GF. 1962. Thalidomide and congenital abnormalities. Lancet 2:912–3.
Sonek M, Bibbo M, Wied GL. 1976. Colposcopic findings in offspring of DES-treated mothers as related to onset of therapy. J Reprod Med 6:65–71.
Soprano DR, Soprano KJ. 1995. Retinoids as teratogens. Ann Rev Nutr 15:111–32.
South MA, Sever JL. 1985. The congenital rubella syndrome. Teratology 31:297–307.
Speidel BD, Meadow SR. 1972. Maternal epilepsy and abnormalities of the fetus and newborn. Lancet 2:839–43.
Spiers PS, Pietyrzyk JJ, Piper JM, et al. 1973. Human potato consumption and neural-tube defects. Teratology 10:125–8.
Spohr HL, Willms J, Steinhausen HC. 1993. Prenatal alcohol exposure and long-term developmental consequences. Lancet 341:907–10.
Spooner EW, Hook EB, Farina MA, et al. 1988. Evaluation of a temporal increase in ventricular septal defects: estimated prevalence and severity in northeastern New York, 1970–1983. Teratology 37:21–8.
Spranger J, Benirschke K, Hall JG, et al. 1982. Errors of morphogenesis: concepts and terms. Recommendations of an international working group. J Pediatr 100:160–5.
Stagno S. 1995. Cytomegalovirus. In: Remington JS, Klein JO eds. Infectious diseases of the fetus and newborn infant. Philadelphia, Saunders, pp 312–53.
Steele EJ, Bellett AJ, McCullagh PJ, et al. 1990. Reappraisal of the findings on agent Orange by the Australian Royal Commission. Toxicol Lett 51:261–8.
Stein Z, Kline J, Kharrazi M. 1984. What is a teratogen: epidemiological criteria? In: Kalter H ed. Issues and reviews in teratology, vol. 2. New York, Plenum Press, pp 23–66.
Steinhausen HC, Nestler V, Spohr HL. 1982. Development and psychopathology of children with the fetal alcohol syndrome. Dev Behav Pediatr 3:49–54.
Stellman SD, Stellman JM, Sommer JF. 1988. Health and reproductive outcomes among American Legionaires in relation to combat and herbicide exposure in Vietnam. Environ Res 47:150–74.
Stern RS. 1989. When a uniquely effective drug is teratogenic: the case of tretinoin. N Engl J Med 320:1007–9.

Sterz H, Nothdurft H, Lexa P, et al. 1987. Teratologic studies on the Himalayan rabbit: new aspects of thalidomide teratogenesis. Arch Toxicol 60:376–81.

Steuerwald U, Weihe P, Jørgensen PJ, et al. 2000. Maternal seafood diet, methylmercury exposure, and neonatal neurological function. J Pediatr 36:599–605.

Stevenson AC, Johnston HA, Stewart MI, et al. 1966. Congenital malformations: a report of a study of series of consecutive births in 24 centres. Bull World Health Org 34(suppl):1–127.

Stevenson RE, Allen WP, Pai GS, et al. 2000. Decline in prevalence of neural tube defects in a high-risk region of the United States. Pediatrics 106:677–83.

Stevenson RE, Huntley CC. 1967. Congenital malformations in offspring of phenylketonuric mothers. Pediatrics 40:33–45.

Stevenson SS, Worcester J, Rice RG. 1950. 677 congenitally malformed infants and associated gestational characteristics. I. General considerations. Pediatrics 6:37–50.

Stockard CR. 1921. Developmental rate and structural expression: an experimental study of twins, 'double monsters' and single deformities, and the interaction among embryonic organs during their origin and development. Am J Anat 28:115–227.

Stoler JM, Huntington KS, Peterson CM, et al. 1998. The prenatal detection of significant alcohol exposure with maternal blood markers. J Pediatr 133:346–52.

Stone DH. 1980. Possible prevention of neural-tube defects by periconceptional vitamin supplementation. Lancet 1:647.

Stone DH. 1987. The declining prevalence of anencephalus and spina bifida: its nature, causes and implications. Dev Med Child Neurol 29:541–9.

Stone DH, Smalls MJ, Rosenberg K, et al. 1988. Screening for congenital neural tube defects in a high-risk area: a epidemiological perspective. J Epidemiol Community Health 42: 271–3.

Strassburg MA, Greenland S, Portigal LD, et al. 1983. A population-based case-control study of anencephaly and spina bifida in a low risk area. Dev Med Child Neurol 25:632–41.

Streissguth AP, Barr HM, Martin DC, et al. 1980. Effects of maternal alcohol, nicotine, and caffeine use during pregnancy on infant mental and motor development at eight months. Alcohol Clin Exp Res 4:152–64.

Streissguth AP, Barr HM, Olson HC, et al. 1994b. Drinking during pregnancy decreases word attack and arithmetic scores on standardized tests: adolescent data from a population-based prospective study. Alcohol Clin Exp Res 18:248–5.

Streissguth AP, Barr HM, Sampson PD, et al. 1989. IQ at age 4 in relation to maternal alcohol use and smoking during pregnancy. Dev Psychol 25:3–11.

Streissguth AP, Barr HM, Sampson PD. 1990. Moderate prenatal alcohol exposure: effects on child IQ and learning problems at age 7-1/2 years. Alcohol Clin Exp Res 14:662–9.

Streissguth AP, Barr HM, Sampson PD, et al. 1994a. Prenatal alcohol and offspring development: the first fourteen years. Drug Alcohol Depend 36:89–99.

Streissguth AP, Bookstein FL, Barr HM, et al. 2004. Risk factors for adverse life outcomes in fetal alcohol syndrome and fetal alcohol effects. J Dev Behav Pediatr 25:228–38.

Streissguth AP, Martin DC, Martin JC, et al. 1981. The Seattle longitudinal prospective study on alcohol and pregnancy. Neurobehav Toxicol Teratol 3:223–33.

Streissguth AP, Sampson PD, Olson HC, et al. 1994c. Maternal drinking during pregnancy: attention and short-term memory in 14-year-old offspring – a longitudinal prospective study. Alcohol Clin Exp Res 18:202–18.

Suarez L, Hendricks KA, Cooper SP, et al. 2000. Neural tube defects among Mexican Americans living on the US-Mexican border: effects of folic acid and dietary folate. Am J Epidemiol 152:1017–23.

Suhonen L, Hiilesmaa V, Teramo K. 2000. Glycaemic control during pregnancy and fetal malformations in women with type 1 diabetes. Diabetologia 43:79–82.

Sulaiman N, Florey CD, Taylor DJ, et al. 1988. Alcohol connsumption in Dundee and its effects on outcome of pregnancy. BMJ 296:1500–3.

Sullivan FM, Barlow SM. 1985. The relevance for man of animal data on reproductive toxicity of industrial chemicals. Prog Clin Biol Res 163B:301–5.

Sullivan WC. 1899. A note on the influence of maternal inebriety on the offspring. J Ment Sci 45:489–503.

Sutow WW, West E. 1955. Studies on Nagasaki (Japan) children exposed *in utero* to the atomic bomb: a roentenographic survey of the skeletal system. Am J Roentgen 74:493–9.

Swan C, Tostevin AL. 1946. Congenital abnormalities in infants following infectious diseases during pregnancy, with special reference to rubella: a third series of cases. Med J Aust 1: 645–59.

Swan C, Tostevin AL, Moore B, et al. 1943. Congenital defects in infants following infectious diseases during pregnancy: with special reference to the relationship between German measles and cataract, deaf-mutism, heart disease and microcephaly, and to the period of pregnancy in which the occurrence of rubella is followed by congenital abnormalities. Med J Aust 2: 201–10.

Swinyard CA. ed. 1969 Limb development and deformity: problems of evaluation and rehabilitation. Springfield, IL, Thomas.

Swinyard CA, Chaube S. 1973. Are potatoes teratogenic for experimental animals? Teratology 8:349–58.

Szeinberg A, Cohen BE, Boichis H, et al. 1963. Phenylketonuria among Jews. In: Goldschmidt E ed. The genetics of migrant and isolate populations. Proceedings of a conference on human population genetics in Israel held at the Hebrew University, Jerusalem. New York, Williams and Wilkins, pp 296–7.

Tamilarasan KP, Kolluru GK, Rajaram M, et al. 2006. Thalidomide attenuates nitric oxide mediated angiogenesis by blocking migration of endothelial cells. BMC Cell Biol 7:17.

Tanaka K. 1963. Differences between Caucasians and Japanese in the incidence of certain abnormalities. In: Goldschmidt E ed. Genetics of migrant and isolate populations. Proceedings of a conference on human population genetics in Israel held at the Hebrew University, Jerusalem. New York, Williams and Wilkins, pp 148–51.

Taruffi C. 1881–1894. Storia della teratologia. Bologna, Regia Tipografia.

Taussig HB. 1962. A study of the German outbreak of phocomelia: the thalidomide syndrome. JAMA 180:80–8.

Temple R, Aldridge V, Murphy HR. 2006. Prepregnancy care and pregnancy outcomes in women with type 1 diabetes. Diab Care 29:1744–9.

Tennes K, Blackard C. 1980. Maternal alcohol consumption, birth weight, and minor physical anomalies. Am J Obstet Gynecol 138:774–80.

Therapontos C, Erskine L, Gardner ER, et al. 2009. Thalidomide induces limb defects by preventing angiogenic outgrowth during early limb formation. Proc Natl Acad Sci USA 106:8573–8.

Thiersch JB. 1952. Therapeutic abortions with a folic acid antagonist 4-aminopteroylglutamic acid (4-amino P.G.A.) administered by the oral route. Am J Obstet Gynecol 63:1298–304.

Thiersch JB. 1956. The control of reproduction in rats with the aid of antimetabolites and early experiences with antimetabolites as abortifacient agents in man. Acta Endocr Suppl 28:37–45.

Thiersch JB. 1960. Discussion. In: Wolstenholme GE, O'Connor CM eds. Ciba Foundation symposium on congenital malformations. Boston, MA, Little, Brown, pp 152–4.

Thiersch JB, Philips FS. 1950. Effects of 4-amino-pteroylglutamic acid (aminopterin) on early pregnancy. Proc Soc Exp Biol Med 74:204–8.

Thom H, Campbell AG, Farr V, et al. 1985. The impact of maternal serum alpha fetoprotein screening on open neural tube defect births in north-east Scotland. Prenat Diag 5:15–19.

Thompson JR, Gerald PF, Willoughby ML, et al. 2001. Maternal folate supplementation in pregnancy and protection against acute lymphoblastic leukaemia in childhood: a case-control study. Lancet 358:935–40.

Tilley BC, Barnes AR, Bergstrahl E, et al. 1985. A comparison of pregnancy history recall and medical records. Am J Epidemiol 21:269–81.

Tolarova M. 1982. Periconceptional supplementation with vitamins and folic acid to prevent recurrence of cleft lip. Lancet 2:217.

Tolarova M, Harris J. 1995. Reduced recurrence of orofacial clefts after periconceptional supplementation with high-dose folic acid and multivitamins. Teratology 51:71–8.

Tookey PA, Peckham CS. 1999. Surveillance of congenital rubella in Great Britain, 1971–96. BMJ 318:769–70.

Townsend JC, Bodner KM, Van Peenen PF, et al. 1982. Survey of reproductive events of wives of employees exposed to chlorinated dioxins. Am J Epidemiol 115:695–712.

Trasler DG, Fraser FC. 1977. Time—position relationships: with particular reference to cleft lip and cleft palate. In: Wilson JG, Fraser FC eds. Handbook of teratology, vol. 2. New York, Plenum Press, pp 271–92.

Trasler JM, Doerkson T. 1999. Teratogen update: paternal exposures – reproductive risks. Teratology 60:161–72.

Treffers PE, Hanselaar AG, Helmerhorst TJ, et al. 2001. [Consequences of diethylstilbestrol during pregnancy: 50 years later still a significant problem.] In Dutch. Ned Tidschr Geneeskd 145:675–80.

Tsubaki T, Irukayama K. eds. 1977. Congenital Minamata disease: intrauterine methylmercury poisoning. Tokyo, Kodansha.

Tuchmann-Duplessis H. 1972. Teratogenic drug screening: present procedures and requirements. Teratology 5:271–85.

Turnbull CD, Scurletis TD, Danielson AH. 1977. Anencephalus in North Carolina: a 25-year experience. N Carol Med J 38:713–18.

Turton JA, Willars GB, Haselden JN, et al. 1992. Comparative teratogenicity of nine retinoids in the rat. Int J Exp Pathol 73:551–63.

Tyl RW, Price CJ, Kimmel CA. 1988. Developmental toxicity evaluation of Bendectin in CD rats. Teratology 37:539–52.

Tzimas G, Nau H. 2001. The role of metabolism and toxicokinetics in retinoid teratogenesis. Curr Pharmaceut Res 7:803–31.

Ueka K, Nishida Y, Oshima K, et al. 1979. Congenital rubella syndrome: correlation of gestational time at time of maternal rubella with type of defect. J Pediatr 94:763–5.

Uhl K, Cox E, Rogan R, et al. 2006. Thalidomide use in the US: experience with pregnancy testing in the S.T.E.P.S. programme. Drug Saf 29:321–9.

Ulleland CN. 1972. The offspring of alcoholic mothers. Ann N Y Acad Sci 197:167–9.

U.S. Bureau of the Census. 1960. Historical statistics of the United States: colonial times to 1957. Washington, DC, U.S. Department of Commerce.

Utiger RD. 1999. Maternal hypothyroidism and fetal development. N Engl J Med 341:601–2.

Van Allen MI, Boyle E, Theissen P, et al. 2006. The impact of prenatal diagnosis on neural tube defect (NTD) pregnancy versus birth incidence in British Columbia. J Appl Genet 47: 151–8.

Van Allen MI, Fraser FC, Dallaire L, et al. 1993. Recommendations on the use of folic acid supplementation to prevent recurrence of neural tube defects. Clinical Teratology Committee, Canadian College of Medicsl Geneticists. CMAJ 149:1239–43.

Van Driel D, Wesseling J, Sauer PJ, et al. 2002. Teratogen update: fetal effects after in utero exposure to coumarins: overview of cases, follow-up findings, and pathogenesis. Teratology 66:127–40.

Van Dyke DC, Hodge SE, Heide F, et al. 1988. Family studies in fetal phenytoin exposure. J Pediatr 113:301–6.

Vargesson N. 2009. Thalidomide-induced limb defects: resolving a 50-year-old puzzle. BioEssays 31:1327–36.

Vekemans M. 1982. The genetical difference between DBA/2 and C57BL/6 strains in cortisone-induced cleft palate. Can J Genet Cytol 24:797–806.

Velby G, Daltveit AK, Engelsen BA, et al. 2009. Pregnancy, delivery, and outcome for the child in maternal epilepsy. Epilepsia 50:2130–9.

Velie EM, Shaw GM. 1996. Impact of prenatal diagnosis and elective termination on prevalence and risk estimates of neural tube defects in California, 1989–1991. Am J Epidemiol 144:473–9.
Vergel RG, Sanchez LR, Heredero BL, et al. 1990. Primary prevention of neural tube defects with folic acid supplementation: Cuban experience. Prenat Diag 10:149–52.
Veurink M, Koster M, Berg LT. 2005. The history of DES, lessons to be learned. Pharm World Sci 27:139–43.
Vianna NJ. 1980. Adverse pregnancy outcomes – potential endpoints of human toxicity in the Love Canal preliminary results. In: Porter IH, Hook EB eds. Human embryonic and fetal death. New York, Academic Press, pp 165–8.
Viberti G. 1995. A glycemic threshhold for diabetic complications? N Engl J Med 332:292–3.
Viljoen D, Craig P, Hymbaugh K, et al. 2003. Fetal alcohol syndrome – South Africa, 2001. MMWR Morb Mortal Wkly Rep 52:660–2.
Vitéz M, Korányi G, Gönczy E, et al. 1984. A semiquantitative score system for epidemiologic studies of fetal alcohol syndrome. Am J Epidemiol 119:301–8.
Volpe EP. 1986. Is Down syndrome a modern disease? Perspect Biol Med 29:423–36.
Vrijheid M, Dolk H, Stone D, et al. 2000. Socioeconomic inequalities in risk of congenital anomaly. Arch Dis Child 82:349–52.
Waddington CH. 1957. The strategy of the genes. London, Allen and Unwin.
Waisbren SE, Doherty LB, Bailey IV, et al. 1988. The New England Maternal PKU Project: identification of at-risk women. Am J Public Health 78:789–92.
Waisbren SE, Hanley W, Levy HL, et al. 2000. Outcome at age 4 years of offspring of women with maternal phenylketonuria. JAMA 283:756–62.
Wald NJ, Bower C. 1995. Folic acid and the prevention of neural tube defects. BMJ 310:1019–20.
Wald NJ, Brock DJ, Bonnar J. 1974. Prenatal diagnosis of spina bifida and anencephaly by maternal serum-alpha-fetoprotein: a controlled study. Lancet 1:765–7.
Wald NJ, Hackshaw A. 1997. Folic acid and prevention of neural-tube defects. Lancet 350:665.
Wald NJ, Law M, Jordan R. 1998. Folic acid fortification to prevent neural tube defects. Lancet 351:834.
Wald NJ, Law MR, Morris JK, et al. 2001. Quantifying the effect of folic acid. Lancet 358:2069–73.
Wald NJ, Polani PE. 1984. Neural-tube defects and vitamins: the need for a randomized clinical trial. Br J Obstet Gynecol 91:516–23.
Walker BE. 1984. Transplacental exposure to diethylstilbestrol. In: Kalter H ed. Issues and reviews in teratology, vol. 2. New York, Plenum Press, pp 157–87.
Walker BE, Fraser FC. 1956. Closure of the secondary palate in three strains of mice. J Embryol Exp Morphol 4:176–89.
Walker BE, Fraser FC. 1957. The embryology of cortisone-induced cleft palate. J Embryol Exp Morphol 5:201–9.
Walker J. 1960. Discussion. In: Wolstenholme GE, O'Connor CM eds. Ciba foundation symposium on congenital malformations. Boston, MA, Little, Brown, pp 156–7.
Walkowiak J, Wiener JA, Fastabend A, et al. 2001. Environmental exposure to polychlorinated biphenyls and quality of the home environment: effects on psychodevelopment in early childhood. Lancet 358:1602–7.
Wallace AR. 1874. Review of "The object and method of zoological nomenclature," by David Sharp. Nature 9:258–60.
Walpole I, Zubrick S, Pontré J. 1989. Confounding variables in studying the effects of alcohol consumption before and during pregnancy. J Epidemiol Community Health 43:153–61.
Walpole I, Zubrick S, Pontré J. 1990. Is there a fetal effect with low to moderate alcohol use before or during pregnancy? J Epidemiol Community Health 44:297–301.
Warburton D, Fraser FC. 1964. Spontaneous abortion risks in man: data from reproductive histories gathered in a medical genetics unit. Am J Hum Genet 16:1–25.
Warburton D, Stein Z, Kline J, et al. 1980. Chromosome abnormalities in spontaneous abortion: data from the New York City data. In: Porter IH, Hook EB eds. Human embryonic and fetal death. New York, Academic Press, pp 261–98.

Warkany J. 1944. Congenital malformations induced by maternal nutritional deficiency. J Pediatr 25:476–80.
Warkany J. 1947. Etiology of congenital malformations. Adv Pediatr 2:1–63.
Warkany J. 1954. Disturbance of embryonic development by maternal vitamin deficiencies. J Cell Comp Physiol 43:207–36.
Warkany J. ed. 1957 Congenital malformations: papers presented at a conference on Teratology. Pediatrics 19:719–92.
Warkany J. 1960. Etiology of mongolism. J Pediatr 56:412–19.
Warkany J. 1971. Congenital malformations: notes and comments. Chicago, IL, Year Book Medical.
Warkany J. 1975. Etiology of down's syndrome. In: Koch R, de la Cruz FF eds. Down's syndrome (mongolism): research, prevention, and management. Philadelphia, PA, Saunders, pp 279–302.
Warkany J. 1978a. Aminopterin and methotrexate: folic acid deficiency. Teratology 7:353–7.
Warkany J. 1978b. Terathanasia. Teratology 17:87–92.
Warkany J. 1986. Teratogen update: hyperthermia. Teratology 33:365–71.
Warkany J. 1988a. Story of a teratologist. In: Kalter H ed. Issues and reviews in teratology, vol. 4. New York, Plenum Press, pp 1–79.
Warkany J. 1988b. Teratogen update: lithium. Teratology 38:593–6.
Warkany J, Beaudry PH, Hornstein S. 1959. Attempted abortion with aminopterin (4-aminopteroylglutamic acid). Am J Dis Child 97:274–81.
Warkany J, Kalter H. 1961. Congenital malformations. N Engl J Med 265:993–1001.
Warkany J, Nelson RC. 1940. Appearance of skeletal abnormalities in the offspring of rats reared on a deficient diet. Science 92:383–4.
Warkany J, Nelson RC. 1941. Skeletal abnormalities in the offspring of rats reared on deficient diets. Anat Rec 79:83–100.
Warkany J, Nelson RC, Scraffenberger E. 1942. Congenital malformations induced in rats by maternal nutritional deficiency. II. Use of varied diets and of different strains of rat. Am J Dis Child 64:860–6.
Warkany J, Roth CB, Wilson JG. 1948. Multiple congenital malformations: a consideration of etiologic factors. Pediatrics 1:462–71.
Warkany J, Schraffenberger E. 1943. Congenital malformations induced in rats by maternal nutritional deficiency. V. Effects of a purified diet lacking riboflavin. Proc Soc Exp Biol Med 54:92–4.
Warkany J, Schraffenberger E. 1944a. Congenital malformations induced by maternal nutritional deficiency. VI. The preventive factor. J Nutr 27:477–84.
Warkany J, Schraffenberger E. 1944b. Congenital malformations of the eyes induced in rats by maternal vitamin A deficiency. Proc Soc Exp Biol Med 57:49–52.
Warkany J, Schraffenberger E. 1946. Congenital malformations induced in rats by maternal vitamin A deficiency. Arch Ophthal 35:50–69.
Watkins ML, Edmonds L, McClearn A, et al. 1996. The surveillance of birth defects: the usefulness of the revised US Standard Birth Certificate. Am J Public Health 86:731–4.
Watson D, Pow M, Ellam A, et al. 1983. Prevention of neural tube defects in an urban health district. J Epidemiol Community Health 37:221–5.
Webster WS. 1998. Teratogen update: congenital rubella. Teratology 58:13–23.
Wegerle H. 1962. Beitrag zur Häufung von kongenitalen Missbildugen der Gliedmassen und ihre mögliche Ätiologie. Med Klin 58:877–83.
Wei PY, Chen YP. 1965. Congenital malformations, especially anencephalus, in Taiwan. Am J Obstet Gynecol 91:870–6.
Weicker H. 1963. Klinik und Epidemiologie der Thalidomid-Embryopathie. Bull Soc R Belg Gynecol Obstet 33:21–7.
Weicker H. 1965. 100 children with thalidomide embryopathy Tokyo. Proceedings of the XIth international congress of pediatrics.

Weicker H. 1969. Epidemiological and etiological considerations in the increase of limb malformations in Germany. In: Swinyard CA ed. Limb development and deformity: problems of evaluation and rehabilitation. Springfield, IL, Thomas, pp 225–32.
Weicker H, Bachmann KD, Pfeiffer RA, et al. 1962. Thalidomid-Embryopathie: vorkommen innerhalb und ausserhalb Deutschlands. Dtsch Med Wochenschrs 87:992–1002.
Weihe P, Hansen JC, Murata K, et al. 2002. Neurobehavioral performance of Inuit children with increased prenatal exposure to methylmercury. Int J Circumpolar Health 61:41–9.
Weinstein JB. 1999. Expert witness testimony: a trial judge's perspective. Neurol Clin 17:355–62.
Weinstein MR. 1979. Lithium teratogenesis: the Register of Lithium babies. In: Cooper TB, Gershon S, Kline NS, et al. eds. Lithium: controversies and unsolved issues. Amsterdam, Excerpta Medica, pp 432–46.
Weinstein MR, Goldfield MD. 1975. Cardiovascular malformations with lithium use during pregnancy. Am J Psychiatr 132:529–31.
Wells C. 1964. Bones, bodies and disease: evidence of disease and abnormality in early man. New York, Praeger.
Wells PG, Bhuller Y, Chen CS, et al. 2005. Molecular and biochemical mechanisms in teratogenesis involving reactive oxygen species. Toxicol Appl Pharmacol 207(2 Suppl):354–66.
Wenstrom KD, Johanning GL, Owen J, et al. 2000. Role of amniotic fluid homocysteine level and of fetal 5,10-methylenetetrahydrofolate reductase genotype in the etiology of neural tube defects. Am J Med Genet 90:12–16.
Werler MM, Hayes C, Louik C, et al. 1999. Multivitamin supplementation and risk of birth defects. Am J Epidemiol 150:675–82.
Werler MM, Lammer EJ, Rosenberg L, et al. 1990. Maternal vitamin A supplementation in relation to selected birth defects. Teratology 42:497–503.
Werler MM, Lammer EJ, Rosenberg L, et al. 1991. Maternal alcohol use in relation to selected birth defects. Am J Epidemiol 134:691–8.
Werler MM, Mitchell AA. 1992. Case-control study of vitamin supplementation and neural tube defects: consideration of potential confounding by lifestyle factors. Ann N Y Acad Sci 678:276–83.
Werler MM, Shapiro S, Mitchell AA. 1993. Periconceptional folic acid exposure and risk of neural tube defects. JAMA 269:257–61.
Wesselhoeft C. 1947. Rubella (German measles). N Engl J Med 236:943–50.
WHO. 1992. ICD-10: international statistical classification of diseases and related health problems, vol. 1. Tabular list, 10th rev ed. Geneva, World Health Organization.
Wichman CL, Moore KM, Lang TR, et al. 2009. Congenital heart disease associated with selective serotonin reuptake inhibitor use during pregnancy. Mayo Clin Proc 84:23–7.
Wide K, Winbladh B, Tomson T, et al. 2000. Body dimensions of infants exposed to antiepileptic drugs in utero: observations spanning 25 years. Epilepsia 41:854–61.
Wiedemann HR. 1961. Hinweis auf eine derzeitige Häufung hypo- und aplastischer Fehlbildungen der Gleidmassen. Med Welt 37:863–6.
Wiedemann HR. 1964. Klinische Bemerkungen zur pharmakogenen Teratogenese. Bull Schweiz Akad Med Wiss 20:544–64.
Wiersinga WM. 2001. Thyroid cancer in children and adolescents – consequences in later life. J Pediatr Endocrinol Metab 14(suppl 5):1297–8.
Wilcox AJ, Maxey J, Herbst AL. 1992. Prenatal diethylstilbestrol exposure and performance on college entrance examinations. Horm Behav 26:433–9.
Wilcox AJ, Weinberg CR, O'Connor JF, et al. 1988. Incidence of early loss of pregnancy. N Engl J Med 319:189–94.
Wild J, Schorah CJ, Sheldon TA, et al. 1993. Investigation of factors influencing folate status in women who have had a neural tube defect-affected child. Br J Obstet Gynecol 100:546–9.
Williams LJ, Mai CT, Edmonds LD, et al. 2002. Prevalence of spina bifida and anencephaly during the transition to manditory folic acid fortification in the United States. Teratology 66:33–9.

Williams LJ, Rasmussen SA, Flores A, et al. 2005. Decline in prevalence of spina bifida and anencephaly by race/ethnicity: 1995–2002. Pediatrics 116:580–6.
Williams RA, Mamotte CD, Burnett JR. 2008. Phenylketonuria: an inborn error of phenylalanine metabolism. Clin Biochem Rev 29:31–41.
Wilsnack SC, Klassen AD, Wilsnack RW. 1984. Drinking and reproductive dysfunction among women in a 1981 national survey. Alcohol Clin Exp Res 8:451–8.
Wilson JG. 1954. Differentiation and reaction of rats to irradiation. J Cell Comp Physiol 43(suppl):11–37.
Wilson JG. 1957. Is there specificity of action in experimental teratogenesis? Pediatrics 19:755–63.
Wilson JG. 1959. Experimental studies on congenital malformations. J Chron Dis 10:111–30.
Wilson JG. 1969. Teratological and reproductive studies in non-human primates. In: Nishimura H, Miller JR eds. Methods for teratological studies in experimental animals and man. Tokyo, Igaku Shoin, pp 16–33.
Wilson JG. 1973. Environment and birth defects. New York, Academic Press.
Wilson JG. 1977. Current status of teratology: general principles and mechanisms derived from animal studies. In: Wilson JG, Fraser FC eds. Handbook of teratology, vol. 1. New York, Plenum Press, pp 47–74.
Wilson JG. 1979. The evolution of teratological testing. Teratology 20:205–11.
Wilson JG, Beaudoin AR, Free HJ. 1959. Studies on the mechanism of teratogenic action of trypan blue. Anat Rec 133:115–28.
Wilson JG, Brent RL. 1981. Are female sex hormones teratogenic? Am J Obstet Gynecol 141:567–80.
Wilson JG, Roth CB, Warkany J. 1953. An analysis of the syndrome of malformations induced by maternal vitamin A deficiency. Effects of restoration of vitamin A at various times during gestation. Am J Anat 92:189–217.
Wilson JG, Warkany J eds. 1965. Teratology: principles and techniques. Chicago, IL, University of Chicago.
Wilson JG, Warkany J. 1985. The history of organized teratology in North America. Teratology 31:285–96.
Windham GC, Fenster L, Swan SH. 1992. Moderate maternal and paternal alcohol consumption and the risk of spontaneous abortion. Epidemiology 3:364–70.
Windham GC, Shaw GM, Todoroff K, et al. 2000. Miscarriage and use of multi-vitamins or folic acid. Am J Med Genet 90:261–2.
Windham GC, Von Behren J, Fenster J, et al. 1997. Moderate alcohol consumption and risk of spontaneous abortion. Epidemiology 8:509–14.
Winter RN, Donnai D, Burn J, et al. 1987. Fetal valproate syndrome: is there a recognizable phrenotype? J Med Genet 24:692–5.
Wiseman RA, Dodds-Smith IC. 1984. Cardiovascular birth defects and antenatal exposure to sex hormones: a reevaluation of some base data. Teratology 30:359–70.
Wogelius P, Nørgaard M, Gislum M, et al. 2006. Maternal use of selective serotonin reuptake inhibitors and risk of congenital malformations. Epidemiology 17:701–4.
Wolfe WH, Michalek JE, Miner JC, et al. 1995. Paternal serum dioxin and reproductive outcomes among veterans of operation Ranch Hand. Epidemiology 6:17–22.
Wolstenholme GE, O'Connor CM eds. 1960. Ciba foundation symposium on congenital malformations. Boston, MA, Little, Brown.
Wood JW, Johnson KG, Omori Y. 1967. In utero exposure to the Hiroshima atomic bomb: an evaluation of head size and mental retardation: twenty years later. Pediatrics 39:385–92.
Wren C, Birrell G, Hawthorne G. 2003. Cardiovascular malformations in infants of diabetic mothers. Heart 89:1217–20.
Wren C, Richmond S, Donaldson L. 1999. Presentation of congenital heart disease in infancy: implication for routine examination. Arch Dis Child Fetal Neonatal Ed 80:F49–53.
Wright JT, Waterson EJ, Barrison IG, et al. 1982. Alcohol consumption, pregnancy, and low birthweight. Lancet 1:663–5.

Wright S. 1934. An analysis of variability in number of digits in an inbred strain of guinea pigs. Genetics 19:506–36.
Wright S, Eaton ON. 1923. Factors which determine otocephaly in guinea pigs. J Agr Res 26: 161–82.
Wylie BR, Kong J, Kozak SE, et al. 2002. Normal perinatal mortality in type 1 diabetes mellitus in a series of 300 consecutive pregnancy outcomes. Am J Perinatol 19:169–76.
Xiao KZ, Zhang ZY, Su YM, et al. 1990. Central nervous system congenital malformations, especially neural tube defects in 29 provinces, metropolitan cities and autonomous regions of China: Chinese Birth Defects Monitoring system. Int J Epidemiol 19:978–82.
Yacobi S, Ornoy A. 2008. Is lithium a real teratogen? What can we conclude from the prosepctive versus retrospective studies? Isr J Psychiatry Relat Sci 45:95–106.
Yamazaki JN, Wright SW, Wright PM. 1954. Outcome of pregnancy in women exposed to the atomic bomb in Nagasaki. Am J Dis Child 87:448–63.
Yang Q, Khoury MJ, James LM, et al. 1997. The return of thalidomide: are birth defects surveillance systems ready? Am J Med Genet 73:251–8.
Yates JR, Ferguson-Smith MA, Shenkin A, et al. 1987. Is disordered folate metabolism the basis for the genetic predisposition to neural tube defects? Clin Genet 3 1:279–87.
Yen IH, Khoury MJ, Erickson D, et al. 1992. The changing epidemiology of neural tube defects: United States, 1968–1989. Am J Dis Child 146:857–61.
Yerby MS. 1992. Risks of pregnancy in women with epilepsy. Epilepsia 33(suppl 1):S23–7.
Yerby MS, Leavitt AM, Erickson DM, et al. 1992. Antiepileptics and the development of congenital anomalies. Neurology 42:32–40.
Yerushalmy J, Milkovitch L. 1965. Evaluation of the teratogenic effect of meclizine in man. Am J Obstet Gynecol 93:553–62.
Young AL. 2002. Vietnam and Agent Orange revisited. Environ Sci Pollut Res 9:158–61.
Young KR. 1983. Effect on birth weight of smoking in pregnancy. J R Army Med Corp 129:101–3.
Zalzstein E, Koren G, Einarson T, et al. 1990. A case-control study on the association between first trimester exposure to lithium and Ebstein's anomaly. Am J Cardiol 65:817–8.
Zhou XP, Cui L, Cui YL, et al. 2008. [Prevalence trend of neural tube defect in Zhengzhou city, 1996–2005.] In Chinese. Zhonghua Yu Fang Yi Xue Za Zhi 42:321–3.
Zilva SS, Golding J, Drummond JC, et al. 1921. The relation of the fat-soluble factor to rickets and growth in pigs. Biochem J 5:427–37.
Zimmermann MB. 2009. Iodine deficiency. Endocr Rev 30:376–408.

The manufacturer's authorised representative in the EU is Springer Nature Customer Service Centre GmbH, Europaplatz 3, 69115 Heidelberg, Germany. If you have any concerns regarding our products, please contact ProductSafety@springernature.com

Printed and bound by CPI Group (UK) Ltd, Croydon, CR0 4YY
30/09/2026
02225510-0006